PAKISTAN

By the same author

South Asia
Place, People and Work in Japan (with Phyllis Reichl)
Bangladesh
India: Resources and Development

PAKISTAN

B. L. C. JOHNSON

Professor of Geography
Australian National University, Canberra

HEINEMANN
LONDON AND EXETER, NEW HAMPSHIRE

To R. O. B.
A very perfect, gentle, man;
sage, teacher and friend.

Published in Great Britain 1979 by
Heinemann Educational Books Ltd
22 Bedford Square, London WC1B 3HH

Published in the U.S.A. 1979 by
Heinemann Educational Books Inc.
4 Front Street, Exeter, New Hampshire 03833

First published 1979

British Library C.I.P. Data

Johnson, Basil Leonard Clyde
1. Pakistan – Description and travel
1. Title
915.49'1 DS377

ISBN 0 435 35484 1

Library of Congress C.I.P. Data

Johnson, Basil Leonard Clyde.
Pakistan.

(Heinemann educational books)
Includes index.
1. Pakistan – Economic conditions. 2. Agriculture – Economic aspects – Pakistan. 3. Pakistan – Population. I. Title.
HC440.5.J56 330.9'549'105 79-10749

ISBN 0-435-35484-1

Filmset in Lumitype Times Roman
and printed by Interprint Limited, Malta

PREFACE

When a new Pakistan emerged in 1971, following the secession of Bangladesh, it was the first time in history that the peoples of the Indus valley had achieved political unification as a nation state. The cultural antecedents of Pakistan may possibly extend back to the third millenium BC, to the Indus civilization flourishing at Mohenjo-Daro and Harappa and contemporaneously with the great valley-based urban cultures of the Nile and Mesopotamia. There is more probability that traces of the civilization that evolved in Gandhara in the northwest from the sixth century BC onwards fusing Persian, Greek and Buddhist influences, may be present in the cultures of modern Pakistan. With greater certainty, modern Pakistan owes much to the invaders from southwest Asia who from the eleventh century onwards carried their religion, their architecture and their administrative skills deep into the sub-continent. Muslim political power reached its zenith under the Moghuls in the sixteenth and seventeenth centuries, though its focus was more often outside the Indus valley at Delhi or Agra, and was superseded by British colonial rule lasting until the partition of their Indian Empire in 1947.

Migration by many Muslims from western India to West Pakistan, and of non-Muslims in the opposite direction, reinforced the sense of religious unity in the Indus valley that had been the main basis for partition. East Pakistan, created by the partition of Bengal, was numerically comparable to West Pakistan, but while the majority of its people were Muslim, they were linguistically, racially and to a large extent culturally quite distinct from the West Pakistanis and geographically remote from them. While the separation of Bangladesh left the new Pakistan with its economy disrupted and its political and social organisms severely bruised, at the same time it opened the way for a more single-minded sense of nationhood to develop among the peoples of the Indus valley and its mountainous margins.

This book explores the basis for unity in the new Pakistan, and the paths for economic recovery and further development for its people. These are examined against the background of the constraints exercised by the country's environments. Man has however done much to adapt these environments to his advantage. Pakistan's greatest asset, the Indus Rivers system, has been harnessed progressively and its waters manipulated to generate power and to provide the irrigation that has wrought miracles of agricultural development in the plains. There has been an impressive degree of modernization in agriculture, and for wheat the country is an outstanding advertisement for the success of the 'green revolution'. Growth in the industrial sector of the economy has tended to lag, and while the rate of increase in the population shows little sign of diminishing, there is an acute and expanding problem of creating the jobs which might raise real living standards.

Political reorientation since 1971 has led Pakistan out of the British Commonwealth and into closer association with the oil-rich Muslim states of southwest Asia which have become important investors in Pakistan's development. Internationally Pakistan remains a 'non-aligned' country, accepting aid from nations of every political complexion. In internal politics stability has yet to be achieved; the pendulum has swung between democratic government and military dictatorship several times since independence in 1947.

ACKNOWLEDGEMENTS

In the many years since I first travelled in the Indus valley – as a soldier in 1941 and subsequently as an academic – innumerable people have contributed consciously and inadvertently towards my understanding of Pakistan, its peoples and its problems. I can name only a few of the many who have made the writing of this book possible. From the time I decided to write the book within the country, Mr Riaz Piracha, Pakistan Ambassador in Australia, was unstinting in his help and encouragement. Through his good offices and those of the Ministry of Education in Islamabad, the University of the Punjab, Lahore invited me as a visitor and extended to me the hospitality of the New Campus. There my friends of long-standing, Professors Mariam Elahi and K. U. Kureshy of the Department of Geography, and their colleagues, made me feel welcome, while my day-to-day life on campus was greatly enriched by the friendship and culinary ingenuity of M. Jacques Nieuviart.

In my own Australian National University I am grateful to Council for the study-leave which made my absence from Canberra possible; to my secretaries who so ably handled my manuscript when I was at the other end of the world; to Mr Kevin Cowan and Mrs Val Lyon who translated into maps the data and scribbled sketches I sent them; to Mrs Margaret Scrivenor for checking with eagle eye most of the figures; to Mrs Anne Coutts who assisted in the research and Miss Sue Gairns who compiled the index; and to my wife who succeeded in removing some of the inelegancies in the text. All the remaining errors of commission and omission are, of course, mine alone.

For permission to reproduce illustrative material my thanks are due to the following: Dr Samuel V. Noe, for Fig. 13.9 of the Old City, Lahore; the Government of Pakistan, Department of Films and Publications, for photographs, 2, 14, 15, 22, 26, 28, 34, 40, 41, 44, 52, 53, 55, 57, 58, 59, 60, 64, 66, 73, 74, 75, 80, 83; Pakistan Tourist Development Corporation for 77; Water and Power Development Authority for 20, 23; Mrs Margaret Scrivenor for 48, 62, 67; other photographs are by the author.

Lahore, Pakistan
February 1978.
B.L.C.J.

CONTENTS

LIST OF TABLES

LIST OF FIGURES

CHAPTER ONE

THE UNITY OF PAKISTAN

SOME PRE-HISTORY AND HISTORY

With the dissolution of the already spatially divided state of Pakistan in 1971 by the secession of its eastern wing to become Bangladesh, a new Pakistan came into being which achieved for its people a measure of physical identity hitherto unknown.

Pakistan occupies the easternmost basin of the three great rivers that traverse the steppe-deserts of the Old World: the Nile, the Tigris-Euphrates and the Indus. That these basins were the cradles of early civilisation gives to the Indus a distinctiveness lacking in other river basins in South Asia.

Pakistan lies at the eastern limits of the subtropical steppe-desert belt that extends westwards through Iran and Arabia to the Atlantic coast of the Sahara. In the heart of this arid zone Islam developed as a faith uniting its peoples in a vigorous culture and a political organisation. The Pakistanis tend to see their cultural affinities in this Islamic world, to a degree not shared by their co-religionists in Bangladesh or in the Muslim nations of Southeast Asia.

In more recent times Pakistan is the successor to a western frontier established in the nineteenth century by the British for their Indian Empire, a frontier albeit in rugged, sparsely-peopled territory that tends to divide like from like rather than to separate the unlike. Its eastern frontier with India is the product of the partition of India in 1947 when Lord Radcliffe had to arbitrate the separation of a Muslim Pakistan from the rest of the Indian Empire. That process established in the erstwhile West Pakistan the long-sought after sense of political unity. The in-migrations of Muslim refugees from the new India, and the out-migrations of non-Muslims subsequent to partition, accentuated the sense of cultural unity within the Western wing of the new nation, even if it was some way short of homogeneity. Parallel processes were at work in East Pakistan, but at a much lower level of intensity. Ultimately, Islam and a common antagonism towards 'Hindu' India (the two related factors that bound together East and West Pakistan) proved inadequate as a matrix to hold the young nation together against the fissiparous tendencies of cultural differences and physical separation.*

It would be a mistake to see in the new Pakistan the recovery of a political unity that had been lost. The modern state has achieved something that has not existed hitherto. The long history of the region that is now Pakistan, while it culminates in the appearance of a unitary state, is not the chronicle of a single, let alone an homogeneous, people. Much of the diversity present in Pakistan today can be traced to the accidents of human history, in which of course a variegated environment has played its part.

A significant regional dualism based upon contrasting environments may in fact be traced through much of Pakistan's history. The plains of the Indus are increasingly arid towards the centre and south, and have always been a region where permanent settlement is dependent on the rivers. Only in a relatively narrow submontane zone, less than 100 km wide, extending southeastwards from the Vale of Peshawar past Sialkot and Lahore, does the combination of moderate rainfall and a shallow watertable readily accessible by wells permit settlement independent of the rivers.

* For a discussion of Bangladesh's secession from Pakistan, see Johnson, B. L. C., *Bangladesh*, Heinemann Educational Books, London 1975, pp. 1–7.

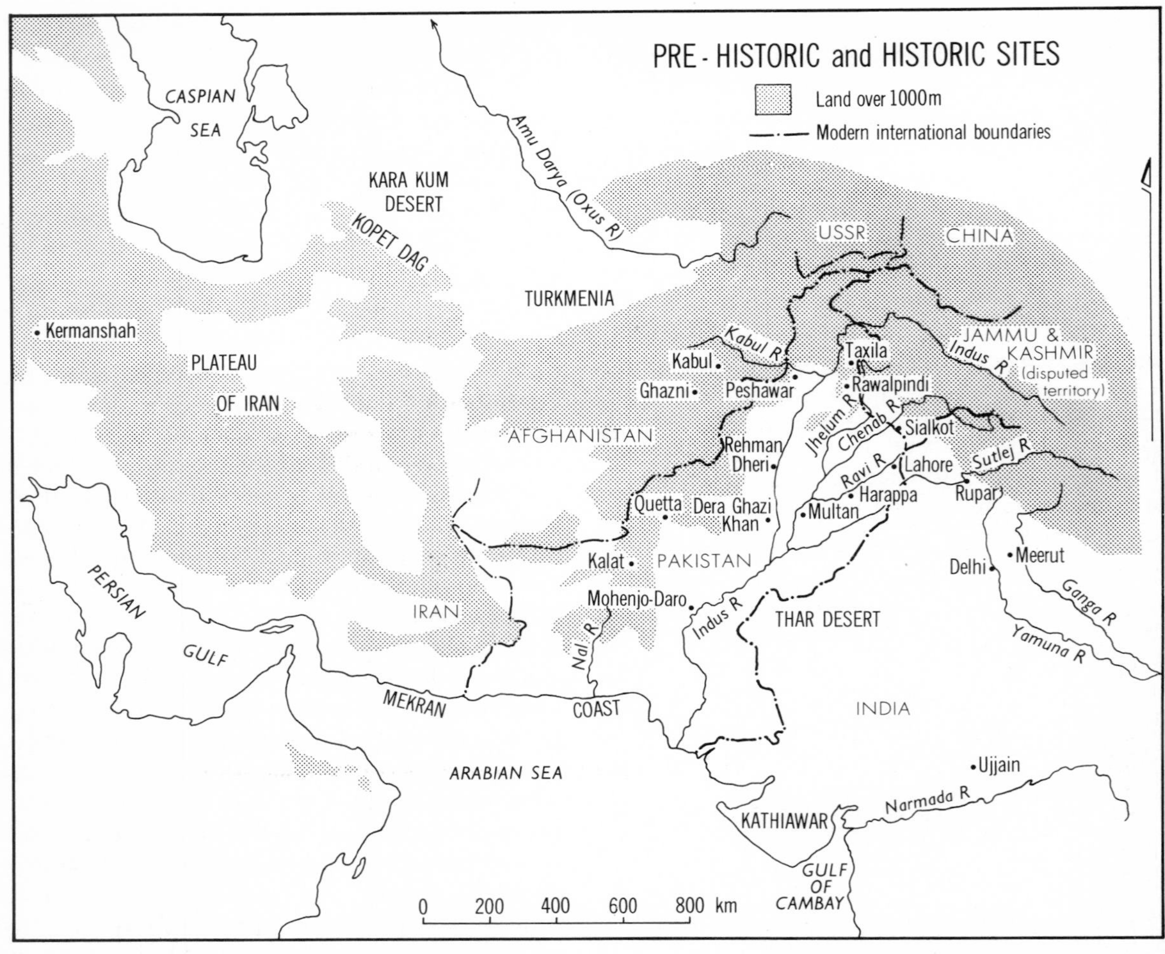

FIG. 1.1

Whether the environments of the Indus plains in the past were any more attractive than now to agricultural settlement using at best only simple systems of irrigation is debatable. The climate may have been slightly more humid, but within a similar regime and variability of rainfall. It was within such environments that early urbanisation took place in the Indus plains. The earliest city so far identified now appears to have been at Rehman Dheri, north of Dera Ismail Khan, lying just west of the Indus. It is thought to have been in existence in 3200 BC, a few hundred years before the cities of the Indus civilisation, and to have been similar in culture to contemporaneous cities at Hissar and elsewhere in Iran, at Mundi Gjak in Afghanistan, and sites in Turkmenia (USSR). The finds at Rehman Dheri suggest an agriculture similar in its crops and livestock to that of the present day: wheat, barley, mustard, and chillies were grown, sheep, goats and cattle seem to have been domesticated.

The Indus civilisation dating from 2500–1700 BC is described by Sir Mortimer Wheeler as the most extensive of Bronze Age civilisations. It was characterised by a pair of large city sites at Mohenjo-Daro (on the western flank of the Indus flood plain 240 km north of Karachi), and Harappa 570 km further northeast, on the banks of the River Ravi (Fig. 1.1). Upwards of a dozen town sites in the lower Indus south of Mohenjo-Daro represent this civilisation, which has outliers westwards, along the Makran coast and in the Nal valley. Southeast of the Indus delta there were several centres in the Kathiawar Peninsula and around the head of the Gulf of Cambay, all now in the Indian state of Gujarat. A gap of some 270 km separates the sites close to Mohenjo-Daro from the nearest of a string of towns starting in a cluster in Bahawalpur District, to the northeast, along the course of the now vanished river Sarasvati or Ghaggar. Eastwards the sites continue into India, the most distant being near Meerut beyond Delhi, and the most northerly on the River Sutlej close to Rupar.

The riverain settlements must have been supported by grain farming in the active flood plains, probably watered by seasonal channel diversions. Crops common in the region today were grown; wheat, barley, peas, sesamum and cotton. Cattle, buffalo and dogs were kept, and probably camels, horses and asses.

Working on contemporary settlements in the Baluchistan Hills, Raikes* concludes that the climate has not changed appreciably over the past 9000 years. Village sites showing more varied cultures, at later stages, related to that of the Indus plains, are common around Quetta and in the valleys turning south and west from Kalat. That climatic conditions were similar to those obtaining today is evidenced by the ruins of numerous 'garabands' for diverting and holding silt and water from intermittent torrents.

The villagers probably engaged, as they do today, in transhumance, wintering their flocks in the plains and returning to the high country for the summer. That the hill environments appear less able to support population nowadays is attributed by Raikes to the activities of man himself, overgrazing the slopes with his flocks and herds with resultant deterioration in the natural herbage, accelerated runoff and the reduction of the arable land through soil erosion and sand deposition.

From about 3000 BC villages with similar pre-Indus cultures producing fine pottery flourished also in the northwest, adjacent to the site of the much later Gandharian civilisation. No links have been established so far, between this village culture and the Indus urban civilisation and one can only conjecture that continuous settlement has persisted in the region.

The sudden disappearance of the Indus civilisation at about 1700 BC has not been satisfactorily explained. Recent climatological research suggests that a more arid climate than previously may have set in about this time, to which an increase in atmospheric dust, perhaps due to man's activities, may have contributed.† As a result, a tantalising hiatus separates the prehistory of Pakistan up to that period, from the dawn of its classical history. This we may take as the conquest of Gandhara centring on Peshawar and Rawalpindi Districts in the humid northwest in the sixth century BC by Cyrus, King of Persia. In the subsequent century the Persians annexed the Indus valley from the sea, having earlier sailed down the river from Kabul on its western tributary. It might be claimed that Darius I controlled as an eastern extension of his Persian Empire an area approximating to the present Pakistan. The Greek expedition under Alexander the Great in 327 BC marked out a somewhat similar area of influence, extending eastward to the River Beas, and including the lower Indus and Makran. From the third century BC these latter areas which are now Sind and Baluchistan slip from the limelight of history for nine hundred years, Sind to become at best the far western adjunct of the Kingdom of the Western Satraps (second century AD) whose capital was Ujjain (now in Madhya Pradesh in central India) or of the Gupta Empire that spread further eastwards to Bihar. The centre of the stage is taken by kingdoms of Indo–Greek origin nestling in the valleys and piedmont of the northern mountains. Thus around 200 BC, Demetrius had his Graeco–Bactrian capital at Taxila, a thriving centre of fusion of Greek and Buddhist cultures.

This northwest corner of Pakistan was to become through the following centuries, a frequent point of entry by groups originating in Central Asia – Scythians, Kushans, Huns and later others – whose capitals included Peshawar in the west and Sialkot in the east of the sub-montane zone. Their spheres of influence probably included the plains of the Punjab rivers south to their confluence with the Indus (near present-day Multan). The Thar Desert has always set an eastern limit to sedentary occupation south of the Punjab, and has forced invaders from the northwest into the con-

* Raikes, Robert, *Water, Weather and Pre-history*, John Baker, London 1967.

† Bryson, R. A., and Murray, T. J., *Climates of Hunger*, Wisconsin 1977, p. 109.

stricted watershed between the Ghaggar (a tributary of the lost Sarasvati River) and the Yamuna, at the head of the Ganga plain. The focus of attraction for these invaders from the harsher environments of Central Asia moved progressively eastwards into the better watered, more heavily populated and culturally increasingly more sophisticated region of the Ganga plain which became the core area of powerful empires that in turn exerted their influence westwards into the Indus Basin.

LANGUAGES

How substantial a contribution these early times made to the cultures of the peoples of present-day Pakistan is arguable. Language and tribal affiliations and antipathies can survive despite the later overlay of strong cultural elements such as religion. Thus Brahui, spoken in parts of Baluchistan (Fig. 1.8) belongs to the Dravidian language family now mainly restricted to South India, and is seemingly a relict of a population migrating into the subcontinent in prehistoric times. Similar but later pockets of distinctive peoples and culture are found in remote valleys of Himalayan Pakistan. The Chitralis in the far north show Central Asian affinities, while some of their non-Muslim Kafir neighbours may be refugees from the threat of forced conversions to Islam in Afghanistan.

The mountains are occupied by speakers of a variety of languages. The Iranian sub-family of Indo-European languages is represented by Balochi (Fig. 1.7) which dominates in Baluchistan (with an inlier of Dravidian Brahui mentioned above) while in the Northwest Frontier Province Pushto speakers straddle the frontier with Afghanistan (Fig. 1.6).

1 *Left* The excavations at Mohenjo-Daro, a city dating from about 2500 BC. In the foreground are baths lined with bricks fired with wood.

2 *Below left* Statuary of the Graeco-Buddhist Gandhara culture in Swat, North West Frontier Province.

3 *Below right* Ruined Buddhist *stupa* at Taxila, near Rawalpindi, a centre of Graeco-Buddhist culture from the third century BC.

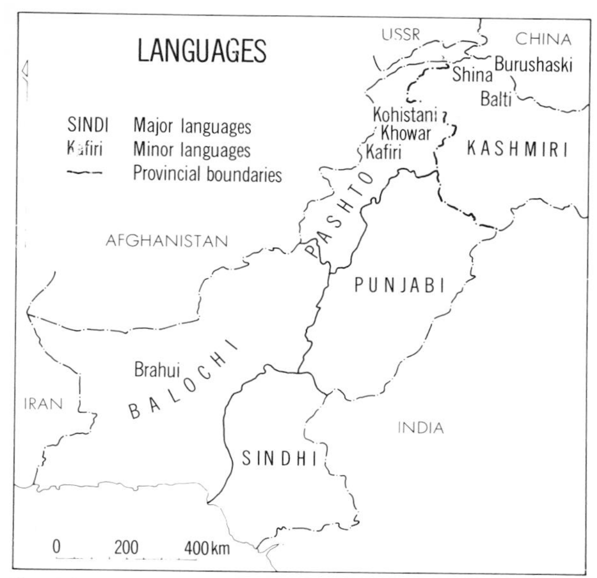

FIG. 1.2

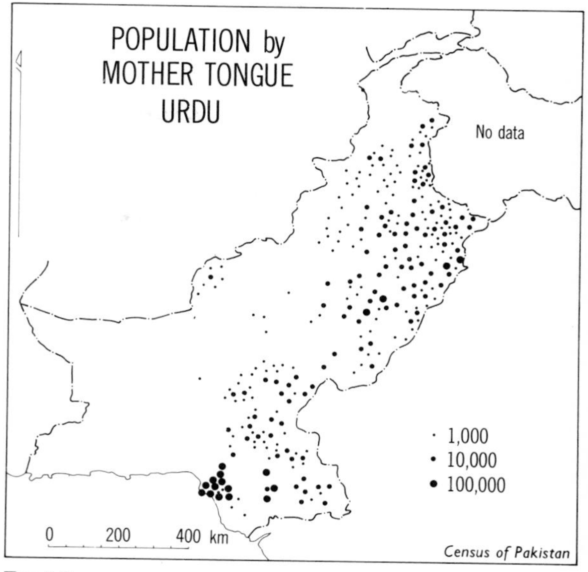

FIG. 1.3

Dardic languages – Kafiri, Kohistani, Khowar (Chitral), Shina (Gilgit) and Kashmiri – are found in the extreme north (Fig. 1.2). The linguistic ancestry of Burushaski, spoken in the Hunza Valley, is unknown. Balti spoken in inner western Kashmir is a Sino-Tibetan language. The linguistic diversity here revealed is further greatly complicated by the multiplicity of dialects that tribal populations have evolved in isolated valleys.

The plains show a much simpler pattern of Indo–Aryan languages, Sindhi in the south (Fig. 1.5), Punjabi in Punjab (Fig. 1.4). The national language, Urdu, (Fig. 1.3) hardly ranks as a regional mother-tongue. It derives from the Persianised Hindi of the Moghul court, is written in an Arabic script, and contains words from Turkish and from the languages of the European mercantile and imperial powers – Portuguese, Dutch, French and English.

Although it is the official national language, Urdu is the 'mother tongue' of only 8 per cent of the population, according to the 1961 Census. It is claimed as a 'language of speech' by 15 per cent however, and in fact a rudimentary knowledge of Urdu as a lingua franca is probably possessed by a majority of Pakistanis. Punjabi, the language of the greatest number is the mother tongue of 66 per cent and is spoken by 68 per cent. In structure and vocabulary it is similar to Urdu.

As the language of the Moghul rulers, Urdu is still most strongly represented in urban areas. The fact that many Urdu-speaking refugees coming into Pakistan from Delhi, U.P. and the eastern part of the former Punjab at the time of partition settled in towns accentuated this tendency. Fig. 1.3 shows Urdu in numerical strength particularly in Lahore, Multan, Hyderabad and Karachi, but only in the latter district is it the language of the majority (54 per cent). In Hyderabad, 24 per cent claim Urdu as their mother tongue.

This relative unimportance of Urdu in most districts has been claimed as one major advantage it has as a unifying factor in Pakistan, since it is not a major regional language and so cannot be said to favour one province above the rest.

Punjabi (Fig. 1.4) strongly dominates throughout the Punjab, no district showing less than 89 per cent of its population claiming it as their mother tongue. Urdu makes up most of the difference, supported by five per cent or less of Pushto speakers in Attock and Mianwali which border on the NWFP. On the other hand Punjabi 'invades' the adjacent Frontier districts of Hazara and Dera Ismail Khan to the extent of 86 and 74 per cent respectively. Unlike other languages in Pakistan, Punjabi has its speakers in some strength in every district of the country, and might with some justice be said to be the language of socio-economic dominance.

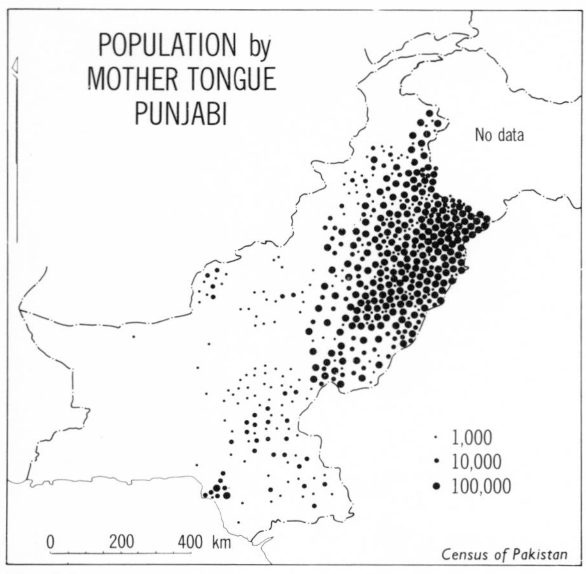

FIG. 1.4

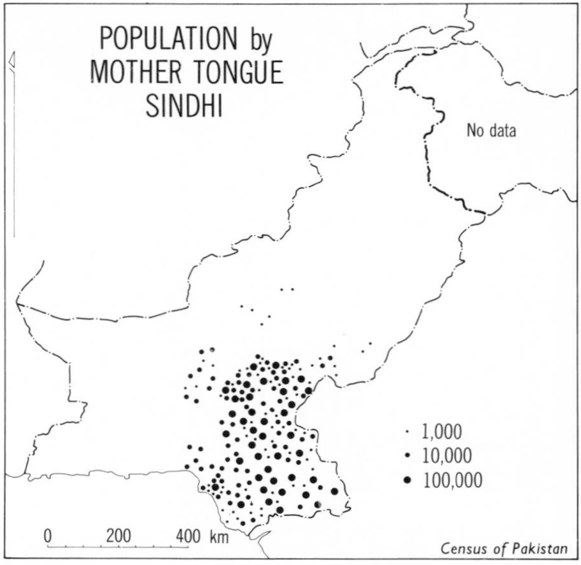

FIG. 1.5

Sindhi, spoken as mother tongue by 13 per cent and altogether by 14 per cent, is more restricted in its distribution (Fig. 1.5). It is a negligible language in the Punjab (except in the border district of Rahimyar Khan) and NWFP, but has relative (though not absolute) strength in the Baluchistan districts of Las Bela and Kalat bordering on Sind. It is a matter of some bitterness that the tide of development and the current of refugees from India have reduced to 9 per cent the Sindhi speakers in the provincial capital, Karachi. Generally the districts of Sind show a greater variety of tongues than are found in other provinces. Reflecting Sind's long attachment to the British Indian Bombay Presidency, Gujrati is still spoken by substantial numbers in Karachi (152,000), Hyderabad (39,000) and Tharparkar (31,000), altogether by nearly 3 per cent of the province, many of them in commerce. Rajasthani finds 41,000 speakers in Sanghar, 29,000 in Tharparkar and 24,000 in Hyderabad, mainly among the farming population. Balochi is second only to Sindhi in Jacobabad (32,000) and Dadu (11,000) which lie adjacent to areas where that language is dominant.

Pushto, mother tongue for 8 per cent of the population is highly concentrated in the NWFP except in Hazara (9 per cent only) and Dera Ismail Khan (24 per cent) (Fig. 1.6). It also dominates the northern districts of Zhob, Loralai and Quetta

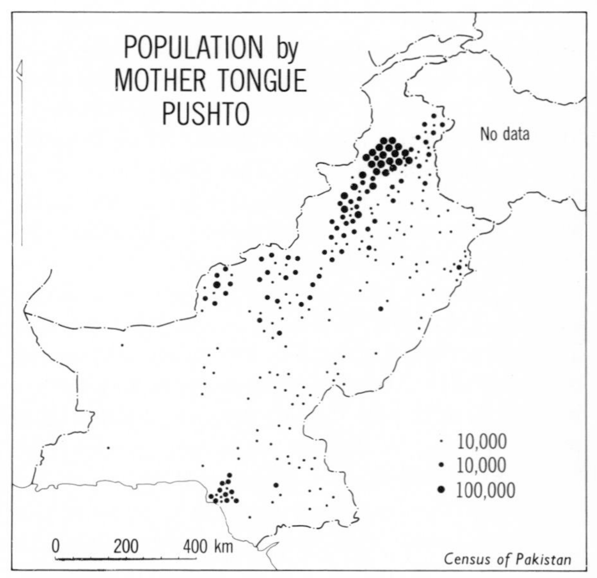

FIG. 1.6

in Baluchistan. The predeliction, indeed the necessity to travel in search of a living that characterises the Pathan, takes him (often without his family) far from home to wherever a job can be found. This explains the wide scatter of Pushto speakers throughout Pakistan, with 105,000 in Karachi alone.

In national terms Balochi (2 per cent) and Brahui (1 per cent) are unimportant, but they are the lan-

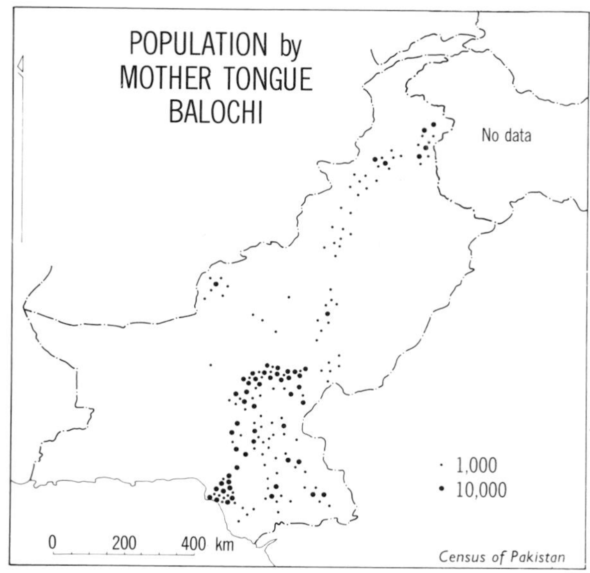

FIG. 1.7

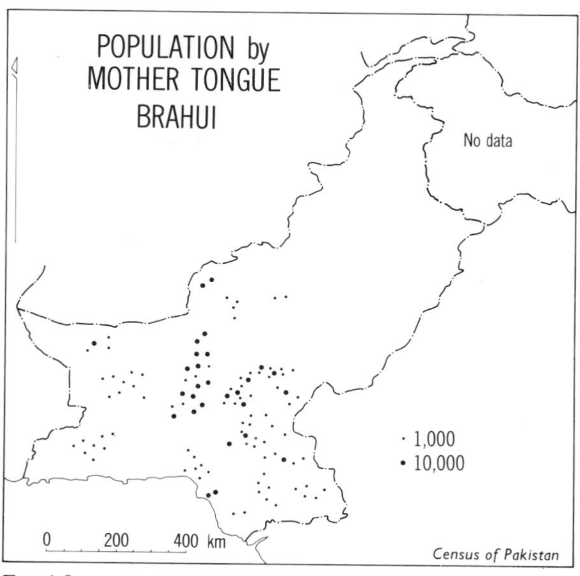

FIG. 1.8

guages of the more backward groups in Baluchistan. Balochi spoken also in eastern Iran is the more important, dominating in Makran where it is spoken by 94 per cent, Brahui by the remainder. In Kharan the scales are still loaded in favour of Balochi (78 per cent) against Brahui (21 per cent), as also in Chajai (60 per cent against 31 per cent). Kalat shows Brahui on top with 39 per cent, with Balochi (32 per cent) and Sindhi (25 per cent) in strength. In Sibi, Balochi (65 per cent) finds Pushto (22 per cent) in competition. Figs. 1.7 and 1.8 show how interwoven with each other and with neighbouring tongues are the Balochi and Brahui speaking realms. Beyond Quetta and Loralai they are hardly represented except for Balochi in the far western districts of the Punjab.

RELIGION

The most powerful factor in unifying the peoples of Pakistan is neither race nor language, but the common heritage of their Islamic religion.

In the 1961 Census, 97.17 per cent of the population were returned as Muslim. In only two districts, Sanghar (85 per cent) and Tharparkar (60 per cent) does the proportion fall below 94 per cent due to the presence of substantial Hindu groups of mainly scheduled castes, representing the western fringe of the Rajput peoples of modern Rajasthan, and the

4 The domes of the Badshahi Mosque, with the city of Lahore beyond, viewed from the northern minaret. The mosque is built in pink sandstone and marble.

Gujaratis. Christianity however is the major minority religion, having in 1961, 583,884 adherents (1.36 per cent of the total population) mainly in the Lahore and Sargodha Divisions of the Punjab.

Islam, originating in Arabia in the seventh century AD was first introduced to the Indus basin in 711 AD by Mohammed ben Kasim, Governor of Basra who arrived by sea to suppress piracy on Arab shipping, and established control of the Indus valley as far north as Multan. More vigorous proselytisation of the non-Muslim population was carried out by invaders from the northwest. Early in the eleventh century the Punjab became the eastern province of an empire ruled from Ghazni in Afghanistan and extending through the plateau of Iran to Kermanshah near the borders of modern Iraq and north to the Amu Darya (Oxus River) and the Kara Kum Desert (Fig. 1.1). Mahmud (998–1030) was the greatest of the Ghaznaavids. They had Lahore as their regional capital in the Punjab; much of Baluchistan was either directly ruled by, or tributary to Ghazni, as was Sind from time to time. In 1206 the centre of Muslim power in the sub-continent shifted to Delhi, inaugurating three centuries of rule by the Delhi Sultans, and a separation of their territories from Ghazni. It ended with the conquest by Babur coming from Kabul in 1525–26 to found the Moghul empire in Delhi and to re-establish for a time the link with Afghanistan. Under the Moghuls, Muslim power spread almost throughout the Indian sub-continent, reaching its zenith in the seventeenth century but declining thereafter. Sind, for example, was lost for a time to Persia in 1739, and internally the empire became fragmented.

The disruption of the Moghul empire from the eighteenth century, and efforts of the British to extend their control, had far reaching implications for modern Pakistan. In the early nineteenth century, the Sikhs, a militant religious group deriving from Hinduism in the sixteenth century, had seized power in the Punjab. Their leader Ranjit Singh, from his capital in Lahore, widened his control of territory to include most of Kashmir and lands west of the Indus, right into the Pathan-occupied hill country that overlooks the plains from the Vale of Peshawar, to Dera Ghazi Khan. This enabled him to push back and contain Afghan influence, which through the Pathan tribes, had usually reached as far as the Indus.

After Ranjit Singh's death the Sikhs came into conflict with the British. Following the First Sikh War, Kashmir was surrendered and sold by the British to a Hindu Raja thus placing its mainly Muslim population under non-Muslim rule, a situation which persists to this day as the major bone of contention between Pakistan and India. Renewed hostilities in the Second Sikh War (1848–49) resulted in the British succeeding to the Sikh's demesne and thus to a frontier with Afghanistan lying athwart the northwestern mountains.

An attempt to settle a frontier demarcating British and Afghan spheres of influence in this area was embodied in the Durand Line of 1893, but clearly failed permanently to meet the aspirations of the Afghans who have subsequently espoused the cause of Pakhtunistan. This latter is the concept of nationhood for the speakers of Pushto, a third of whom form a substantial minority in Afghanistan, and two-thirds live on the Pakistan side of the frontier. While a potential rallying point when leaders in NWFP seek to lever concessions from a Punjabi-dominated government, Pakhtunistan is unlikely to command serious support among Pakistan's Pathans who would have much to lose and little to gain from union with Afghanistan. In Sind, a British adventure under Napier in 1843 led to its annexation from its Baluchi rulers and attachment to the Presidency of Bombay, an event that demonstrated the continuing negative influence of the Thar Desert in separating the lower Indus basin from political developments in the Punjab plains. Sind achieved separate provincial status within British India in 1936 with Karachi as its capital.

PARTITION

When Britain transferred sovereignty over its Indian Empire to the new Dominions of India and Pakistan, the latter inherited territories with a variety of administrations. Sind, with those parts of Punjab acquired by partition and the more settled parts of NWFP within the 'inner line' of administration, were under direct British control, but had governments increasingly representative of the local population. Bahawalpur and Khairpur were princely states on the Punjab and Sind margins of the Thar Desert respectively. In NWFP, Dir, Swat and Chitral, and in Baluchistan, the Baluchistan States Union (including Kalat and Las Bela), were similarly princely states under indirect rule. The immediate Baluchistan frontier with Afghanistan in Chagai District was under direct British control. In NWFP the frontier ran partly in Tribal territory, partly on the western borders of Dir and Chitral.

Independent Pakistan has absorbed the princely states as administrative districts under direct control. The peculiar problems of the tribal groups in some areas of Baluchistan and NWFP are recognized in special local government arrangements. Economic and social development is relied upon gradually to bring these formerly extremely independent and sometimes fanatically fractious peoples into harmony with the more advanced and advantaged plainsmen. A major problem is to wean some tribes in Baluchistan from their feudal traditions under the Sardari system in which members owed absolute loyalty to their chieftain, towards having confidence in the less visible and impersonal authority of the state. The transition can be difficult for all concerned.

Some of the problems that followed the arbitration of the Indo-Pakistan border in 1947 are discussed in subsequent chapters. The most urgent was the disruption of the supply of water to parts of Pakistan formerly fed from barrages now in Indian territory. Under the Indus Waters Agreement of 1960 and with colossal financial and technical assistance from the World Bank this problem has been satisfactorily resolved.

The antecedents of the Kashmir issue have been touched on above. It is not appropriate here to rehearse the rights and wrongs of the case, real and imaginary. The pity is that the British, on planning to quit India, failed to provide more adequate machinery for the self-determination by its people, of the future of Jammu and Kashmir or even for its partition on principles similar to those that were followed in the Punjab.

Jammu and Kashmir was a princely state with Muslims forming 77 per cent of its population in 1941. Only in the extreme east and south around Jammu were non-Muslims the community in the majority. Under the instruments of partition the princely rulers of states were 'free to accede either to India or to Pakistan, or to remain independent. They were advised, however, to accede to the contiguous dominion, bearing in mind geographical and ethnic considerations.'*

The Hindu ruler of Jammu and Kashmir vacillated. Having earlier come to an understanding with Pakistan he gave way to Indian pressure, and fearing tribal warfare from within and without the state, he agreed to join India, which accepted provisionally 'pending a free and impartial plebiscite'. Pakistani 'tribesmen' attempted to force the issue in 1948 but Indian military forces were brought in and expelled them from the Vale of Kashmir. A 'cease-fire' line now referred to as the 'line of control' under United Nations' supervision divides Pakistan-controlled 'Azad' (Free) Kashmir from the Indian-controlled larger part which is administered as a fully-fledged state of the Indian Union. For the foreseeable future the dispute appears to be in deadlock.

Pakistan continues to lay claim to two small enclaves in the Kathiawar Peninsula of Gujarat State in India, whose Muslim rulers wished to join their co-religionists, but which India absorbed. They are shown as Pakistan territory entitled Junagadh and Manavadar on official maps of Pakistan.

INTERNAL PROBLEMS

Unlike in India where very considerable re-organization of constituent states took place in the early years of independence in response to pressures for linguistic self-determination, in Pakistan conservative traditions have preserved the provincial map to a large extent in the form inherited at partition. What changes have occurred have been largely subdivisions of existing districts to create more manageable administrative units, and the absorption of the former princely states.

The status of the four provinces has been some-

* Aziz Ahmad, 'India and Pakistan', in Holt, P. M., Lambton, A. K. S. and Lewis B. (Eds.) *The Cambridge History of Islam*, Cambridge University Press, 1970, p. 115.

thing of a political football. In 'combined' Pakistan they were for a time absorbed into a single West Pakistan in order to placate East Pakistan sensibilities at being one province among five although exceeding all the others together in terms of population. Provincial administration was restored however even before the secession of Bangladesh.

The provinces, divisions and districts of Pakistan are shown in Fig. 1.9 as at the beginning of 1978. Districts are further divided into *tehsils*. The basic structure of administration still bears the imprint of British rule. Much day-to-day responsibility rests with the Deputy Commissioner at district level, and at the next level above, the Commissioner of the division. The provinces have Governors appointed by the Federal Government, but are normally ruled by elected assemblies headed by the Chief Minister. At the centre there is a National Assembly.

A further relic of British administration is the

FIG. 1.9

retention of the Tribal Territories – Mohmand, Khyber, Kurram, North and South Waziristan – as centrally administered areas within the 'Inner Line'. Adjacent to these, in Peshawar, Kohat, Bannu, Dera Ismail Khan and Dera Ghazi Khan, there are small tribal areas administered by the districts. Gilgit Agency in the extreme north is also centrally administered and is not regarded as part of 'Azad' Kashmir, although under British rule it was part of the responsibilities of the British Resident in Jammu and Kashmir, while not being part of the Maharajah's territory.

Democratic government has had a sorry record in Pakistan ever since 1947. Without previous experience of working together as a unit, provincial and national leaders and administrators were faced with a plethora of problems at home and abroad. The collapse of newly created democratic institutions under the strain was not perhaps surprising, and for the more traditional and feudally minded of West Pakistan, military administrations were, for thirteen out of twenty-three years, before the loss of Bangladesh, accepted as the lesser of evils.

PAKISTAN SINCE 1970

General Yahya Khan, the second of the military rulers, held general elections in December 1970 as a result of which Sheikh Mujibur Rahman's Awami League in East Pakistan gained an overall majority of seats in the National Assembly. On the strength of this the Sheikh demanded a certain measure of autonomy for the East Wing. It was the refusal by Zulfikar Ali Bhutto, leader of the Pakistan People's Party with a majority in West Pakistan, to concede these demands that ultimately led to the proclamation of an independent Bangladesh, and to the nearly ten months of civil strife that culminated in India's intervention in December 1971.

Mr Bhutto assumed the leadership of the new Pakistan, with the task of restoring what the geographer, K. U. Kureshy describes as a 'shattered morale, a crumbling economy, and a low prestige in the community of nations'.

Until General Zia proclaimed the country under martial law in 1977, Bhutto had been pursuing a policy popular with the urban masses and with the rural poor. He nationalized basic industries, banks and insurance and attempted to democratize the civil service and to bring specialist experts into government. In all these areas of reform he could be seen to be attacking the strongholds of the entrenched power of Punjabi semi-feudal landlords who had held effective political and economic control. Later in 1976, the abolition of Sardari rights in Baluchistan struck at another aspect of feudalism and potentially, at least, opened the way for modernization in that backward province.

More positive social policies concerned education and health. Far-reaching reforms were introduced to accelerate literacy and to promote secondary and tertiary studies, and a comprehensive programme was begun to bring basic health services to the villages. A major economic success was to devalue in 1972, removing what has been described as the corrupting effect of an over-valued currency, which discourages exports, encourages imports, especially of luxuries for the elite, and drives capital out of the country.*

On the international stage several changes in Pakistan's position have occurred since 1970. When Australia, Britain and New Zealand recognized Bangladesh, Bhutto took Pakistan out of the British Commonwealth. Even before the separation, other changes in alignment had been taking place with the object of forming ties with China, as a means of balancing India as a threat. One result has been the opening of the frontier to China with the Karakoram Highway, and some technical and economic aid. More significant economically has been the new look westwards to the Islamic World, to the wealthy oil-producing states in particular.† Despite some cooling of relations with the Soviet Union over the latter's support of Bangladesh's independence, Pakistan has accepted Soviet sponsorship of the Karachi Steel Mill now under construction. Aid and trade with the USA

* For some shrewd studies, seen in retrospect, of Bhutto's early policies, see Korson, J. Henry (Ed.), *Contemporary Problems of Pakistan*, E. J. Brill, Leiden 1974.

† A straw in the wind of change towards Islam and the Islamic world is the decision to introduce Arabic as a compulsory subject in schools from Class 8 onwards, a move which could open the schools to an influx of teachers with little knowledge other than of the Holy Koran (which is in Arabic).

continues, so it can be said that Pakistan is being non-committal about foreign ties. Some of these matters are discussed further in appropriate chapters.

In its efforts to discover its identity Pakistan appears, from within at any rate, to be stressing its adherence to Islam, and minimising its British colonial heritage. Lyallpur and Campbellpur, the last vestiges of British-named districts in the Punjab have become Faisalabad (to commemorate the visit by the Saudi Arabian King) and Attock respectively. With more reason, Urdu is to become the language of instruction up to B.A. level, leaving English as the requirement for M.A. studies. The efforts to put meaning into the country's title as an Islamic Republic lead to much retrospection in the newspapers at the contribution of Muslims to this and that aspect of science or philosophy, and a general tendency by aspiring politicians and administrators to 'jump on the bandwagon' proclaiming the need to restore Nizam-i-Mustafa, traditional Islamic law. Even Bhutto, it is said, accepted that the Islamic Council could veto any law passed by the National Assembly which it considered contrary to the tenets of Islam, and more recently the Chief Martial Law Administrator, General Zia, announced that any existing laws found incompatible with Nizam-i-Mustafa could be challenged in the courts. While the wisdom and the logic of these trends for a nation seeking to develop and to modernize are not without critics, the present mood of the country does seem to be away from secular rationalism and in the direction of orthodox religion becoming a strong force in social and political affairs, a direction which Gunnar Myrdal, the outstanding scholar of South Asian socio-economic development would regard as retrograde and irrational.

He would still no doubt classify Pakistan as a 'soft state', a state in which assumedly responsible bureaucrats and politicans practise self-delusion as a matter of course.* An illustration of this is given in J. Henry Korson's edited volume cited above, where it is claimed that the economic planner responsible for the Third Five Year Plan admitted that the Plan had used a population increase rate of 2.7 per cent per annum at a time when clear evidence indicated the rate to be in excess of 3 per cent, saying that the lower rate had been adopted 'to avoid pessimism' in the Plan!

The present outlook for Pakistan is problematical. Survival as a nation is not seriously in doubt, but what kind of a society will emerge by the turn of the century is hard to tell. The chapters that follow assess the country's potentialities and point up the constraints, human and environmental, that must form part of any appreciation of Pakistan's present and future.

* Myrdal, G., *Asian Drama: an Inquiry into the Poverty of Nations*, Pantheon, New York 1968.

CHAPTER TWO

THE NUMBER AND CONDITION OF THE PEOPLE

SUMMARY

This chapter surveys the demographic characteristics of the population of Pakistan: its number, its growth and areal spread over time. It concludes with an attempt to draw together some facts about the social and economic condition of the people in order to arrive at an 'index of development' by which the relative position of each of the forty-five districts may be judged. While the average level of development of the people is undoubtedly low by any standard, some areas show much higher levels than others, and this is found to be a consistent tendency over a number of criteria. The level of development can thus usefully be examined in this light as a framework for investigation, in later chapters, of the factors involved.

HOW MANY PAKISTANIS?

By the end of 1978 or early 1979 the population living in Pakistan will be over 80 million, with another million of its citizens living abroad.

Several factors conspire to make it difficult to estimate population numbers accurately in Pakistan. The country is not alone in finding it impossible to conduct a wholly reliable census. Conservatism, suspicion of all forms of governmental enumeration, and mass illiteracy are common enough in the under-developed world.*

*The author shares with Gunnar Myrdal a distaste for the euphemisms in common use to avoid branding nations with assumedly pejorative adjectives. 'Developing' countries or 'less developed' countries, may sound and seem nicer than 'under-developed', 'backward' or 'underdeveloped', but the width of the gap between 'developed' and 'less developed' is such as to suggest a sharper distinction should be made, at least at the extremes.

To these Pakistan must add the further handicap to the census-taker, the practice among orthodox Muslims of keeping womenfolk and babies in seclusion, making a direct head count of a household impracticable. Furthermore, within the territories of former British India, and continuing since Pakistan and India gained their independence, it has been necessary to recognize that censuses are open to manipulation for political purposes. Dating at least from the Census of India in 1931, under- or over-enumeration has been common, depending on the political objective. In 1931 the civil disobedience movement led by Mahatma Gandhi and the Indian Congress Party as a demonstration to promote the cause of independence from Britain, registered its feelings by understating the numbers in households. The 1941 census, coming at a time when communal feeling between Hindus and Muslims was running high and the idea of a possible partition was in the air, resulted in over-enumeration, each community seeking to exaggerate its size.

Pakistan's first census of 1951 was modest compared with that of 1961, which for many purposes remains the best source of information over a wide range of topics. The latter is however considered to have been under-enumerated by at least 7.5 per cent and possibly as much as 9 per cent. The census planned for 1971 had to be postponed because of the dislocations due to the crisis which led to the secession of Bangladesh, and the enumeration made towards the end of 1972 was on the basis of a very limited range of questions. Its figures, especially for NWFP and Baluchistan where the Pakistan People's Party was keen to enhance numbers in order to increase the number of seats it might command in the

National Assembly, are accepted as inflated. A possible excess of 400,000 is inferred from earlier forecasts. While this is not a gross error overall, it was perhaps reached through larger local anomalies. Another factor explaining the unexpectedly high figure was the increased enumeration of women, who in Punjab and Sind 'came out of hiding' to quote on study.* Table 2.1 below gives an estimate of the population of the present area of Pakistan by decennial periods since 1901. Up to 1961 the data are those quoted in official sources, viz. the *Economic Survey of 1975–76*. The figure for 1961 is the census total increased by 7.5 per cent which is the officially accepted rate by which it is estimated to be understated (unofficial estimates give 9 per cent).† The 1972 census figure is accepted as approximating to an average growth rate of 2.9 per cent annually since 1961. Thereafter the estimate for 1978 uses the 3.2 per cent annual increase rate suggested by demographers since the 1972 census.

TABLE 2.1
Population of Pakistan, 1901–78

	Million	*Average intercensal increase per cent*
1901	16.6	
1911	19.4	0.7
1921	21.1	0.9
1931	23.5	1.1
1941	28.3	1.8
1951	33.8	1.7
1961	42.9[a]	2.4
1972	64.98[c]	2.9[b]
1978	78.5[d]	
1979	81.0[d]	

[a]46.1 if adjusted for under-enumeration.
[b]from the adjusted total for 1961.
[c]Census taken at end of year.
[d]Estimates for end of year at 3.2 per cent increase.

Forward projections of population over several decades are fraught with uncertainty as to whether birth and death rates will remain constant. Assuming these constant, Burki suggests a population of 170 million by the turn of the century, and would arrive at a total of 77.0 million in 1978–79. (Figures used in the Monthly Statistical Bulletin use 3 per cent increase in 1978–79.) He further projects that if the birth rate can be reduced by 1995 to the level required for replacement while death rates slightly fall, it would be 2075 before a state of zero population growth could be achieved. In terms of the theory of the demographic transition, Pakistan is still in the situation where crude birth rates remain high and fairly constant at around 50 per thousand (with a low estimate of 45). It is probable that little or no impact has yet been made by birth control measures. The death rate at around 20 per thousand (low estimate 15) can be expected to fall through the improvement in general health and nutrition. Thus the rate of natural increase could still expand, until the death rate reaches a minimum level, after which a reduction in the birth rate will be needed if the rate of increase is to be lowered.

FERTILITY

While the crude death rate responds to secular changes in the society brought about without the individual taking decisions on his own account, the birth rate is not directly susceptible to such changes. High fertility is endemic in a society in which marriage is practically universal and takes place at an early stage of the reproductive age. (The 1961 Family Law set the minimum age of marriage at 16 for girls and 18 for boys, but even this is frequently flouted.) Infant mortality is high, running at one child in eight dead by the age of one, and one in six by 14. The figures relate to a survey in 1963–65 and could well have improved somewhat. The prevailing attitude that children, particularly sons, are needed to support parents in old age, is unlikely to alter rapidly and in any case not until alternative guarantees of welfare have been introduced, a distant hope at present.

It is difficult to avoid the conclusion that government is lukewarm at best towards its avowed

*Krotki, Karol J. and Parveen, Khalida, 'Population Size and Growth in Pakistan Based on Early Reports of 1972 Census', *Pakistan Development Review*, XV, 1976, pp. 290–318.

†Burki, Shahed Javed, 'Rapid Population Growth and Urbanization: the Case of Pakistan', *Pakistan Economic and Social Review*, 11, (3) 1973, pp. 239–276.

policy to promote family planning. Religious extremists are against birth control, and the more backward provinces may see central government policy as 'neo-colonialist', aimed at preventing them gaining strength through numbers and the ballot box.* The responsibility for the promotion of family planning has passed to the provinces, who while they may be closer to the people are less able than federal authorities to convey the need at the national level. As in other fields of modernization endeavour, reduction of resources and dilution of the few experienced personnel can soon lead to apathy and stagnation. The Annual Plan for 1976–77 showed the birth control achievement of the previous year to be as follows:

Conventional contraceptives	148.2 million
Pills (cycles)	5.2 million
I.U.D. (insertions)	.255 million
Sterilizations	13,835

At the accepted rate needed to avoid one birth, these contributed to a saving of 698,000 births, which out of a possible number of some 3,685,000 births (at the CB Rate of 50 per 1,000) would indicate a reduction of the CBR to 40.5. Only the next census can say whether these projections are reasonable. If they are, and if the family planning movement can be sustained, a reduction in the rate of natural increase from 30 to 20 per 1000 may be a realistic hope,. but the meaningful decisions will be taken in private by a still largely illiterate population among whom women have low status, and experience elsewhere in the underdeveloped world warns against excessive optimism.

MIGRATION

Births and deaths are not the only factors leading to population change. Migration has also played an important role in Pakistan's demographic history. We have no way of telling what was the scale or character of the migrant streams entering the region from the West during historic time up to the first comprehensive census in 1881. With the rapid development by the British of perennial irrigation in the Punjab at the end of the nineteenth century, however, the demographic record becomes available. The 1901 census records at least 443,500 immigrants, 56 per cent of a total of 792,000 in the Chenab Colony, many of them coming from districts now in the Indian state of Punjab and among these numerous Sikhs from Amritsar, Jullunder and Gurdaspur. These streams from the eastern Punjab had been flowing from mid-century when following the disbandment by the British of Ranjit Singh's Sikh army, homes and livelihoods for many were found in the developing canal colonies of the Punjab.

Sadly the flow was reversed in the agony of the partition of the Punjab between Pakistan and India in 1947 when West Pakistan lost 5.5 million Hindus and Sikhs, to gain 7.5 million Muslim refugees from India. In recent years the migrant flow has been outward. For a while entry into the UK for permanent residence was reasonably easy, and many thousands settled in the industrial cities of Yorkshire and the West Midlands particularly. Of late the movement has been more temporary and less distant, to the oil-rich states of the Persian Gulf, where semi-skilled labour has been scarce.

In the population of 1961, fourteen years after partition, 5.87 million, almost 15 per cent of the total, had been born outside what was then West Pakistan. The distribution of this group is shown in Fig. 2.1. While many were absorbed ultimately into the agricultural areas vacated by their India-bound counterparts, the bulk had to be accommodated in the cities, to which as is the habit of migrants, they brought the spirit of enterprise and innovation induced by their transplantation. The districts close to the border of the Indian Punjab stand out, with more than 55 per cent of the total migrants being in the ten easternmost districts from Sialkot through Gujranwala, Sheikhupura, Lahore, Faisalabad, Sahiwal, Multan, Bahawalpur, Bahawalnagar and Rahimyar Khan. Important outliers were metropolitan Karachi with 918,000 (16 per cent of the total) and the newly opened canal colony in the northern Thal, notably Mianwali District which took 657,000 (11 per cent). Refugees are more readily assimilated, housed, fed and found work in cities than in rural areas. Burki (*op. cit.*) estimates that in the 19 major cities in 1951, the census year nearest the event, over

*Bean, Lee L. and Bhatti, A. D. Pakistan's Population in the 1970s: in Korson, J. Henry (Ed.), *op. cit.* Problems and Prospects pp. 99–118.

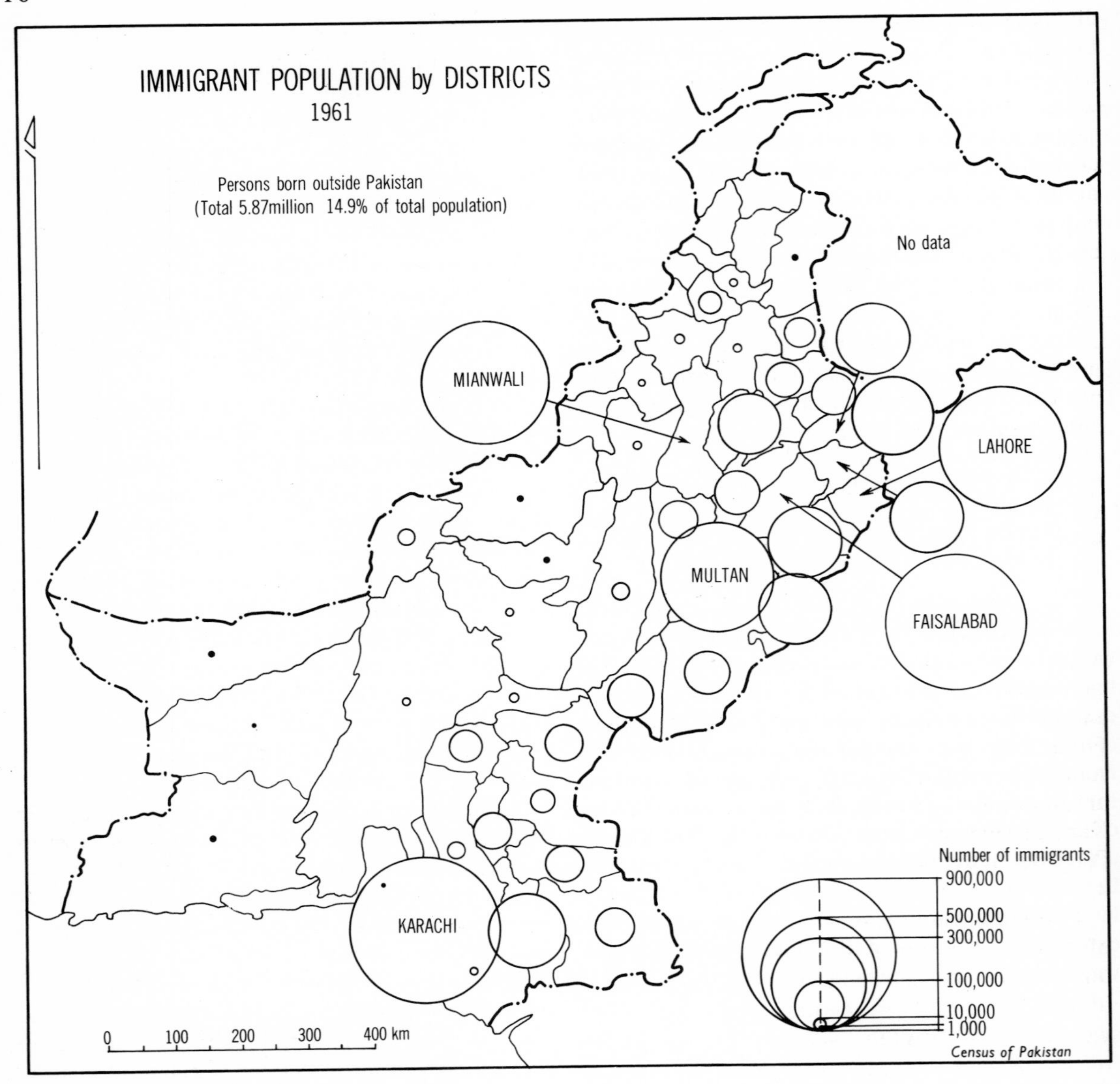

FIG. 2.1

46 per cent of the population were refugees. Faisalabad (then Lyallpur) had 70.4 per cent, Sargodha 67.9, Hyderabad 64.5, Karachi 57.1, Sukkur 54.5 and Gujranwala 51.2 per cent. He comments that such cities became quite suddenly divorced from the social milieu of their surrounding tributary area. The heavy influx of adult non-Sindhis into Sind, and particularly Karachi, threw the traditional political system out of balance.

Migration among the home-born Pakistanis has been on a smaller scale. Nearly 85 per cent of the native-born originated in the districts where the census enumerators found them in 1961. Only 1.9 million were born in non-contiguous districts of Pakistan, and 919,000 in contiguous districts. Fig. 2.2 summarises the situation by plotting the percentage of population born in the district where enumerated. In most western districts more than 90 per cent remained in their home district. Those with less than 80 per cent are the major recipients of refugees. Karachi, for the first few years the national capital, the largest city and the most diverse industrially held only 42 per cent born in the district. Of the remainder, 74 per cent were from outside Pakistan and 26 per cent from 'elsewhere' (beyond the contiguous districts which supplied a negligible proportion). Its polyglot character confirms the cosmopolitan background of its population: 62 per cent spoke Urdu (the mother tongue of most Indian refugees;) 15 per cent Punjabi, the nation's majority group; only 10 per cent spoke Sindhi, the language of the provincial majority.

Balochi, the language of the closely adjacent province, claimed 6 per cent, and Pushto, spoken by the most inveterate migrant, the Pathans, another 6 per cent. Karachi's low sex ratio, 808 females per 1,000 males in 1972, is a further reflection of its strongly migratory population. In the age groups 35–60 the ratio falls to less than 700, with 565 at the 45–49 age level (see page 21).

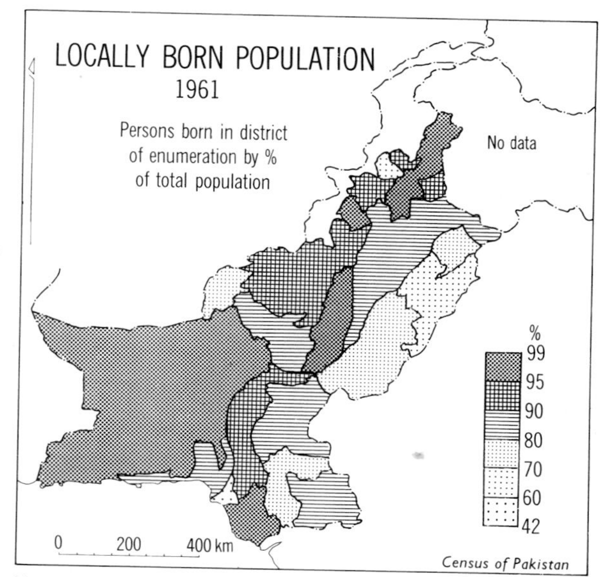

FIG. 2.2

AGE AND SEX STRUCTURE

Age – sex 'pyramids' are a convenient means to summarise the population structure of a country and of its constituent elements, rural and urban. Figures illustrate the age-sex structures for Pakistan as a whole (Fig. 2.3), and for its rural and urban populations (Fig. 2.4), for its largest cities, Karachi (Fig. 2.5) and Lahore, (Fig. 2.6) and for Peshawar District, urban (Fig. 2.7) and rural (Fig. 2.8).

Two factors combine to make accurate analysis of age – sex pyramids for Pakistan a hazardous undertaking. In any country with a high level of illiteracy, particularly in rural areas and among women, it is difficult to ensure an accurate answer to the questions posed by the census enumerator, 'How old are you, and how old are the members of your household?' Understandably these questions are more readily answered in respect of the younger age groups but among the elderly, unless a great deal of cross questioning takes place to arrive at age by reference to memories of outstanding events, it may only be possible to approximate the answer. It is for this reason that in the age – sex pyramids and tables, the five-year age groups recorded in the census of 1972 have been 'smoothed out' above the 50 year level, to avoid the irregular pattern that would otherwise occur, with the

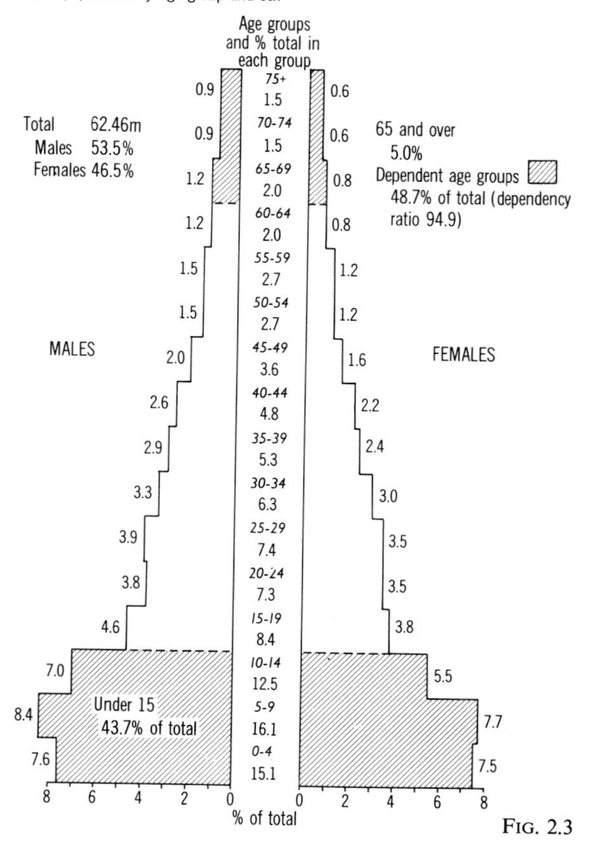

FIG. 2.3

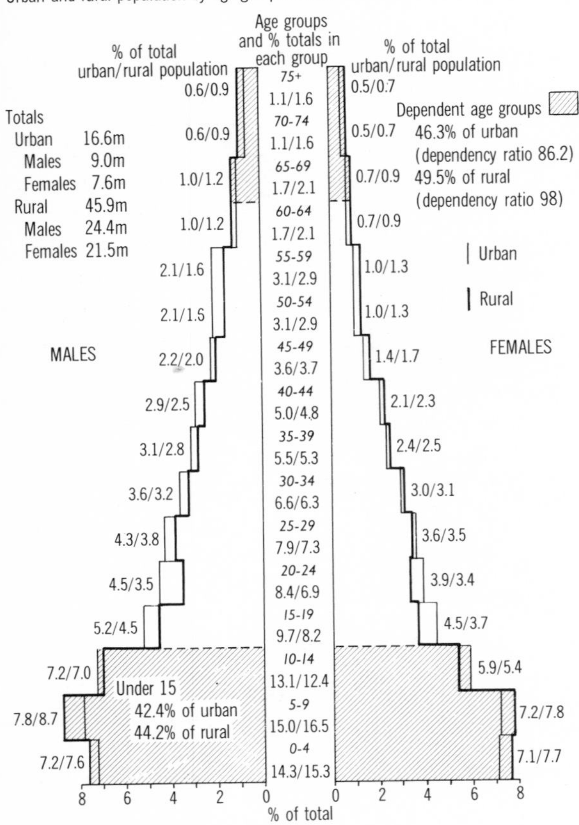

FIG. 2.4

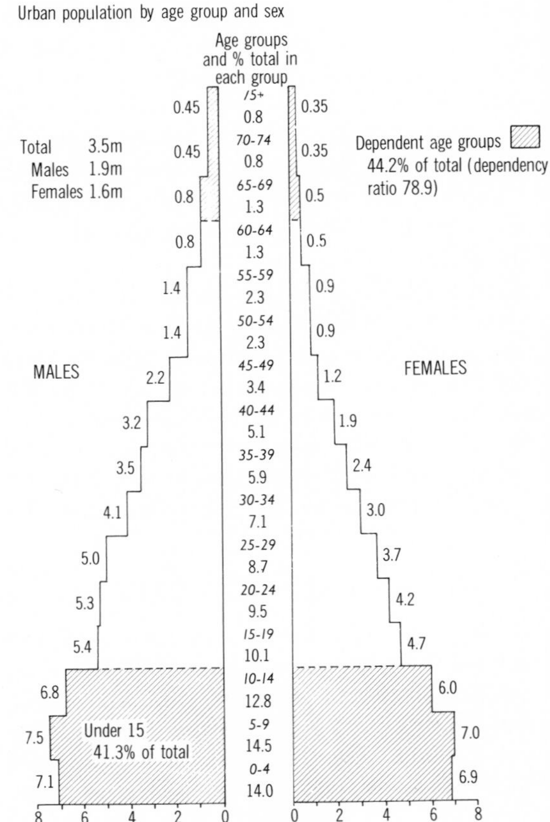

FIG. 2.5

50–54 and 60–64, groups gaining those who return their age as 'about 50' or 'about 60', leaving 55–59 and 65–69 relatively depleted. This modification does not remove this problem entirely. The pyramid for rural Peshawar (Fig. 2.8) indicates that the problem exists there even in the 35–39 age group as compared with that for 30–34. Here the explanation almost certainly lies in the character of local traditional mores, in which the household under a patriarchal head, is a more or less closed group as far as the outsider is concerned.

The tendency to secretiveness about family matters reaches its extreme where the number and age of women in the household are concerned. No constant factor can, however, be applied to correct for the under-enumeration of women. It is likely to be worse in rural than in urban areas, and in the more 'backward' regions generally or in those, like Peshawar, more strongly traditional in social matters. 'Purdah', the practice of veiling women in public and generally of keeping them out of sight, operates to a large extent everywhere, though it is less strictly observed in the cities among the educated classes, and by some in parts of the rural Punjab. On visiting any Pakistani home care is taken to enable those members of the family in purdah to retreat into the inner-rooms unless the visitor is female or is a close male relative.

Again, because they are less likely to be visible to the census enumerator, the infants in the age group 0–4 are under-reported, while those 5–9 may be over-reported. This helps to explain the otherwise biological improbability of the 5–9 age group quite markedly exceeding its immediate predecessor. A sudden adoption of family planning has not occurred!

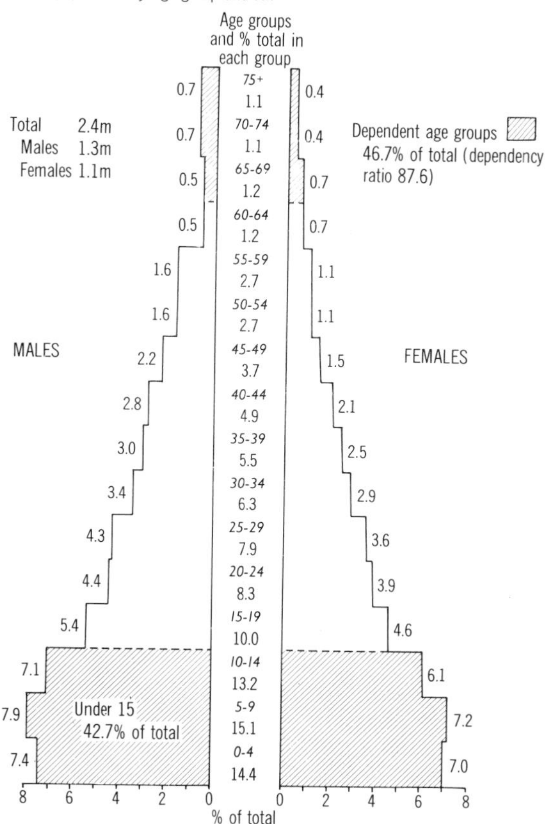

FIG. 2.6

DEPENDENCY

The age – sex structure, apart from these Pakistani peculiarities, has the shape characteristic of countries in the underdeveloped world: a very broad base, due to high fertility, a rapid tapering on account of a high mortality rate among the young, and an early narrowing apex reflecting the relatively low life expectancy. In each diagram, the percentage of population in each age group, male, female and total, is given, and the percentage in the dependent age groups (0–14 plus 65 and over). In the country as a whole 48.7 per cent were dependent by this definition in 1972 giving a dependency ratio of 94.9*. For the rural population the figure is 49.5 per cent (ratio 98) indicating that the very young and the aged tend to remain in or return to their villages, while the urban population's share of the economically active age group increases relatively. A slightly more marked difference between urban and rural in this context is seen in the pyramids for Peshawar District, where dependency in the rural villages reaches 50.1 per cent in a population of 1.25 million (a ratio of 100.4 dependents for every 100 in the active age groups), compared with 44.5 per cent (ratio 80.2) in the city with a total of 272,700. Karachi (Fig. 2.5) has the lowest dependency ratio among these examples, 78.9, and relatively fewer in the over 65 age groups, as would be expected of a city that has grown rapidly by in-migration over a comparatively recent time.

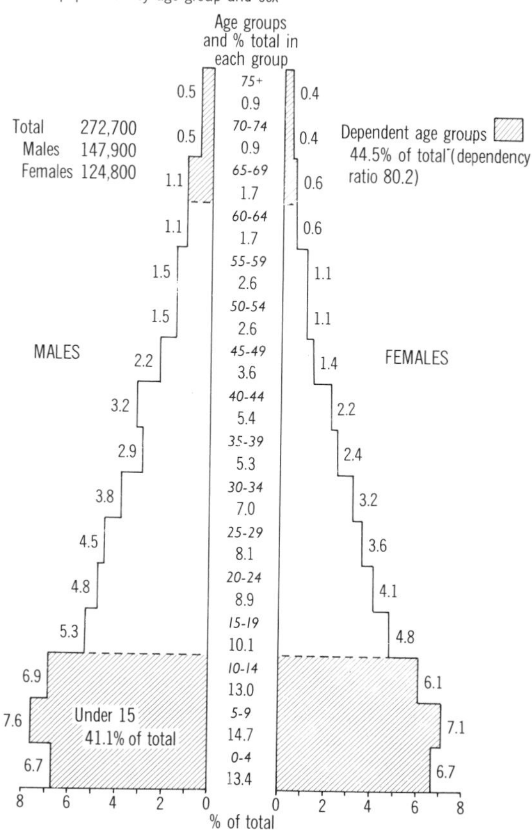

FIG. 2.7

*Dependency ratio: $\frac{\text{population } 0\text{–}14 + \gg 65}{\text{population } 15\text{–}64} \times 100$

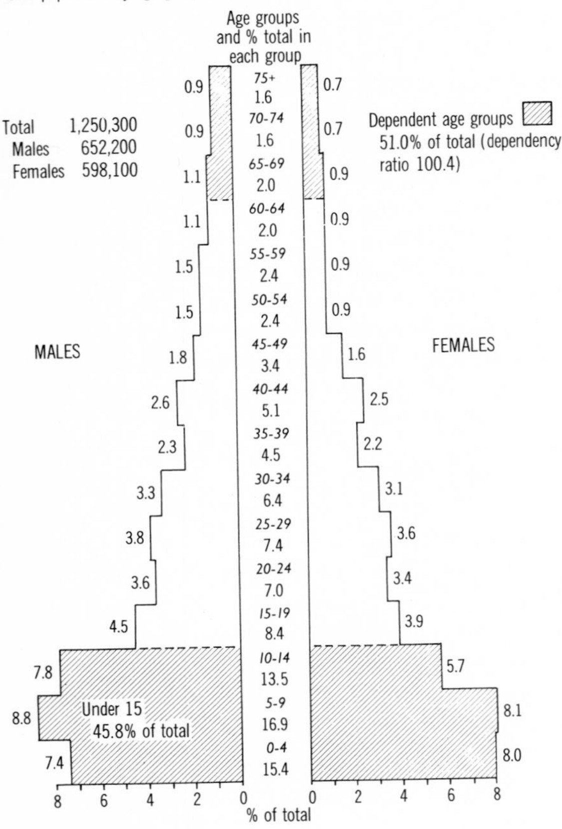

FIG. 2.8

The concept of a dependency ratio as expressed in the manner used above is grossly simplified. In Pakistan's society there is no universal compulsory education, and as will be shown below, a minority of the age groups 5–14 which in a developed country would be at school, are being educated. But they are far from idle, as a rule. It is customary for children to help their parents or near relatives in their daily occupations as soon as they can be useful. Boys work around vehicle workshops, tea shops, markets and so forth, and girls help their mothers at home, minding the babies, spinning cotton, knitting, and weaving, both for the family and for sale. In the villages children spend much time helping collect fodder and fuel. So a strict comparison with the dependency ratios of developed countries can be misleading. The young pull their weight in the socio-economic system. The economically inactive are found among the womenfolk of the affluent and the slightly affluent, whose men can afford the luxury of employing others to do what the average village housewife has to do. As education absorbs more of the young and as the expectation of life extends, the real dependency ratio is likely to increase rather than otherwise, but it is to be hoped that it will do so in an age distribution less heavily weighted than at present in the 0–14 range.

Closer study of the proportions of the population in the (conventional) active age groups indicates the tendency for urban migration to affect the active age groups and to be stronger among the males than the females. This is seen in the direct comparisons made easy in the case of the Pakistan urban/rural pyramid (Fig. 2.4) in which the steps of urban population extend beyond those of the rural even slightly at the age group 10–14, and thereafter more substantially, up to group 55–59. It is impossible to correct the female side of these pyramids for the general under-enumeration of women, particularly in the nubile age groups, but there does seem to be a diminution of the excess of urban as compared with rural in the case of women the 20–24 age group. Perhaps a factor here is the return of marrying girls to their home villages where even urban dwelling parents may prefer to find husbands for them.

Confining attention mainly to the males, the differences between Karachi, Lahore and Peshawar say something about the character of those cities. Lahore is the most mature of the three, demographically speaking, as evidenced by the high dependency among both young and old, suggesting that people are tending to be born, to work and to retire as city dwellers. The cases of Karachi and Peshawar have an apparent similarity at these age levels. Both cities receive a higher proportion than does Lahore of migrants in the active age range. In the case of the older city, Peshawar, many return to their rural villages to retire while Karachi has had too short a time for its workforce to reach retirement in substantial numbers. Since many of these were migrants from outside Pakistan, they have no ancestral home to retire to and perforce will remain in Karachi to help thicken the apex of its age pyramid.

SEX RATIOS

Table 2.2 below shows the sex ratios for these three cities, for Pakistan (total urban and rural), for Peshawar District (rural) and for Sanghar district in Sind which has no large urban centre within or close to it.

For each age-group with a sex ratio of fewer than 750 females per thousand males the figure has been outlined in the table to bring out the pattern visibly. Certain general observations are possible. There is a deficiency of females, no doubt due to under-enumeration, in the age groups 10–14 and 15–19 in every instance. In most cases there is an increasing deficiency towards old age; this is clearly seen in the overall figures for Pakistan, and for the rural population which in any case dominates with almost 75 per cent of the total. Under-enumeration cannot be discounted, and indeed it is to be suspected when Karachi with its high proportion of migrants and presumably a higher level of feminine emancipation than elsewhere shows a pronounced upturn in the ratio in the groups from 50 onwards. Maybe migrant women are tougher for the experience, and so survive better. Certainly the lot of village women is hard in the extreme, and their status the lowest.

Deficiency of females among the economically most active age groups in urban areas is most marked in Karachi where it touches less than 750 from age 25 onwards till 69. In Lahore and Peshawar Urban, and in the urban population of Pakistan as a whole this level is reached later, at 40, and persists to the end in Lahore and in Pakistan Urban.

OCCUPATIONS OF THE POPULATION

Table 2.3 overleaf (and Fig. 2.9) show the estimated breakdown of the population into major groups for 1976–77. Largely due to the high level of dependency (even if in some degree illusory – but working children helping parents are not counted separately in the workforce) the participation rate is less than 30 per cent of the population. Agriculture occupies the majority, over 54 per cent, with less than a quarter of that figure in manufacturing industry. Almost 27 per cent are in the tertiary sector, in trade, commerce, transport and service activities.

Women are a significant element in agriculture, making up over 10 per cent of the workforce, with many more working as an extension of their domestic duties on subsistence holdings, or as casual

TABLE 2.2
Sex ratios by age group, urban and rural, 1972
(Females per 1000 males)

Age group	*Pakistan total*	*Pakistan*		*Karachi urban*	*Lahore urban*	*Peshawar*		*Sanghar*
		Rural	*Urban*			*Urban*	*Rural*	
0 – 4	994	1004	959	968	947	994	1082	1083
5 – 9	905	900	921	936	911	942	919	885
10 – 14	787	768	840	887	859	873	801	726
15 – 19	833	821	859	873	839	903	858	791
20 – 24	941	974	867	788	880	854	966	948
25 – 29	896	927	822	740	838	798	939	880
30 – 34	926	965	828	735	839	826	963	825
35 – 39	860	892	779	692	821	813	965	841
40 – 44	861	917	725	609	746	693	948	841
45 – 49	814	866	678	565	703	622	889	825
50 – 54, 55 – 59 *	783	817	688	633	679	682	865	732
60 – 64, 65 – 69 *	723	731	695	717	651	573	803	708
70 – 74, 75 + *	743	742	744	820	665	759	757	787
Total	870	882	838	808	839	844	917	866

**Note*: These pairs of age groups have been averaged to avoid skew.

TABLE 2.3
Major occupation groups in the workforce, 1976–77

Estimated population	73,425,000
Participation rate	29.62 per cent
Labour force	21,700,000
	%
Agriculture, forestry, fishing	54.3
Mining	0.14
Manufacturing	13.25
Electricity and gas	0.51
Construction	4.97
Trade, commerce	11.23
Transport, communications	5.02
Banking, insurance	0.66
Community, social, personal services	9.62
Other	0.32

Source: Annual Plan, 1976–77.

labour assisting their husbands. In the non-agricultural sector they formed 6 per cent of the total in 1961 mostly in service occupations. To the extent that they are employed in manufacturing industry it is mainly in the small-scale cottage industry sector where they can work within a family group. It could be said that the womenfolk of the agricultural villages have a greater opportunity for social and economic participation, if within a more limited environment, than do their urban sisters. In Pakistan society urbanisation may mean greater freedom of choice for men in what they do and how they live. For women a move to the city can mean closer confinement than in their home villages.

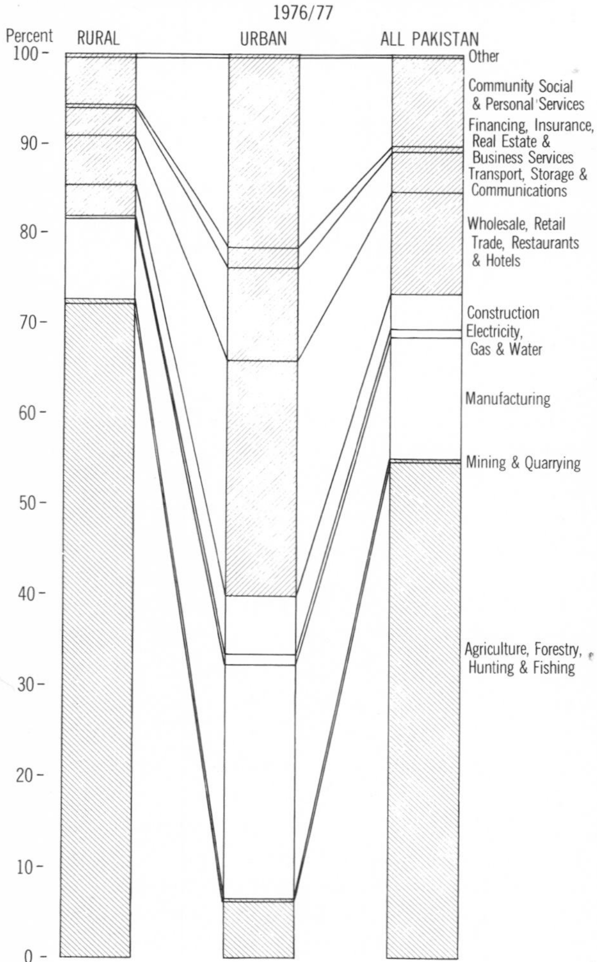

FIG. 2.9

5 These girls in a refugee colony in Karachi are unlikely to appear in the occupational statistics. They are making paper bags for groceries out of old cement bags and glue.

WOMEN'S ROLE IN THE RURAL COMMUNITY

Since it is difficult for the 'westerner' to envisage what life is like for the average rural Pakistani, particularly for the women, it may be helpful to interpolate here some points derived from a recent study of the activities of rural women.* The wife of the average small farmer is up before dawn to answer the call of nature under the cover of darkness in the fields near the village. After prayers she churns the milk boiled the day before, and milks the buffaloes which practically share the house with the family. Water for washing up and watering the cattle has to be brought from the tube well in the village or hand pumped from a shallow well nearer the home. The fresh milk has to be boiled, and then early breakfast is taken, consisting of the previous day's left overs – chapatti roti (unleavened bread the shape of a large pancake), butter, gur (unrefined sugar) and lassi (yoghurt). Rarely will tea be used.

The first task after breakfast is to knead some flour into a dough (to be left for baking later), after which she cleans the house, puts away the bedding and the beds (charpoys), in summer time off the roof where they sleep for coolness.

Bread is baked in the home in winter, but in summer she takes the dough to a woman with a tandoor oven who bakes for her at the cost of a tenth of the flour. She brings the roti back home for the main breakfast – lunch between 8.30–9.30 am, consisting of roti, onion, chillies, gur and lassi. Rarely is meat or fish taken, once or twice a year by the poor, once or twice a week by the rich. If the men are working in the fields she or one of the children takes their meal to them.

During the morning, helped by the children, she collects cow dung and makes dung cakes which are plastered onto the wall of the compound to dry. From each buffalo 12 dung cakes are collected daily and any surplus to the household fuel needs are sold. A stock is stored in a conical heap plastered with mud to keep it dry.

*Khan S.A. and Bilquees F., 'The Environment, Attitudes and Activities of Rural Women: a case Study of a village in the Punjab', *Pakistan Development Review*, XV, 1976, pp. 237–271.

6 Among the traditional service occupations, the water carrier performs an important role. Here he is seen with his goat-skin bag slung over his shoulder having just filled it from an ornamental pond in the Shalimar Gardens, Lahore. Stone-masons are at work in the background.

The animals spend most of the time in the compound, where they have to be watered and fed with grass from a stack which she keeps replenished by cutting and carrying from the fields until about 1.30 pm. Early afternoon she spends washing clothes at the tube well using soap and a stick to beat them clean. The morning visit to the tandoor and this afternoon trip to the tube well are her main opportunities to gossip with her neighbours. If there is work to be done in the fields for payment in cash or in kind (like sugar cane trimming for the sake of the green leaf fodder) this task has to be fitted in with her regular duties.

An afternoon meal of roti and lassi may be followed by a brief period of 'leisure', spent mending clothes, embroidering or crocheting. More fodder has then to be collected and chopped on a chaff cutting wheel, after which she fetches more water, grinds pulses and the wheat (or takes them to be ground at the oxen-turned mill at a fee of one-fortieth the quantity), and takes the animals to the tube well to wash and home again to feed them. The children may share in these jobs.

The evening meal is prepared at about 4 pm. She makes rotis and a vegetable curry or dhal (from pulses). Milking, milk boiling and curd-making come round again, and after tying up the stock for the night she gets ready the men's huqqa (hubble-bubble, communal smoking pipe) and makes the beds. After washing up she is in bed by 8.30 pm. having spent about 15 hours at work. For women there is little rest!

Such is the daily round. The march of the seasons brings some variety and occasional excitement into her life. Chief among its highlights is the big show held every year in March, in the lull in field activities immediately prior to the wheat harvest. This is the season of entertaining, buying or making new clothes, sweets and toys for the children, and for the housewife repairing the house by plastering with mud already brought in from the fields. She uses some of the mud to make a parola in which the grain will be stored and to repair the mangers where the cattle eat. Chicken cages have to be made or repaired too.

7 Women making cow-dung balls for fuel, mixing in straw as binding material. They find a ready market in Lahore nearby.

During the wheat harvest everybody works hard, and she most of all, preparing and carrying meals to the fields and helping to winnow the grain, first that of the landlord's share and then their own. By early June the harvest is over, with enough in store for 9 to 10 months. Some grain is traded at the village shop for necessities like kerosene or thread.

8 Flour mills powered by water are a common sight in the foothills, as here near Abbotabad, NWFP.

9 Inside a flour mill. The grind-stone is at the top end of a vertical shaft which has propellor-like wooden blades at the other end, driven by water. Maize is being fed in at the top, and the flour is scraped away below.

At the height of the hot weather in June – July little work can be done beyond essentials, but time is found to make mango achar (chutney) and to help sow rice in readiness for the rains when it has to be transplanted in another busy period in July – August. The rain promotes growth of every kind, and she has to build up stocks of fodder for the lean months of the dry winter. Indoor jobs occupy spare moments in the rains: the women are expert basket weavers and plaiters of straw, and pride themselves in their skill making jullians, (padded quilts made from rags).

As the rainy season slackens and the sun comes out for more reliable spells, she has to take the wheat out of store to dry it and put it back into store with salt to reduce damp and fungus. September is another slack time in the fields awaiting the next round of harvesting of rice and then maize. Meanwhile house repairs are undertaken where the rains have damaged the mud plaster, and all spare hands can be used scaring birds off the ripening crops.

October–November is again busy with harvesting, threshing and storing grain. The women have to dry the maize cobs and then beat out the grain. Cotton picking for cash wages is an opportunity not to be missed in November–December. The housewife will obtain kapas (new cotton) as wages in kind perhaps, which she gins herself to remove the seed before spinning into yarn or using to fill quilted jullians. She may be involved in picking chillies in December. When the cold weather sets in, the grass grows slowly or not at all, giving her some respite from its daily cutting and carrying. Then perhaps she can sit in the calm soft sunshine to spin, to make chillie achar and to pop corn. February brings in the round of visiting again, for which clothes must be made or refurbished and gifts selected.

This record is somewhat idealized, busy as it is nonetheless. To all these activities the housewife has to add the bearing and rearing of children in regular uncontrolled succession, and the suffering and survival of all the associated health problems. Many of the aids to welfare that are taken for granted in the developed world are quite remote from village experience, and one suspects that the basic morning prayer must subconsciously at least be for strength and good health to carry on.

10 Punjabi women buying semi-processed ginned cotton for spinning into thread at home. They will sell the skeins of thread back to the 'factor' who dyes them ready for weaving. The women are dressed in winter clothing: trousers, shirt, woollen cardigan and cotton shawl.

11 A Punjabi girl winding cotton thread onto bobbins, checking the twist of the thread at the same time.

NUTRITION

To appreciate the importance of the various crops grown by the Pakistani farmer for his own family's consumption and for sale in the market, it is necessary to know something about the people's diet. Table 2.4 below sets out the annual per capita availability of the principal items of food consumed in Pakistan. Some data on per capita consumption in affluent overfed Australia are given for comparison. The salient features of the average person's diet have been touched on in the section immediately above. Wheat flour, coarse ground and wholemeal, is made into roti as the staple item of a meal. Rice may be taken at one meal instead or in addition to roti. For second-class protein, apart from wheat, the major source is pulses. Animal protein is scarce and expensive; meat is a rare treat, though eggs may be commonly used. Fats and oils are from milk and vegetable sources. Sugar is, as often as not, taken as gur, a very palatable form of crudely crystallised sugar resembling hard fudge. Tea (and coffee even more so) tend to be urban drinks and are rarely taken by the rural poor.

There has been much debate in recent years about the nutritional needs and deficiencies of Third World populations, and sweeping claims have been made that one-half, or two-thirds or whatever large proportion of the world goes hungry to bed each night. It seems that the dieticians who calculate the tables of 'minimum needs' set those well above the levels required for spartan but continued survival. While one can accept that many people could and would eat more if they could afford it, and that deficiency diseases may be widespread, the fact is that starvation due to under-nutrition is extremely rare. *Homo sapiens* one suspects is a far tougher animal for survival than the western nutritionists would have us believe.

This is not to deny that better nutritional levels, and in particular, better balanced diets should be the aim in every country, and especially for everybody in that country. A major problem hidden in Table 2.4 of average available food, is how that available food is distributed through the population.

The basic food needs of an individual are usually expressed in calories per day, with an indication of the weight of protein required as shown in Table 2.5.*

TABLE 2.4
Per capita available food, Pakistan 1975–76 and average consumption, Australia 1974–75 (in kg per year)

	Pakistan (kg)		*Australia (kg)*	
Wheat	113.8	138.3	87.0	139.5
Rice	24.5			
Potatoes	?		52.5	
Pulses	7.7		?	
Eggs (no.)	–		219	
Cheese	–		5.2	
Sugar, refined gur	28.9		49.1	
Ghee/butter veg.oil/margarine	5.4		18.6	
Meat	7.3		108.8	
Milk (litres)	60.4		113.0	
Tea	0.8		2.0	
Coffee	–		1.2	
Vegetables	30.6		66.3	
Fruit	?		84.3	

Source: Annual Plan, 1975–76, and Pocket Compendium of Australian Statistics, 1977.

TABLE 2.5
Calorie and protein needs

	Calorie need per day	*Protein need (grammes)*
Child 0–9 years	1,744	29.82
Male 10 years and above	2,512	52.06
Female 10 years and above	2,011	45.08

Table 2.6 shows that by these standards the average urban and rural person is undernourished, the poor particularly so. The data, from Alauddin's study, take an average urban income of Rs 300 per month and rural Rs 250 per month as indi-

*Abdul Wasay, 'An Urban Poverty Line Estimate', *Pakistan Development Review*, XVI, 1977, pp. 49–57. An earlier study has also been used here: Talat Alauddin, 'Mass Poverty in Pakistan, a Further Study', *Pakistan Development Review*, XIV, 1975, pp. 431–450. Table 2.7 is derived from this paper which had it from the *Household Income and Expenditure Survey* of 1971–72.

cating poverty.* Although the urban dweller gets less cereal in his diet than the countryman, his average calorie intake is lower both on average and among the poor.

TABLE 2.6
Calorie intake by urban and rural population, average and poor

Year	*Rural*			*Urban*		
	Average	*Poor*	*% in cereals*	*Average*	*Poor*	*% in cereals*
1963–64	1,988	1,897	90	1,731	1,595	86
1968–69	1,974	1,857	87	1,713	1,664	83
1969–70	1,983	1,815	85	1,707	1,691	83
1970–71	1,950	1,810	88	1,734	1,681	85
1971–72	1,898	1,736	86	1,702	1,641	84

In 1971–72 a study was made of daily calorie intake in a sample population, rural and urban, covering a full range of income levels. The results are in Table 2.7.

It is clear that only the wealthy reach the levels claimed as necessary, and there is a wide discrepancy between rich and poor. The rural population consistently gains more of its calories from cereals than does the urban, who can choose from a wider selection of foods in the markets and shops. The farming population eats mainly what it produces. For both groups a major deficiency is animal protein which optimally should constitute half the protein intake. On average Pakistanis probably get at best 15 per cent as animal protein.

TABLE 2.7
Daily per capita calorie intake by income group

Monthly income (Rs)	*Rural*		*Urban*	
	Calories	*% cereal*	*Calories*	*% cereal*
All groups	1,898	83.3	1,702	82.6
50	1,642	86.0	1,585	87.0
100	1,736	86.0	1,676	85.0
150	1,812	85.0	1,614	84.0
200	1,851	83.0	1,672	82.0
250	1,892	83.0	1,730	82.0
300	1,985	83.0	1,658	80.0
400	2,075	81.0	1,705	80.0
500	2,334	81.0	1,691	77.0
750	1,912	79.0	1,835	75.0
1,000	2,384	86.0	1,939	73.0
1,500	5,257	82.0	1,992	68.0
2,000 +	3,871	75.0	2,179	63.0

THE HEALTH OF THE PEOPLE

Man's survival, personal development and well-being are not just a matter of adequate and properly balanced nutrition, important as these are. Physical well-being requires preventive and curative medical services being available in the community. Like any other service medicine is subject to economic constraints on its spatial distribution within a society. It is not due to some perverse accident that doctors are concentrated in urban areas; this phenomenon is common in the developed world also. Fig. 2.10 shows how the total number of people (potential patients, if you like) are related to the number of qualified medical practitioners and dentists, in each of 45 districts.

* Income levels are difficult to conceptualise in the abstract. In 1978 a chowkidar (night watchman, caretaker) received Rs 375 per month (about US $37) as did a sweeper (cleaner) on the Punjab university staff. An advertisement for a Research Assistant, with a Masters degree in social science and a diploma in demography offered Rs 625–1325 (US$62–132); the applicants were required to have a year's research experience and to be 30–35 years old. Jobs in the Airport security force were:

Inspector (Graduate)	Rs 430–830 per month
Sub-Inspector (Intermediate)	Rs 335–575
Assistant Sub-Inspector Matriculation	Rs 290–470
Security Guard I (Class 8)	Rs 270–390
Security Guard II (Class 5)	Rs 260–365

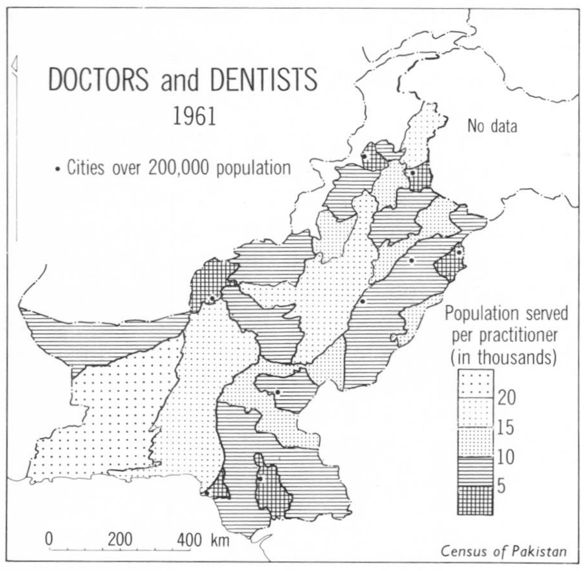

FIG. 2.10

While the major urban centres show one practitioner for less than 2500 people, much less in Quetta (1090 people), Lahore (1367), Multan (1388) and Karachi (1439) for example, the predominantly rural areas are poorly served. In the more closely settled districts a doctor may serve 5–10,000, while in the sparsely populated desert and foothill districts there are upwards of 15,000 and 20,000 for every practitioner. This does not spell an overworked medical profession, but the harsh fact that the bulk of the population never see a doctor from birth to their life's end, medicine being an aspect of modern life quite beyond their purse, and in any case often so remote as to be inaccessible when needed by the seriously afflicted.

Among the more promising developmental reforms initiated in 1975 was the People's Health Scheme which aimed to bring basic health services to the masses, by training para-medical 'Health Guards' for immunisation and preventive medical duties, and by extending midwife, child care, and family planning services. For each village of about 1000 population there would be one male and one female community health worker. To serve several villages, with a total population of about 10,000, there would be a Basic Health Unit, with 4–6 para-medical Health Guards. At the start of 1975 there were 369 such units in operation. Rural Health Centres (136 established by 1975) to provide professional services for several Basic Health Units, would be staffed by two doctors and 6–8 auxiliary para-medicals and auxiliaries, and have up to 10 beds. Above this level would be a Tehsil hospital serving maybe 300,000 with surgical, x-ray and pathological laboratory facilities. District hospitals would be more elaborate, with teaching hospitals, a Provincial responsibility in the capitals and a few major cities.

Not only are medical practitioners heavily concentrated in urban areas. The same applies to hospitals and to maternity and child clinics, despite the strongly rural nature of the population. In 1975–76, 75 per cent of the 548 hospitals were in urban centres, 80 per cent of the 38,033 beds, and 56 per cent of the 715 clinics. While the need for doctors is obvious, the *Economic Survey 1975–76* commented that of 18,000 registered in Pakistan, only 10,000 were available in the country, the rest having migrated to more lucrative employment

12 Village schoolmaster near Kohat, NWFP. He holds a little cane to encourage concentration among his pupils, and a double-barrelled shotgun and a bandolier of ammunition, to discourage attention from feuding enemies. The boys seated on the ground have wooden writing boards. The water jug is needed for washing ink from the boards, for cleaning the blackboard and for slaking teacher's thirst in summer. This picture was taken in winter.

abroad. Among the latter would be 300 provided by government to Iran under an agreement in 1973!

EDUCATION

Education in a curriculum relevant to the needs of the society and for the enlightenment of its individuals is an essential element in development, but one which, like medicine, is difficult and expensive to provide equally to all its members. While the numbers attending schools, colleges and universities is a measure of current commitment to formal education, the level of literacy in the population indicates at a grosser level the cumulative impact of formal and more casual learning. The 1961 Census provides data by age group and sex of school attendance, urban and rural. The numbers, rural and urban, attending schools or colleges, as a percentage of their respective age group, male and female are shown in Table 2.8. Figs. 2.11 and 2.12 show by district the percentage of the 10–14 year age group, male and female respectively, in school. The percentages are highest for this age group.

For boys, the peak values of 58, 54 and 44 per cent are in Jhelum, Rawalpindi and Gujrat districts, and may be explained by the military traditions of these districts which encourage education. Lahore and Karachi with 42 and 41 per cent contain the country's major urban concentrations. Apart from these, fairly high values occur in the belt Sialkot–Gujranwala to Faisalabad, rich agricultural areas with important towns which attracted refugees at the time of partition. It is recognised that transplantation, more often than not, acts as a stimulus to enterprise, and education

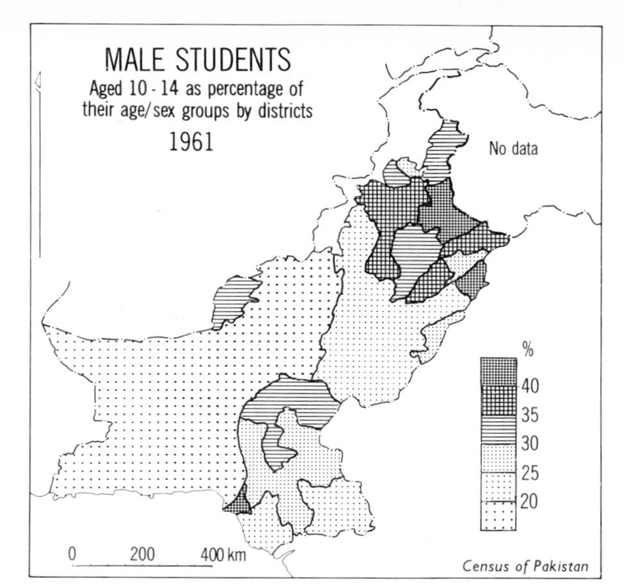

Fig. 2.11

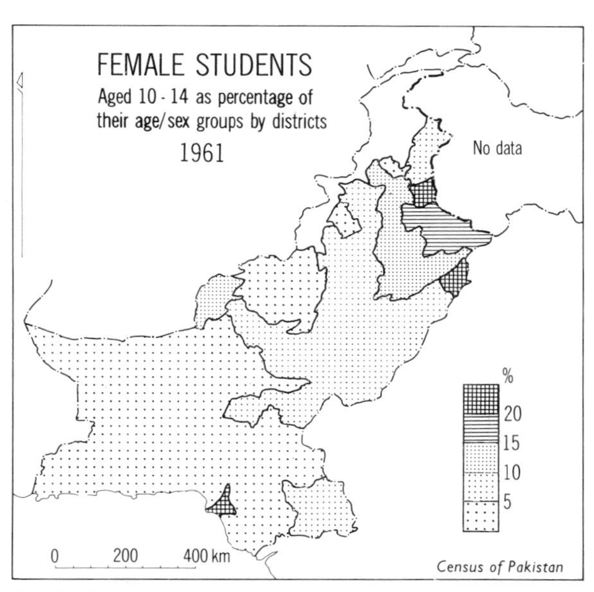

Fig. 2.12

Table 2.8
Number in educational establishments, 1961 (millions)

Age group	*Rural*						*Urban*					
	Total numbers			*% in school*			*Total numbers*			*% in school*		
	Total	*Male*	*Female*	*Total*	*M*	*F*	*T*	*M*	*F*	*T*	*M*	*F*
5–9	5.0	2.7	2.3	12.0	17.0	6.0	1.5	0.8	0.7	26	31.0	22.0
10–14	2.8	1.6	1.3	17.0	25.0	6.0	1.0	0.6	0.5	42	49.0	32.0
15–19	2.6	1.4	1.2	5.0	9.0	0.5	1.0	0.5	0.4	20	26.0	12.0
20–24	2.2	1.1	1.1	0.1	0.3	—	0.9	0.5	0.4	2	2.5	0.8

Source: Census of Pakistan 1961.

for children is seen as an important asset to economic advance. The same applies to Mianwali, where many refugees and other migrants were settled on the newly opened Thal irrigation area. In Kohat and Attock the military tradition may again be responsible.

At the other end of the scale, districts with less than 20 per cent of boys at school (and most of these had 12 per cent or less) cover the whole of backward Baluchistan with the exception of Quetta, whose military cantonment and civil administration for the Province have brought better education in their wake. The 'Dera' districts and Bannu in the Sulaiman Piedmont and a block in the southern Punjab with 20–24 per cent at school stand in some contrast to most of Sind where perhaps modernisation of the canal system following the completion of the Gudu and Ghulam Mohammed barrages and the considerable inflow of refugees have had an influence.

While the picture of education for boys may not be particularly rosy, that for the girls in the same age group is dismal in the extreme, reflecting not only the underdeveloped state of the country's social infrastructure, but also the conservative and masculine attitudes characteristic of Islam. In only three districts, Rawalpindi, Lahore and Karachi does the percentage of girls in school exceed 20. In 31 out of the 45 districts enumerated, the percentage is below 10. These include most of Baluchistan (except Quetta), Sind, NWFP (except Peshawar) and the southern half of Punjab.

The greater probability of urban youth getting to school is seen in Table 2.8 which shows urban and rural school populations as percentages of the respective age-sex groups. Since many of those who eventually go to school start at some point between 5 and 9 years of age, and even more leave soon after they reach the mid-teens, the 10–14 age group is best representative of exposure to formal education. Girls have a better chance of schooling if they live in town, where, if anywhere, more liberal attitudes prevail. In rural areas at most a quarter of the percentage attend school as for boys, while in urban areas girls come closer, to within 60 or 70 per cent of the percentage for boys, in the 5–9 and 10–14 age groups at any rate. From the age of 15, with the onset of puberty, girls become much more sheltered,

13 Scribes outside the General Post Office, Lahore, help illiterate correspondents keep in touch with home, compose petitions and write applications for jobs.

even in towns, and only half as many girls as boys are in educational institutions at 15–19 and 30 per cent at 20–24. In rural areas the falling off of female attendance is even more marked, with a comparable ratio of 6 per cent at 15–19 and a negligible number at 20–24. Understandably there are few educational opportunities at tertiary level for either sex in rural areas.

In 1976 there were 11 university institutions, the most recent being Gomal University at Dera Ismail Khan in NWFP and one at Bahawalpur in the Punjab. Enrollments totalled 27,500 students of whom only 4500 or 16 per cent were women.

LITERACY

Literacy is another aspect of general education with great significance for modernisation, since without the ability to read, the individual is very dependent on the chance diffusion towards him of new ideas, and unless he can write he may be unable to record the necessary details for adopting an innovation. Figs. 2.13 and 2.14 show separately the percentage of the male and female population which is literate. In general the pattern of male literacy follows closely that of male schooling (Fig. 2.11). Data for the major cities is also shown and these march 13 to 27 per cent above their corresponding district in 9 cases out of the 12. In the case of girls' schooling the major cities are generally not quite so far ahead of their districts. Rawalpindi City with 30 per cent of the 10–14 aged girls literate compares with 13 per cent in the district as a whole. In most cases however the difference is between 7 and 12 per cent. For Karachi, data available from the 1972 Census set out in Table 2.9, shows clearly the better chance urban dwellers have of becoming literate, and the relative levels of female literacy for different age groups, which demonstrates a changing attitude in favour of educating girls, at least in that city.

Overall, 14 per cent of the population were recorded as literate, and in terms of the population aged 5 years and above, 14 per cent were literate in Urdu, 3 per cent in English and 2 per cent in Sindhi.

Newspapers and periodicals are the most immediate uses for mass literacy. In 1975 there were 1584 newspapers and periodicals in regular publication, of which 105 were dailies. Urdu (882 papers) was the language most commonly used, but English with 312 was a strong contender as the national lingua franca of the better educated throughout the bureaucracy and elsewhere. The province of publication reflected fairly closely

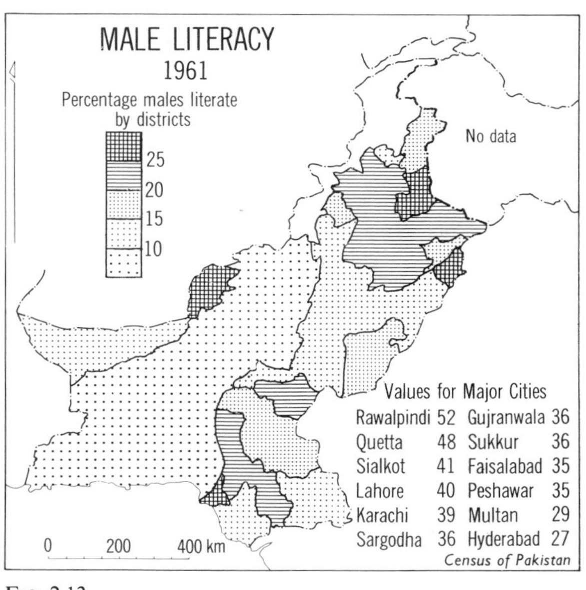

FIG. 2.13

FEMALE LITERACY
1961
Percentage females literate
by districts
25
10
5
3
No data
Values for Major Cities
Rawalpindi 30 Sargodha 17
Sialkot 27 Peshawar 16
Karachi 26 Faisalabad 15
Lahore 24 Sukkur 14
Quetta 19 Hyderabad 13
Gujranwala 19 Multan 12
0 200 400 km
Census of Pakistan

FIG. 2.14

TABLE 2.9
Literacy in Karachi District, 1972

Age group	Urban Total	Urban Male	Urban Female	Urban % F/M ratio	Rural Total	Rural Male	Rural Female	Rural % F/M ratio
Total 10+	52	56	46	82	28	35	19	54
10–14	56	58	54	93	35	43	24	56
15–19	63	65	60	92	37	44	28	64
20–24	57	61	52	85	33	42	23	55
25–34	50	55	43	78	28	36	19	53
35–44	47	53	38	72	25	31	17	55
45–54	45	52	32	62	20	26	12	46
55+	37	46	23	50	17	22	11	50

their population disparities: Punjab had 960 publications, Sind 484, NWFP 108 and Baluchistan 32.

LEVEL OF DEVELOPMENT

Using some of the measures described above, together with others to be discussed, an attempt can be made to measure the comparative level of development across the country. The factors that have a bearing on the level of development are of such a disparate nature that a measure of the level cannot be achieved in absolute terms. One is faced with the problem of combining economic yard-sticks which can be given monetary values, with social and demographic indices which can only be abstractions. The nine factors used in conbination to arrive at the development indices for districts mapped in Fig. 2.15 are tabulated in Table 2.10.

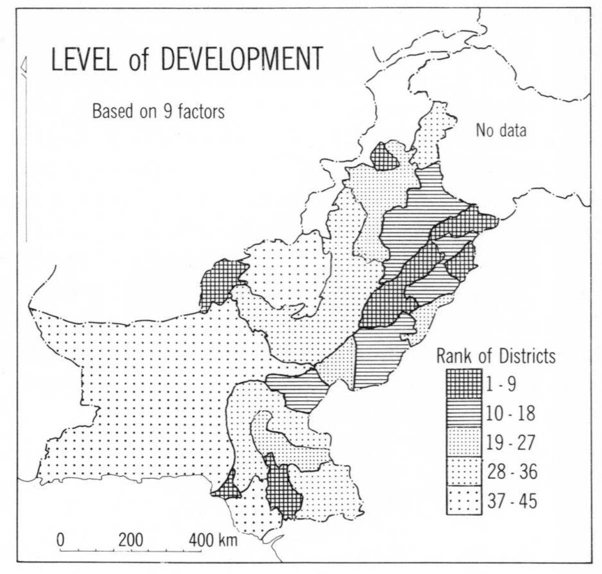

FIG. 2.15

1. A measure of production.
2. Level of urbanisation.
3. & 4. Measures of modernization in agriculture: tractors and fertilisers.
5. & 6. Measures of education: percentage literate and percentage attending school.
7. A measure of social welfare: the population served by each medical practitioner.
8. A measure of social modernization and feminine emancipation: the ratio of females in the school population aged 10–14.
9. Industrialisation: the percentage of the non-agricultural labour engaged in manufacturing industry.

1. *Production*: both agricultural and non-agricultural economic activity were brought into a single index of value per head of population in each district. Agricultural production was assessed by calculating the value of actual production of twelve major crops: wheat, Basmati rice, other rice, sugar cane, jowar, bajra, tobacco, maize, oil seeds (rape and mustard), gram, cotton and fruits. For non-agricultural production a more arbitrary value was arrived at by multiplying Rs 350 per month, i.e. Rs 4200 per annum (an approximation to the average income), by the total non-agricultural labour force as enumerated in the census of 1961.

2. The value for *urbanisation* was taken from the 1961 Census since only in this census were data on employment, education and literacy also available. It is probable that were it possible to use 1972 data, Rawalpindi would change rank with Quetta in the overall development index.

3. & 4. The percentages of farm households using *tractors* and of cropped area to which *fertilizer* is applied were obtained from the 1972 Census of Agriculture.

5. & 6. The percentages of the population *literate*, and of the age group 10–14 attending *school* came from the 1961 Census. The reasons for choosing this particular age group have been discussed above. Literacy covers the whole population and is not necessarily a function of formal schooling.

7. The 1961 Census provides a breakdown of professions, including *medical practitioners, dentists and specialists* under the one head. This figure was used to arrive at a comparative measure of provision of medical services to the population.

8. The traditional Muslim *attitude to women* must be regarded as a handicap to development. The proportion of females in the 10–14 age group attending school was taken as an indicator of relative emancipation and enlightenment, though in itself school attendance is no guarantee of either.

9. The numbers in *manufacturing industry* are

TABLE 2.10
Levels of development – rank ordering of districts

DISTRICT	*Combined index*	*Production per head*	*Urbanisation*	*Tractors*	*Fertilisers*	*Literacy*	*Schooling*	*Medical welfare*	*Females at school*	*Industrialisation*
Gujranwala	1	3	7	3	15	6	6	12	3	4
Faisalabad	2	8	10	3	2	6	8	11	12	2
Lahore	3	15	2	9	13	3	4	3	2	15
Karachi	4	4	1	27	12	1	3	2	1	25
Sialkot	5	21	18	8	18	6	6	8	6	4
Peshawar	6	9	6	6	6	12	16	4	15	34
Multan	7	14	10	10	5	22	27	9	12	6
Hyderabad	8	10	4	27	9	3	19	6	21	19
Quetta	9	18	3	1	31	5	14	1	7	41
Rawalpindi	10	27	5	17	33	2	1	5	4	32
Sheikhupura	11	6	25	11	13	21	17	22	11	13
Gujrat	12	22	25	15	24	6	5	27	7	9
Sargodha	13	12	7	20	26	12	10	18	12	15
Bahawalpur	14	27	12	18	16	22	23	7	9	15
Jhang	15	12	18	27	19	16	14	31	17	1
Sahiwal	16	13	33	5	10	27	31	23	20	3
Jhelum	17	38	23	27	37	3	2	14	9	15
Sukkur	18	17	9	36	24	10	13	15	24	20
Mianwali	19	8	12	23	30	16	10	37	28	9
Sangar	20	18	18	27	1	22	17	13	36	21
Mardan	21	1	25	2	7	32	25	21	37	27
Bahawalnagar	22	23	25	18	23	27	19	30	15	8
Nawabshah	23	20	25	20	2	16	19	24	41	25
Kohat	24	43	12	6	32	16	10	10	32	36
Sibi	25	3	25	20	21	40	39	20	5	27
Attock	26	35	39	11	34	12	8	32	21	11
Rahimya Khan	27	27	33	13	11	27	27	34	18	13
Tharparkar	28	34	25	13	4	32	34	21	26	27
Larkana	29	25	18	40	20	16	17	33	30	24
Khairpur	30	32	39	23	8	22	23	29	37	11
D.I. Khan	31	26	12	23	34	27	31	28	23	22
Muzaffargarh	32	16	42	15	21	32	27	43	28	6
Dadu	33	24	33	40	17	12	25	26	37	27
Hazara	34	44	7	34	26	22	19	40	30	27
D.G. Khan	35	31	25	34	29	37	31	38	26	20
Bannu	36	32	33	33	34	27	34	17	40	36
Jacobabad	37	12	33	36	37	32	36	35	35	34
Zhob	38	37	28	23	40	38	41	25	24	43
Chagai	39	35	23	39	39	38	38	20	32	43
Thatta	40	27	44	27	28	32	36	36	41	38
Loralai	41	42	42	38	40	41	40	39	15	39
Makran	42	41	17	41	42	42	42	44	43	32
Kalat	43	39	41	41	42	42	42	42	32	39
Kharan	44	45	22	41	42	42	42	45	44	43
Lasbela	45	40	45	41	42	42	45	41	44	42

a measure of the movement of the economy away from its traditionally agricultural bias.

The 45 districts for which data is available were ranked for each of the nine indices, low ranking indicating a high level of development. The nine indices were then added to give a generalized development index. In calculating an overall index the problem always exists of how to balance the factors being combined. Should, for example, production be given greater weight than some other factor? Is modernization in agriculture over-represented by the use of both a tractor and a fertilizer index? In practice the total picture changes but little if one or other of the individual factors are weighted by a factor of two. Table 2.10 provides the opportunity for experimenting with a variety of weightings or for making exclusions in the total. The index as mapped in Fig. 2.15 gives equal weight to each of the nine factors.

Arbitrary as its derivation must be, this map presents an entirely consistent picture of the regional inequalities in development in Pakistan. Unfortunately lack of data prevents the inclusion of four districts in the far north of NWFP, of the Tribal Territories, and of the Federal Capital Territory of Islamabad which had not been established at the time of the 1961 Census. Islamabad and, as has been mentioned, Rawalpindi, with which it is closely associated in growth, would almost certainly now be among the top ranking ten districts. In addition to the provincial capitals of Lahore, Karachi, Quetta and Peshawar, the nine districts of the upper quintile include the major cities of Sialkot, Gujranwala, Faisalabad, Multan and Hyderabad. Since seven of the nine factors tend to be associated with processes of urbanisation such a result is to be expected.

The second group of nine districts form, with the single exception of Sukkur, a continuous block with the main core of Punjab districts in the highest group. Rawalpindi is in this rank because of its importance as the temporary capital (after 1959) while Islamabad was being constructed, and as Pakistan's major military centre. Jhelum district also contains military training establishments, and provides large numbers of army recruits, a long term consequence of which tends to be a concern for education and welfare. For the remainder, the districts constitute with the rural portions of the top ranking groups, the best irrigated lands in the country, and the principal basis for its economy. Sukkur forms an outlier of this long-irrigated tract, and furthermore has an industrial centre of some importance.

The middle group of districts include the areas in Mianwali District, of recent canal colonisation in the northern Thal tract; Kohat and Mardan in NWFP, each with a considerable core of irrigated agriculture surrounded by impoverished hills, and Attock in the Potwar Plateau, where high technology industrial centres help compensate for agriculture of mediocre productivity. In southern Punjab Bahawalnagar and Rahimyar Khan are predominantly rural, and somewhat marginal to the major perennial canal systems, while Nawabshah and Sanghar in central Sind, well developed as they are by the standards of that Province, suffer environmental constraints which are less critical in the Punjab. The relatively lower level of development in Sind generally as compared with Punjab is clearly shown by this index. Indeed two of its districts fall in the lowest category: Jacobabad on the fringe of the irrigated area and experiencing probably the most extremely hot climate in the world, and Thatta in the Indus delta, where tidal incursions vie with the water problems characteristic of the tail-end of an irrigation system to make life difficult for the cultivator.

Five districts form a block in the fourth ranking group. They are Muzaffagarh (southern Thal), Bannu, Dera Ghazi Khan and Dera Ismail Khan in the trans-Indus piedmont of NWFP, and Sibi in the Baluchistan piedmont. Dependent to a large extent on 'rod kohi' (torrent channel irrigation) productivity is always a gamble with an erratic rainfall, in addition to which general backwardness and traditionalism put a brake on social change on any kind. Hazara, in the north is probably representative of the other mountain districts beyond it for which data are incomplete.

Finally the districts of lowest rank form a near continuous block in Baluchistan, broken only by the accident of Sibi's extended shape. In effect from Zhob to Makran, the wastelands of Baluchistan provide bare subsistence for a sparse population. Only its strategic position and some ingenious indigenous irrigation practices set Quetta apart.

CHAPTER THREE

THE NATIONAL ACCOUNTS AND PLANNING

SUMMARY

The structure of a country's Gross National Product (GNP) provides a good perspective view of the various economic activities of its population, the value of which is not necessarily tied directly to the numbers involved in the different sectors. Changes over time in the contribution to GNP from the various sectors can indicate whether the economic structure is in a static or dynamic state. Divided by the estimated population, GNP allows calculation to be made of per capita income, a grossly simplified concept, but one commonly used especially in a time series to chart the movements in the economy, for better or for worse.

Foreign trade, vital to national prosperity in a country heavily dependent on its exports to earn exchange to pay for imports, is analysed by major commodity groups and by direction. The expanding and politically increasingly uninhibited spread of trading connections is demonstrated and the quantum and origins of foreign capital received in grants, loans and commercial investment concludes the chapter.

The final section reviews the content and stated objectives of the Five Year Plans up to 1970 when the Fourth Plan was abandoned in favour of Annual Plans, and the Fifth Five Year Plan, 1978–83.

GROSS NATIONAL PRODUCT

The composition of Gross National Product is set out in Table 3.1 and Figs. 3.1 and 3.2. The table shows for the eleven sectors constituting Gross Domestic Product, and for the external invisible receipts and expenditures of money which convert GDP to GNP, their share of GNP over three years since the secession of Bangladesh. In Fig. 3.1 the pattern is traced back to the early days of Pakistan's independence, and since the values used are adjusted at constant factor cost, a proper comparison of the size of the economy over the years and of its changing sectoral composition can be made. Distortions arising from inflation and currency devaluation are thus ironed out. Fig. 3.2 shows the changes in relative importance of the major sectors of the economy.

The continuing strong dominance of the agricultural sector is not a cause for great optimism in a poor country seeking development, however defined. This is further underlined by the fact that the small decline in this sector's share between 1971–72 and 1977–78 is not taken up in productive manufacturing, which also loses ground, but rather in construction, defence and administration. These latter are not renowned anywhere for their contribution in themselves to developmental growth. Their function is rather to serve the productive sectors, and in an underdeveloped economy their over expansion is economically unhealthy. The decline in energy and transport reflect the general reduction in buoyancy in the economy due largely to political factors within Pakistan.

Per capita income at constant factor cost and at current value is appended to the table to show how flat is the gradient of development for the abstract 'average citizen'. Recent political unrest was accompanied by much industrial trouble, curfews and loss of production, with a consequent *fall* in the level of per capita income in 1976–77 compared with 1973–74. 1977–78 showed a recovery of 6 per cent over the previous year. The economic growth rate in 1976–77 was reported as 2.8 per cent only, but prospects for 1977–78 were much brighter at 9.2 per cent.

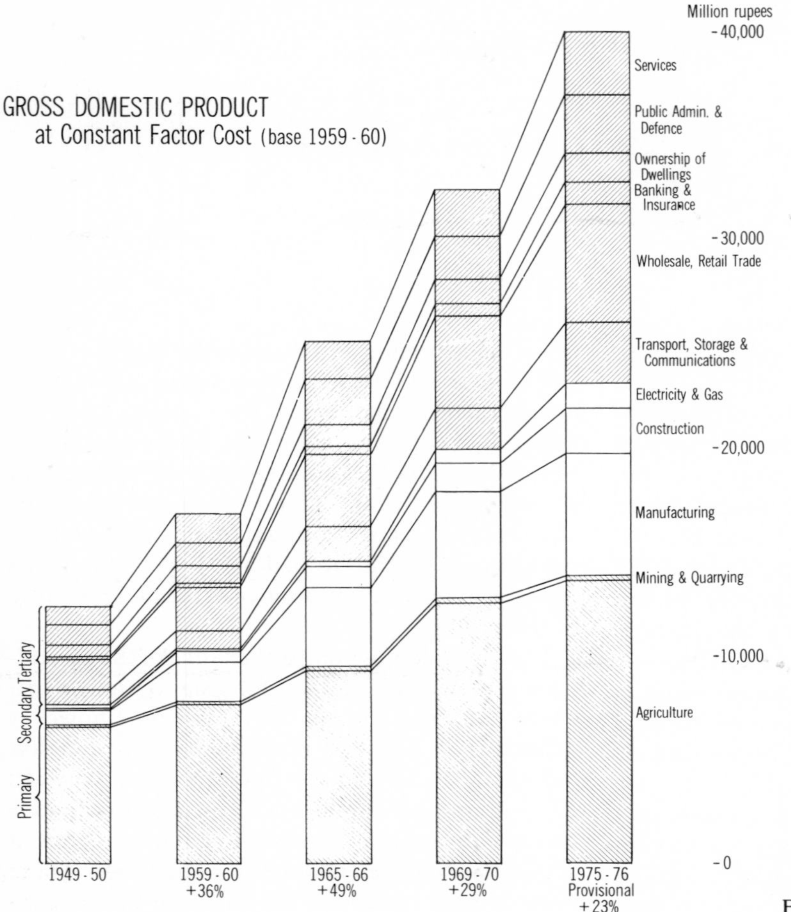

FIG. 3.1

THE CHANGING STRUCTURE OF FOREIGN TRADE

From 1971–72, the present political entity of Pakistan functioned as an independent economic unit. After the partition of British India in 1947 and until Bangladesh, formerly East Pakistan, separated from West Pakistan, the national accounts covered the totality of erstwhile Pakistan. In the data on exports and imports for example, trade between the two wings is regarded as internal, and so does not appear in the separate statements about the trade of West and East Pakistan respectively with the rest of the world. For the earliest years of Pakistan, data on a basis comparable with later years is not available.

At partition, Pakistan inherited an underdeveloped colonial economy, even less industtrialised than that of India at the same time. Although for a year or two the economic links with India persisted, particularly between East Pakistan and Indian West Bengal in the raw jute trade, the political estrangement of the two nations and fiscal difficulties due to India devaluing independently of Pakistan, led to the practical cessation of commercial intercourse. In 1949–50 Pakistan had nothing to sell to the world other than the primary products of agriculture. Apart from raw jute, East Pakistan contributed a little fish and tea, and the nation's export trade was dominated by West Pakistan's raw cotton, raw wool, hides and skins, cotton seed and fish. Cotton made up about 84 per cent of West Pakistan's exports, wool 7.5 per cent, hides almost 7 per cent.

TABLE 3.1

Structure of gross national product at constant factor cost (1959/60) as percentage of total

Sector	*1971–72*	*1973–74*	*1976–77*	*1977–78[a]*
1. Agriculture	38.5	35.6	34.5	32.2
2. Mining & quarrying	0.5	0.5	0.5	0.5
3. Manufacturing	15.3	15.6	14.4	13.4
Large-scale	11.6	12.2	10.9	10.2
Small-scale	3.7	3.4	3.5	3.2
4. Construction	3.6	4.0	5.1	5.0
5. Electricity & gas	2.4	2.8	2.4	2.7
6. Transport, storage, communications	6.2	6.6	6.2	6.5
7. Wholesale & retail trade	13.5	14.3	13.6	13.1
8. Banking & insurance	2.0	2.3	2.	2.6
9. Ownership of dwellings	3.6	3.4	3.5	3.3
10. Public administration & defence	7.0	7.3	8.1	7.7
11. Services	7.3	7.1	7.1	7.4
12. Gross Domestic Product	99.8	99.5	98.2	94.5
13. Net factor income from/to rest of the world	0.2	0.5	1.8	5.5
14. Gross National Product	100.0	100.0	100.0	100.0
15. Per capita income (Rs.)	517.0	559.0	551.0	597.0
16. Per capita income (current values) (Rs.)	774.0	1,116.0	1,730.0	2,165.0

Source: National Accounts, 1973–74 to 1976–77 and *Pakistan Statistical Yearbook*, 1974 (for 1971–72). Provisional figures for 1977–78 from *Pakistan Economic Survey, 1977–78.*

[a]GNP totalled Rs. 45,145 million.

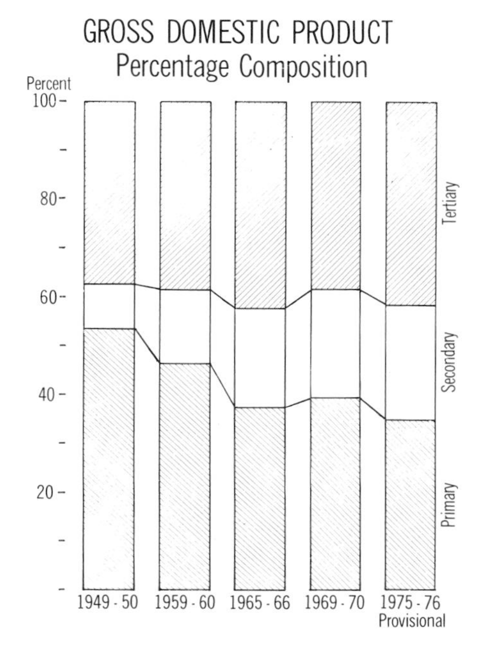

FIG. 3.2

It is impossible to apportion imports between the two wings at this time, but they shared a common lack of manufactured products of all kinds and were not even completely self-sufficient in food stuffs. The major items imported were cotton goods (24 per cent by value), cotton yarn (15 per cent) for handloom industries and the incipient textile industry, machinery and vehicles (12 per cent) metal products (4 per cent), petroleum products (4 per cent) and food (3 per cent).

Among trading partners Britain was still well in the lead in both imports into Pakistan (32 per cent) and exports (27 per cent). India was second with 16 per cent of the imports and 12 per cent of exports. Other substantial providers of imports were USA (11 per cent), Japan (9 per cent), Italy (7.5 per cent) and China (3 per cent). Important buyers of exports were USA (10 per cent), Hong Kong (9 per cent), Japan (9 per cent), USSR (7 per cent) and West Germany and Italy both with 5 per cent. It is likely that some of these percentages

14 Bales of cotton, Pakistan's main raw material export, being loaded at Karachi for shipment overseas.

would be altered were separate figures available for East and West wings, but since both jute and cotton are in demand in most industrialised countries the change might not be significant.

Tables 3.2 and 3.3 show the changing composition of West Pakistan's exports and imports over the thirty-year period of independence of British colonial rule. The predominant role of primary raw or semi-processed commodities persisted, though manufactured cotton, wool and leather increased their share of the total. In 1965–66 more or less raw materials still made up about 60 per cent of the whole. This share declined by 1971–72 to a little under half, where it has remained until 1976–77 when a serious decline in raw cotton exports from 29 per cent in 1971–72 to barely one-tenth of that, reduced raw materials to about 41 per cent of the total. Semi-manufactured goods, such as processed raw materials make up 17 per cent, and fully manufactured goods 42 per cent of the total. Rice exports have increased in importance however, Pakistan's high grade Basmati rice finding a ready market and a high price overseas. Guar is the seed of the field vetch, *Gyamopsis psoralioides*, not unlike soya bean as a source of protein extracts. It is grown widely as a kharif (hot weather) green fodder crop.

TABLE 3.2
Composition of imports, 1965–66, 1971–72 1976–77

Main group	*Percent of total value* 1965–66	1971–72	1976–77
Machinery and transport equipment	45.0	30.0	28.8
Manufactured goods	13.8	16.4	17.1
Food & live animals	9.0	13.9	8.9
Chemicals, fertilizer, etc.	6.0	8.0	11.6
Mineral fuels etc.	2.1	7.3	18.0
Animal & vegetable oils and fats	2.8	2.4	7.2
Total (Rs millions)	2,880.3	3,495.4	23,012.0

TABLE 3.3
Composition of exports, 1965–66, 1971–72, 1976–77

	Percent of total value		
Main group	*1965–66*	*1971–72*	*1976–77*
Raw cotton & waste	24.6	29.1	2.8
Cotton fabrics & yarn	20.8	29.4	24.8
Raw wool	5.0	0.6	0.7
Woollen carpets	1.9	3.2	7.8
Raw hides & leather	7.4	5.7	5.7
Leather products & footwear	0.7	1.2	7.3
Rice	11.0	8.1	21.9
Fish and products	11.0	3.3	3.4
Tobacco and products			1.5
Guar and products	2.2	1.2	1.6
Petroleum products	1.6	1.2	2.4
Sports goods	0.2	1.5	1.8
Surgical instruments	0.6	0.7	1.2
Total (Rs millions)	1,203.6	3,371.4	11,293.0

Any list of exports says as much about the state of the world's markets as it does about the state of the country's productivity. While the bulk of the exports reflect primary and secondary production based on resources of soil, water and climate, the sports goods and surgical instruments categories are the result of business enterprise married to Pakistani craftsmanship, focussed particularly on Sialkot. In an endeavour to increase foreign exchange earnings, Free Trade Zones are being established at Karachi and Lahore. Foreign investors are to be given facilities and a 15 year tax holiday as incentives to develop export oriented industries using Pakistan labour and raw materials as far as possible.

Machinery, vehicles and manufactured goods together heavily dominate imports in each year of the record. Development has created industry within Pakistan, but brings with it an increased demand for high technology products from abroad and for the fuels and lubricants to keep the expanded economy moving. The abrupt rise in oil prices in the early 1970s has added to the nation's problems but is not responsible for the whole of the increase in the value of petroleum imports. Meanwhile the growing population must eat to live, and what food it cannot itself produce must be imported. At the same time efforts towards achieving self-sufficiency in food grains require heavier inputs of fertilizers, an increasingly important item in the country's shopping list.

The apparent jump in the values of trade from 1971–72 to 1976–77 is in part due to inflation and to the devaluation of the currency by 140 per cent in 1972 but also to the low state of the economy at the time of the separation of Bangladesh following the brief but traumatic war with India in 1971.

The effect of some fresh political realities are evident in the line-up of Pakistan's trading partners following the Bangladesh crisis (Table 3.4) While the long-established export links with Britain, USA, Western Europe, Hong Kong and Japan remain (note the share by these countries in 1948–49), their relative importance has changed, and some significant new participants emerge, notably in the Middle East.

As a source of imports Britain is now well behind USA. Perhaps most remarkable about this list is the great number of countries involved in supplying goods and commodities to Pakistan. The list is also noteworthy for the catholicity of its political colouring: West, East and Uncommitted, friends and former foes, the wealthy and the impoverished alike appear. A common device to encourage mutually advantageous trade without risking pressures on the foreign exchange reserves of either country involved is the barter trade agreement. Currently Pakistan has such agreements with USSR and five other Eastern European countries, with China and North Korea, Sweden, Yugoslavia and Iran.

THE BALANCE OF TRADE

A less rosy picture is seen in Table 3.5 which shows the monotonous regularity with which Pakistan's balance of trade runs into deficit; during the mid-1970s the balance of trade deficit was alarmingly severe with exports paying for little more than half the imports. The massive devaluation of 1972 and the redirection to new markets of exports that formerly went to East Pakistan accounted for the short lived credit balance in 1972–73. To try to balance the account Pakistan has had to borrow extensively abroad, and up to 1976 had incurred a foreign debt of US $8255 million, 77 per cent of which has already been disbursed (Table 3.6). The cost of servicing this debt places an increasing

TABLE 3.4

Pakistan's major trading partners, 1948–49, 1971–72 and 1976–77

	Exports			Imports		
	1948–49[a]	*1971–72*	*1976–77*	*1948–49*[a]	*1971–72*	*1976–77*
Total value (million Rupees)	Rs 775	Rs 3371	Rs 11,294	Rs 952	Rs 3495	Rs 23,012
	%	%	%	%	%	%
Canada		0.3	5.7		1.2	2.2
USA	10.0	5.2	5.1	11	20.9	14.9
EEC countries		19.0	22.8		30.9	23.5
UK	27.0	7.7	7.1	32	10.1	8.1
W. Germany	5.0	3.1	5.7		9.8	6.4
Italy	5.0	3.7	3.7	8	5.5	3.1
France		1.8	2.3		2.2	2.2
EFTA countries		0.2	3.0		2.5	3.2
Spain		0.5	2.1		0.1	0.2
E. Europe		11.7	4.4		10.3	4.5
USSR	7.0	4.0	2.2	1	2.4	1.9
Middle East		13.5	31.3		8.0	18.3
RCD Iran & Turkey		0.6	8.4	2	2.1	0.2
Abu Dhabi		0.6	0.3		0.9	3.9
Dubai		1.2	4.8		0.2	0.7
Iraq		1.6	4.4		0.1	0.6
Kuwait		3.1	12.8		1.8	5.1
Saudi Arabia		2.4	4.1		2.6	6.9
China	1.7	4.3	0.9	5	2.8	2.8
Hong Kong	9.0	26.8	6.3		0.7	0.8
Japan	9.0	16.0	8.1	9	10.0	14.3
Bangladesh			0.6			1.1
India	12.0			16		1.0
Sri Lanka		1.9	3.9		3.1	1.9
Afghanistan		1.3	1.1		2.9	1.3
Indonesia		1.5	1.6		0.3	1.7
Malaysia		0.7	0.2		0.4	1.8
Singapore		1.8	1.2		0.7	0.9
Australia	1.4	0.8	0.9		1.0	2.4

Source: *Pakistan Statistical Yearbook*, 1976; *Three Years of Pakistan*; *Monthly Statistical Bulletin*, May–June 1977.

[a]Separate total for West Pakistan not available for 1948–49. The percentages are estimates for undivided Pakistan's trade structure.

TABLE 3.5

Balance of trade (Rs millions)

Year	*Imports*	*Exports*	*Balance*[a]
1950–51	1,167	1,342	+175
1955–56	964	742	−222
1960–61	2,173	540	−1,633
1965–66	2,880	1,203	−1,677
1970–71	1,998	3,602	−1,492
1971–72	3,495	3,371	−72
1972–73	8,393	8,551	+224
1973–74	13,479	10,161	− 3,246
1974–75	20,925	10,286	−10,469
1975–76	20,465	11,253	−9,091
1976–77	23,012	11,294	−11,429

Source: *Pakistan Statistical Year Books* (various dates).

[a]The Balance takes account also of re-imports and re-exports but excludes 'invisibles'.

TABLE 3.6

Foreign debt incurred: 1977

(US $ Millions and per cent of total)[a]
Total: $8255 million of which disbursed $6341.5 million.

		%			%
USA	2,187.3	26.5	Canada	315.2	3.8
IDA	791.2	9.6	China	263.3	3.2
Iran	778.6	9.4	France	261.8	3.2
W. Germany	617.2	7.5	UK	223.6	2.7
IBRD	542.9	6.6	Saudi Arabia	130.7	1.6
USSR	517.6	6.3	Italy	121.7	1.5
Japan	412.3	5.0	Abu Dhabi	113.0	1.4
ADB	392.8	4.8	Netherlands	111.8	1.4

Source: *Pakistan Economic Survey* 1977–78.

[a]Only totals of more than 1% are included.

burden on an economy that needs capital for investment in development, but never seems nearer to being able to generate enough increased production to catch up with the back-log.

By 1975–76 the total burden of debt was assessed at US $5700 million on which, from a variety of rates of interest, a weighted average of 3.7 per cent gives an annual cost of US $211 million *in interest alone*, equivalent to almost 19 per cent of the earnings from current exports. A press report in December 1977 gave 28 per cent of foreign earnings as the cost of servicing debts by then incurred. In addition to loans, Pakistan has been receiving very substantial grants in aid, some of them tied to the Tarbela Dam project.

TABLE 3.7
Foreign grants in aid (US $ millions)

	1971–72	*1972–73*	*1973–74*	*1974–75*	*1975–76*
Consortium countries	27.6	32.4	55.5	82.2	53.3
Indus–Tarbela Fund	33.9	16.4	13.3	11.6	14.7
Islamic countries	—	—	—	7.2	30.5
Total	61.5	48.8	68.8	101.0	98.5

Source: *Pakistan Statistical Yearbook*, 1976.

Table 3.7 shows the grants received since 1971–72 from three groups of donors. The appearance of the Islamic countries from 1974–75 is linked to the economic resurgence of Islam associated with the increase in petroleum prices by OPEC. South Asian countries suffered along with the rest of the developing world when their bill for oil quadrupled; Pakistan enjoys some measure of compensation in the form of grants and investments by OPEC countries in its development.

A small but increasingly important element in the National Account is helping to bring foreign exchange into the country, even though it may be channelled into savings rather than into direct investment. This is the remittances from a million or more Pakistanis living and working abroad, but continuing to help support their relatives back home. In Table 3.1 the 'net factor income from/to the rest of the world' is made up mainly of such money flows, and has increased from 0.2 per cent of the Gross National Product in 1971–72 to 1.8 per cent in 1976–77, a not insignificant sum, amounting to Rs 2995 million of a gross income of Rs 5735 million. Prospects are that the gross total may reach Rs 10,000 million in 1977–78 with a net figure not less than half that amount, the difference being due to 'invisible' expenditure in remittances abroad by expatriate individuals and firms in Pakistan.

For 1975–76 the estimated invisible receipts totalled approximately Rs 5920 million ($592 million) and invisible payments Rs 6480 million made up as follows (in Rs millions):

Receipts		*Expenditure*	
Transport	1170	Transport & insurance	2130
Other services	1340	Technical assistance	530
Remittances	3140	Factor services	1770
Other	270	Other	2050
	5920		6480

THE FIVE YEAR PLANS

Governments in newly independent countries of the Third World have customarily used Five Year Plans, or plans in other time periods, as political manifestos for consumption at home and abroad, and as indications of intent to develop this or that sector of the economy and to further social well-being through education and reforms of various kinds aimed at traditional institutions and customs. In that the implementation of economic planning generally depended heavily on foreign aid to support the capitalization of industry and public works, the plans had to be framed in such a way as to persuade the donors of such aid of the recipient's bona fide intentions. These remarks may serve to remind us that planning documents may be more than merely economic blue prints, and their declarations may need to be read with circumspection. However, taken at their face value, the periodic plans over a span of two decades or more may reveal how firmly or otherwise planners, and the governments they serve, are able to hold to the high hopes and principles that embellish their earliest efforts.

The Five Year Plans of Pakistan relate to the time before the secession of Bangladesh, an event which helped shatter the Fourth Plan even though it was already in disarray when the final separation came. In the terms of reference by the Government when setting up the commission for the First Five Year Plan in 1953 it says, 'The economic and social objectives of Government's policy are well known. They are to develop the resources of the country as rapidly as possible so as to promote the welfare of the people, provide adequate living standards and social services, secure social justice and equality of opportunity and aim at the widest and most equitable distribution of income and property.'

The plan contained discussion of the ideals of giving the land to those who tilled it (see Chapter 10 below) and other visions of a 'brave new world'.

The Plan was to serve the years 1955–60, but did not receive Government approval till 1957, half way through its postulated life. In the event it proved an over-optimistic programme. Aiming to increase gross National Income by 15 per cent, it achieved 11 per cent, and as more rapid population growth occurred than had been anticipated, per capita income rose no more than 3 per cent instead of 7 per cent. The target of self-sufficiency in food could not be reached and aid and earned exchange had to be used to buy food for a predominantly agricultural people – a state of affairs that has persisted ever since, with rare interruptions. It proved harder to earn income from exports as raw material prices fell, and harder to import the capital goods so sorely needed to create alternative jobs to agriculture, and everything took longer than the planners had assumed.

The relative priorities given to the different sectors of the national economy as they affect West Pakistan in different plans are shown in Table 3.8.

The contrast between agricultural and industrial investment implied in the table was not as sharp as it appears since the major share of 'water and power' went to support agriculture; nonetheless, despite the quite fundamental role of agriculture in the domestic economy and in the export trade, by no manipulation of the figures can it be brought to top priority. Planners, politicians and bureaucrats were from urban elites and the fact of their urban bias in planning, to use Michael Lipton's telling phrase, was possibly a quite unconscious one.*

The Second Five Year Plan, 1960–65, appeared on schedule. It is written in optimistic vein, referring to the people 'responding to dynamic and determined leadership . . . with remarkable fervency and confidence.' The failures of the First Plan to reach its targets are blamed on 'political instability, and absence of sustained endeavour.' Adverse climatic accidents and terms of trade beyond Pakistan's control bear part of the blame. In its introduction the Second Plan elaborates a little on the country's evolving political philosophy. Thus: 'No doctrinaire assumptions underlie the Plan, and neither an exclusively capitalist nor an exclusively socialist economy is postulated. The approach throughout is pragmatic. The fundamental problem is how, under severely limiting conditions, to find some way towards the liberation of the people from the crushing burden of poverty.'

* Michael Lipton, *Why Poor People Stay Poor: Urban Bias in World Development* Temple Smith, London and A.N.U. Press, Canberra, 1977.

TABLE 3.8

Sectoral priorities in Five Year Plans for West Pakistan (percentage of total allocated by sector)

Sector	*Pre-Plan 1950–55*	*First Plan 1955–60*	*Second Plan 1960–65*	*Third Plan 1965–70*
Agriculture	6	7	13	15
Industry, fuel, minerals	36	31	28	26
Water & power	13	17	19	15
Transport, communications	14	17	17	18
Physical planning, housing	22	20	15	13
Education, training	5	6	4	5
Health	3	2	1	2
Manpower & welfare	1			1
Works			3	5

Source: *Third Five Year Plan.*

The need is recognized for the economy to grow faster than the population, but population control, along with water, power, education and research is seen as a field for long term investment, while the immediate need is to put investment where quick results may be achieved. Three strands dominated the planners' thinking:

1. The agricultural economy, characterized by the low productivity and an inability to feed even itself.
2. Industrial development, for which much reliance was placed on the private sector, which should be freed as far as possible of restraints and controls, to produce substitutes for imported goods and an exportable surplus.
3. Education at all levels.

Expenditure under the Second Plan was 65 per cent greater than under the First, though inflation reduces this in real terms. 20 per cent growth was looked for during the Plan, as an essential foundation for 25 per cent in the Third and 30 per cent in the Fourth and Fifth. Agriculture, by increasing production by 14 per cent, would become self-sufficient. Industrial production would rise 60 per cent in the large scale and 25 per cent in the small scale and cottage sector. For the first time the concept of a semi-public sector emerged. Of the total Plan cost of Rs 19,000 million, Rs 9750 million went to the public sector, Rs 6000 million to the private sector and Rs 3250 to the semi-public sector. This latter covered various Government corporations set up to promote industrial activity in cooperation with private enterprise, or to operate certain functions in an autonomous fashion, independent of direct ministerial control. Thus the P.I.D.C. (Pakistan Industrial Development Corporation) had the task of establishing joint enterprises between public and local or foreign enterprises which, once running, could become fully private concerns. Other instruments in the semi-public sector were the Small Industries Corporation, Pakistan International Airways Corp., various organisations for the then capital, Karachi: Port Trust, Electricity Supply Corporation, Development Authority, Road Transport Authority, the West Pakistan Road Transport Board and the Lahore Improvement Trust.

Agriculture, with its share of water development, probably went a little ahead of industry, depending on how the fuels item is divided. The breakdown of the allocations in more detail than that in Table 3.8 is not available till the Third Plan. In physical terms, the Plan's targets for West Pakistan included the addition of almost one million hectares of newly irrigated land, 38 per cent increase in cotton production, 40 per cent in the installed electricity generating capacity, 36 per cent in cotton yarn, 186 per cent in cement, and 107 per cent in coal production.

The preamble to the Third Five Year Plan, 1965–70, speaks of the 'intention to move in the direction of a "welfare state" with efficiency and despatch. The term "welfare state" has a well accepted connotation which in our country would be accepted as synonymous and interchangeable with "Islamic Socialism".' The apportioning by sectors of the Rs 14,000 million that was allocated for West Pakistan is given in Table 3.8 above. Within manufacturing there was a shift in emphasis to build up a capital goods industry as a step in the direction of greater industrial self-sufficiency. Heavy outlays continued to be necessary to implement Pakistan's share of the Indus Waters Agreement, which was to be a heavy burden for the years to come. The Third Plan reviewed the progress of planning to date as in Table 3.9 overleaf, from which the disparity between the two wings becomes apparent, a disparity which gave East Pakistan cause to complain of their 'less than equal treatment in the allocation of resources for development'.

The hopes for the Third Plan soon foundered in the brief war with India in 1965 as a result of which US aid was suspended for a while. Drought in 1965–66 and floods in 1966–67 took their toll and in the end the plan was underspent by 17 per cent, public investment falling by 29 per cent. GNP rose 5.7 per cent compared with an expected 6.5 per cent, agriculture 4.5 per cent (5 per cent) and exports by 7 per cent (9.5 per cent).

At the time the Fourth Five Year Plan was prepared for 1970–75 the whole planning process was under attack from critics who claimed that a small privileged class was being created at the expense of the masses. As a consequence the Fourth Plan was abandoned, and there began a series of Annual

TABLE 3.9

GNP and per capita income (at 1959/60 prices)

	1949/50	1954/55	1959/60	1964/65	Annual compound rate of growth 1949/50–1959/60	1959/60–1964/65
GNP at factor cost (Rs million)						
Pakistan	24,466	27,908	31,439	40,525	2.5	5.2
W. Pakistan	12,106		16,494	21,070	3.1	5.0
Per capita income (Rs)						
Pakistan	311	316	318	360	0.2	2.5
E. Pakistan	287		278	318	−0.3	2.7
W. Pakistan	338		366	411	0.8	2.4

Source: *Third Five Year Plan.*

Development Plans which continue today. That for 1970–71 set out to try to redress the balance of development in favour of East Pakistan which had been relatively neglected in earlier plans and by the way the economy functioned. Another objective was 'to pay greater attention to social objectives and to correct some of the imbalances that have developed in the country'.

In the event, 1970 brought disaster to East Pakistan in the form of a devastating cyclone, followed by the elections which provided the stimulus for secession. 'Civil' disturbances throughout 1971 culminated in war with India, and the disruption of Pakistan's economy.

For the 'new' separated Pakistan the First Annual Plan was that for 1972–73 which was introduced after certain reforms had been enacted to provide, it was thought, sounder foundations for economic development. Nationalization of ten basic industries was the major change announced in January 1972 and others followed. The general objectives in the plan were in the fields of living standards, public services, better income distribution and employment opportunities, development of the backward regions of the country, and the achievement of greater self sufficiency through import substitution industries. The comment that 'success in revitalizing the economy hinged on revising people's faith in the equity of the socio-economic system and on creating harmony among various classes' is the last reference one finds to social objectives. Subsequent Annual Plans restrict themselves to matter of fact economic statements, and even the chapters specifically on social welfare are non-committal on the overall objectives of the plan.

In the 1974–75 Annual Plan the planners talk of the two earlier plans as having been 'aimed mainly to attain the short term goal of revival of the economy in a balanced setting of distributional justice, reform of the economic system and price stability.' Reform there certainly was, but its affects on distributional justice and price stability are arguable.

Table 3.10 gives the actual sectoral allocation of the annual plans from 1971–72 to 1975–76 and the planned allocation for 1976–77 and for the Fifth Five Year Plan 1978–83.

With water and power allocations separated it becomes clear that the agricultural sector plus water development lags behind industry if a fair slice of the power investment is on its behalf. In fairness, some share of the huge cost of Tarbela must go to agriculture. The increase in the allocation to industry is accounted for in part by investments in the nationalized sector and in new expensive projects like the Pakistan Steel Corporation's Karachi Steel Mills. Otherwise the year to year changes in percentage share of the total show little in the way of trends. As Tarbela's drain diminished its share was spread widely over other sectors. Fuel costs rose sharply following the world escalation of oil prices in 1974, health got a better, but still minimal share and government interest in housing the masses as reflected in the data, rose for a time to fall again.

In the Fifth Five Year Plan, 1978–83, announced in mid-1978, the agricultural sector's share is increased on that of recent years. Of industry's allocation, half is for the Karachi Steel Mills. In addition to the Rs 148,170 million distributed to the Public Sector, a further Rs 62,000 million is approved in the plan for investment by the Private

Sector in agriculture (18 per cent) industry and mining (31 per cent), transport (18 per cent), housing (21 per cent) and miscellaneous (12 per cent).

TABLE 3.10
Sectoral allocation of annual plans and Fifth Five Year Plan (percentage of total)

	1971–72	*1972–73*	*1973–74*	*1974–75*	*1975–76*	*1976–77*	*1978–83*
Total. (Rs. million)	*2,655*	*3,869*	*6,417*	*11,372*	*14,583*	*17,000*[a]	*148,170*
Agriculture	8.5	14.3	10.7	8.4	7.2	7.3	10.1
Water	7.2	9.1	9.1	9.2	8.5	9.1	11.6
Power	14.2	13.8	14.1	17.7	15.9	13.8	18.8
Industry	2.9	6.4	9.8	13.7	13.6	31.6	14.2
Fuel	1.9	2.7	4.1	4.6	4.6	4.1	3.8
Minerals	0.2	0.2	0.8	0.9	0.5	0.5	1.3
Transport & communications	16.4	16.1	20.0	18.9	16.5	16.3	18.5
Housing, etc.	4.8	6.9	7.4	8.3	8.3	7.3	6.6
Education	4.7	5.4	5.0	4.7	4.8	3.4	6.9
Health	2.2	2.5	2.7	3.2	4.3	4.2	4.5
Population planning	1.0	0.7	1.6	1.2	1.3	1.3	1.2
Social welfare	0.3	0.2	0.2	0.1	0.1	0.13	0.1
Manpower	0.13	0.14	0.6	0.5	0.2	0.14	0.5
Peoples' Works	1.4	3.0	2.3	1.4	1.3	1.2	1.0[b]
Mass media					0.6	0.6	0.5
Other	0.1	0.1				0.2	
Indus basin/Tarbela	34.1	18.6	11.5	7.1	6.3	2.9	
Earthquake relief						0.7	
Sports complex					5.8	0.6	
Tourism							0.4

Source: *Pakistan Economic Survey*, 1976–77 and *Fifth Five Year Plan* 1978–83.
[a]The sectoral percentages are of a preliminary gross total of Rs 18259 million, from which a shortfall of Rs 1259 million is anticipated.
[b]Rural Development.

CHAPTER FOUR

THE ENVIRONMENTAL BASIS OF TRADITIONAL AGRICULTURE

SUMMARY

Pakistan is a country of considerable environmental variety. Its landscapes are boldly drawn on a grand scale, from the Karakoram Range with the greatest concentration of high peaks in the world, through the hundreds of kilometres of rugged colourful rocky ranges of the western margins of the plains and the interior of Baluchistan, to the fluvial morphology of the mighty Indus and its tributaries, and the sand dunes on the fringe of the desert. In these plains the contrasts in elevation of the terraces, in the texture of alluvium and in the degrees of incision of the rivers have significant bearing on man's use of the country.

Climatically Pakistan is mostly desert or near-desert, made habitable by the presence of rivers. The arc of hills in the north traps precipitation to feed these rivers, and in a narrow zone rainfall is sufficient for some unirrigated agriculture. The seasonal rhythms of climate and the extremes of variability that may be experienced from time to time give diversity to the pattern.

THE LANDSCAPE OF PAKISTAN

The simplest delineaments of the landscape of Pakistan are shown in Fig. 4.1. In the north of the country the westernmost ranges of the Himalayan system aligned from southeast to north west in the Karakoram the Pir Panjal and the intervening ranges that culminate in Nanga Parbat, sweep in a curve to taper away almost at right angles into the Hindu Kush in Afghanistan. Lesser ranges strike southward as the Sulaiman and Kirthar Ranges whose pediments slope gently towards the River Indus. The structural trends are better seen in Fig. 4.2 showing solid geology. The sweeping structures around northern Pakistan, flanking the Pamirs in USSR, mark the extension of the great Indian continental 'plate' which has been thrusting north to buckle into folds the sedimentary formations of a former geosyncline. The Sibi re-entrant between the Sulaiman and Kirthar Ranges marks a submerged prong of the same plate. Westwards again the folds resume more latitudinal sweeps, picked out by the rare river valleys, the upper Zhob and Beji, the Rakshan and the Gudhri, and by the intervening ranges. The rugged nature of this intricate landscape cannot be appreciated at the scale of these figures. Portions of Baluchistan are mapped at a larger scale in Figs. 7.8 and 7.9 in Chapter 7 below. Principal ranges are the Siahan, the Chagai Hills and Ras Koh. While rivers in the south and east of Baluchistan find their way ultimately to the sea when they deign to flow, the northwest of the state comprises basins of internal drainage the major one having at its lowest point the playa lake Hamun-i-Mashkel.

Beyond the high Himalaya in Kashmir, the Indus flows parallel to the strike of the ranges in a broad gravelly valley, before it plunges through some of the world's deepest gorges in a series of sharp angular bends in its antecedent valley athwart the rising structures of the mountain folds. In this region of high glaciers and precipitous screes, the mountain spurs fan out southward, each major valley providing in the past the economic base for a feudal state. Such were Chitral and Dir and Swat, now districts of the NWFP. From the River Kabul southwards, the mountain front though lower in elevation, is more difficult to penetrate for lack of major valleys. Fig. 4.2 gives some idea of the

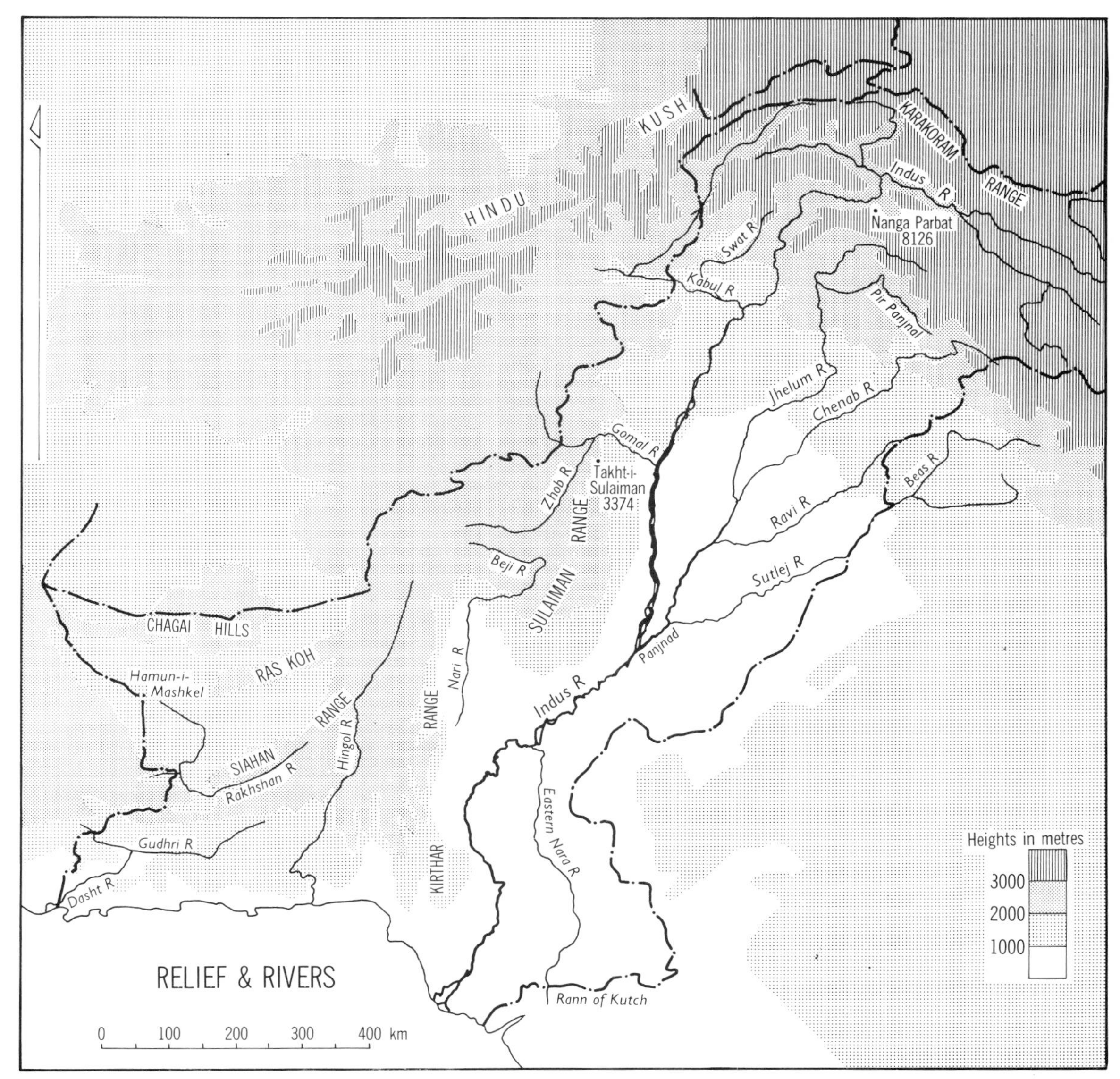

Fig. 4.1

structural complexities in the region north of the Indus plains proper. The alluvial basin of the Vale of Peshawar, geologically not dissimilar to that of the Valley of Kashmir to the east, is cut off from the main Indus plains by the belt of the comparatively low Salt Ranges of steeply folded tertiary sediments, in a few structures of which petroleum is preserved in exploitable quantities. North of the abrupt southward facing scarp that the Salt Ranges present to the Jhelum plain (Plate 16) the Potwar plateau is a confusion of rocky ridges and basin plains, in many areas thickly covered in Pleistocene times, with a mantle of loess which has become dissected into a fantastic 'bad-land' of gullies and loess buttes.

The alluvial geology and landforms of the Indus plains are separately mapped in Figs. 4.3 and 4.4. The morphological system is basically simple, and an understanding of its features is very necessary to a full appreciation of the use that man has made of them. The block diagram (Fig. 4.5) shows simplified portions of the plains from the Salt

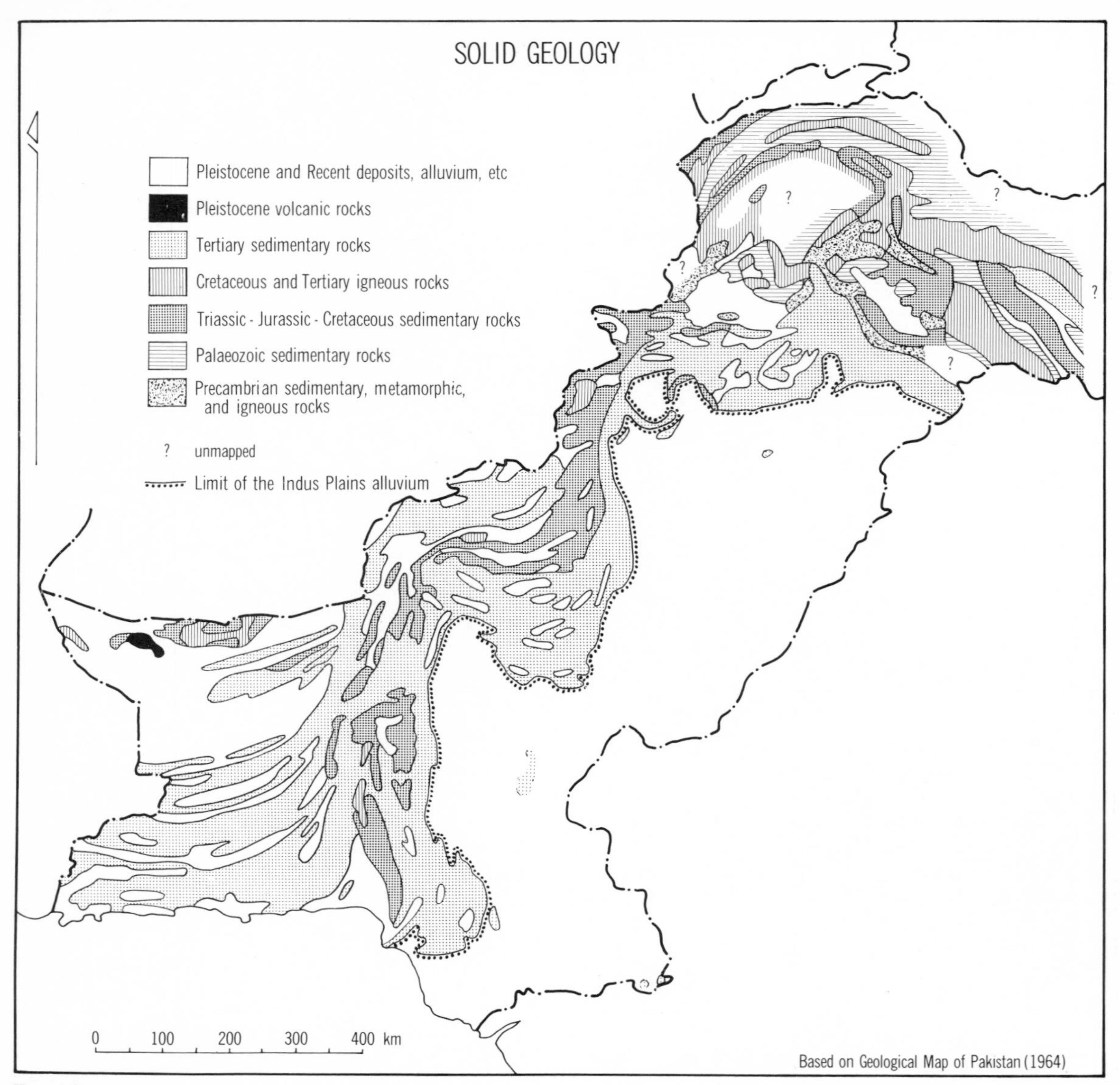

FIG. 4.2

Range in the north across the interfluve between Jhelum and Chenab, and from the Kirthar Range across the Indus just south of Sukkur. Three groups of morphological elements overlap in the Indus plains.

1. The bare bedrock slope of the mountain face is often skirted with fans of very coarse alluvial gravels which merge into the pediment slope or piedmont plain, down which the alluvial material becomes finer, and across which torrents flow intermittently in steep-sided *nalas*. This pattern of features is common around the northern and western rim of the plains.

2. The Indus plains proper are the work of the Indus and its major tributaries, Jhelum, Chenab, Ravi and Sutlej. Their successive terraces form distinctive morphological features.

3. Over parts of the plains, wind-blown deposits mask the fluvial features and give their own distinctive land forms.

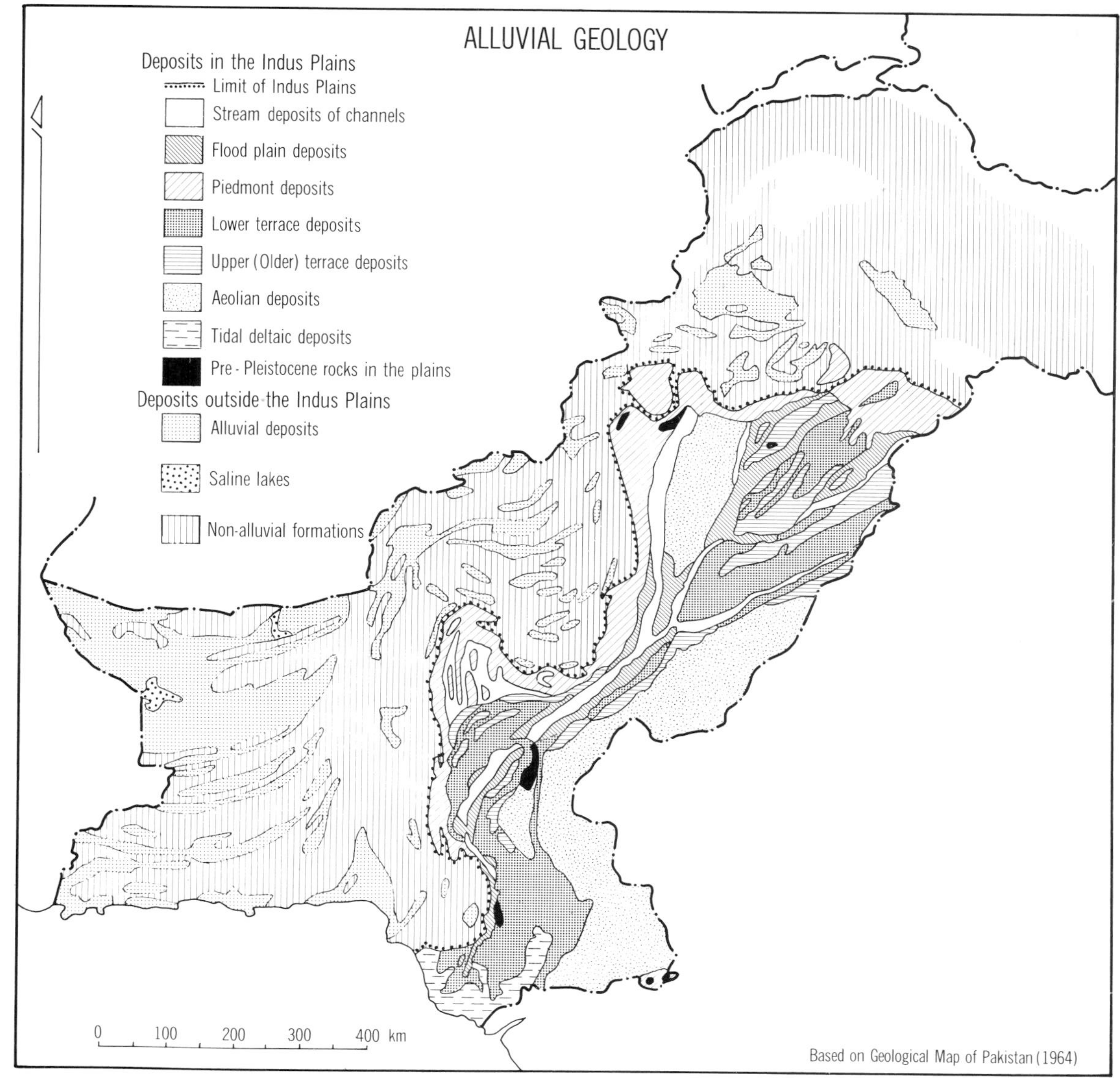

FIG. 4.3

The oldest elements of fluvial origin are the scalloped interfluves, terrace areas of old alluvium now standing high on the interfluve tracts, and so called because of the pattern that distinguishes their edge, where meander scars of later rivers sweeping laterally at a lower level have cut into them. Scalloped interfluves are notable features in the Punjab doabs (as the major interfluves are termed locally), between Jhelum and Chenab and again between Chenab and Ravi. They are too high to be affected by flooding in these rivers. Below Multan this morphological element is missing, and the oldest unit is the cover flood plain, common also to the Punjab. The cover and meander flood plains generally merge and are distinguished by the relative absence or presence of features of past river meanders: relics of oxbows, levees, sand bars, etc. In the cover flood plains, sheet wash has reduced such remnants to the general level of the plain from which they may only now be identified from the texture of the soil. The meander flood plains carry clear evidence of former river work,

15 Gilgit in the mountainous far north has terraced fields climbing into the pine forests. Here the crops are rice, being harvested (note the sheaves) and maize, both kharif crops.

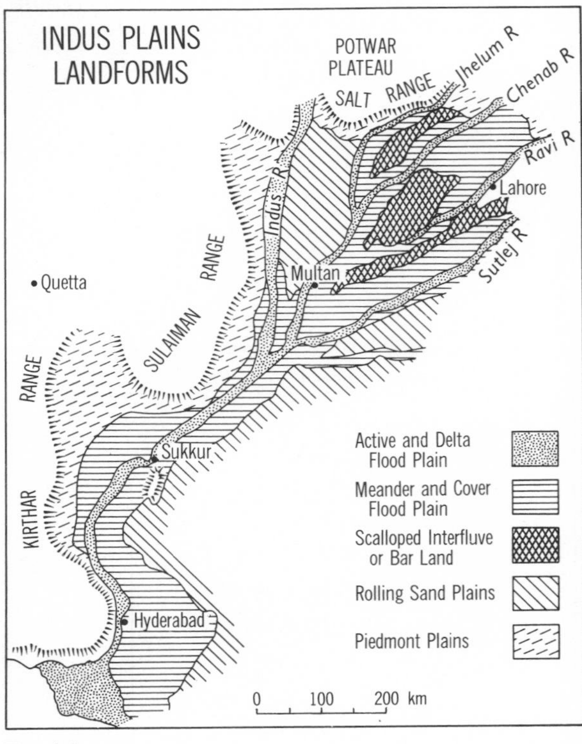

FIG. 4.4

FIG. 4.5

INDUS PLAINS MORPHOLOGY

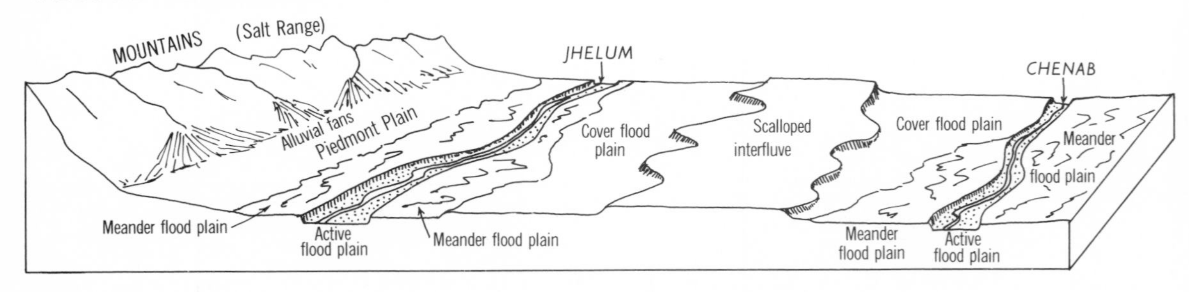

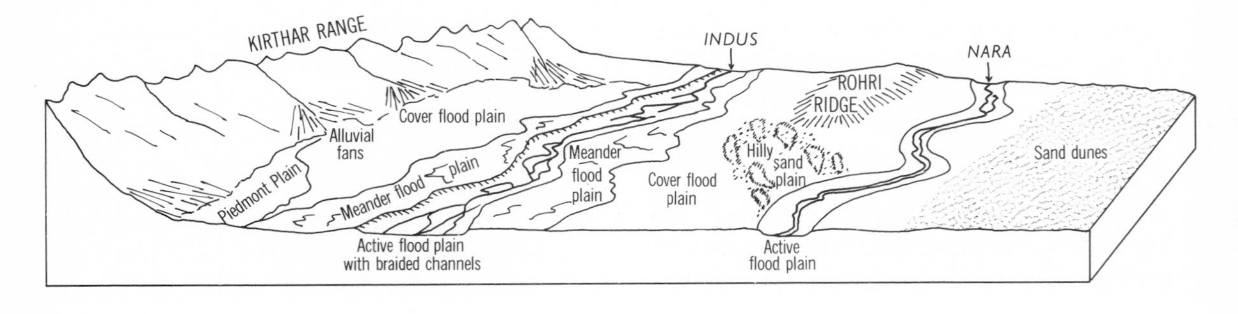

16 A section of the Salt Range scarp with villages at its foot at the junction with the Jhelum plains. The extreme dissection of the Potwar plateau above the scarp is clearly seen.

17 The bridge of boats across one branch of the Indus at Dera Ismail Khan. It is serviceable only during the dry season. When the river rises, a ferry steamer carries the traffic across fifteen kilometres or more of flood waters which completely inundate the island seen in the background.

and indeed represent a younger more recent phase. The lowest fluvial element is the active flood plain in which the present rivers trace their braided changing courses among the sand banks, filling the plain from bank to bank during times of flood.

At the mouth of the Indus the active flood plain becomes a delta, where the distributaries flow on their alluvial ridges between levees, with tidal back swamps and creeks carrying saline water when the river discharge is low.

From the Punjab plains to the delta the character of the flood plains changes, and from being incised into the meander flood plain the rivers flow between shallower banks. This fact was crucial in the days of irrigation by inundation canals. In the Upper Punjab the flood waters could not be led far from the river because of the height of the banks, while in the lower reaches downstream from Multan, the flatter profile of the plain made flood irrigation easier.

Another important change from north to south is the increasing fineness of the alluvium towards the mouth of the Indus not only at the surface but through the subsurface profile also. Because of this fact, tube wells are less effective as a means of tapping groundwater in lower Sind than in Punjab, since water will flow only slowly in clays, which are thus poor aquifers compared with coarse silt and sand.

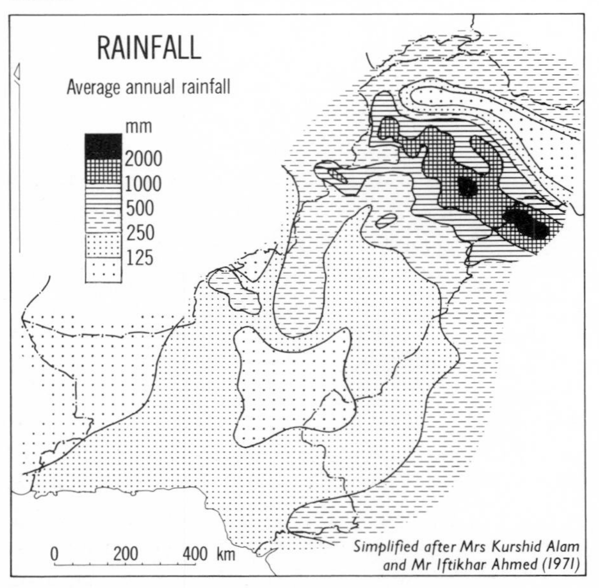

Fig. 4.6

Eolian, or windblown, surficial deposits cover extensive areas in the southeast, on the borders of the Thar Desert, where there are still active dunes. Another large tract of windblown sand is in Thal, the interfluve between Indus and Jhelum. Generally the sand is in hummocky dunes separated by swales of finer sediments where water occasionally collects. Canadian geomorphologists gave the terms 'rolling sand plain' and 'hilly sand plain' to describe different degrees of unevenness in the surface. In northern Thal levelling of the dunes has been undertaken over a wide area where canal irrigation has made the former desert bloom.

THE CLIMATES OF PAKISTAN

Aridity and Rainfall

The outstanding characteristic of Pakistan's climate is dryness, as the map of mean annual rainfall (Fig. 4.6) clearly shows. More than three-quarters of the country has less than 250 mm annually, and the small portion with more than 500 mm amounts to about 7 per cent of the area, and most of that is on mountain slopes. About 20 per cent of the total area has less than 125 mm and it should not be forgotten that the lower the total rainfall the higher its variability as a rule.

Over the Indus plains more than half the rainfall comes in the three months of the summer monsoon, July, August and September (Fig. 4.7). North and west of the mountain fringe less and less rain comes from the monsoon, and the major source is the travelling depressions passing from the Mediterranean under the path of the jet stream in winter. Since temperatures over the plains are extremely high in summer, generally averaging over 32°C with daily maxima often in the 40's, evaporation is high, greatly reducing the effectiveness of the summer rain. In Fig. 4.8 the annual evaporation rate is compared with average annual rainfall which in the plains consistently lags far behind. This high rate of evaporation has to be borne in mind when considering what happens to irrigation water in the summer. Rivers and canals flowing through the Indus plains are everywhere subjected to intense evaporation, far in excess of any contribution they may receive from local rain-

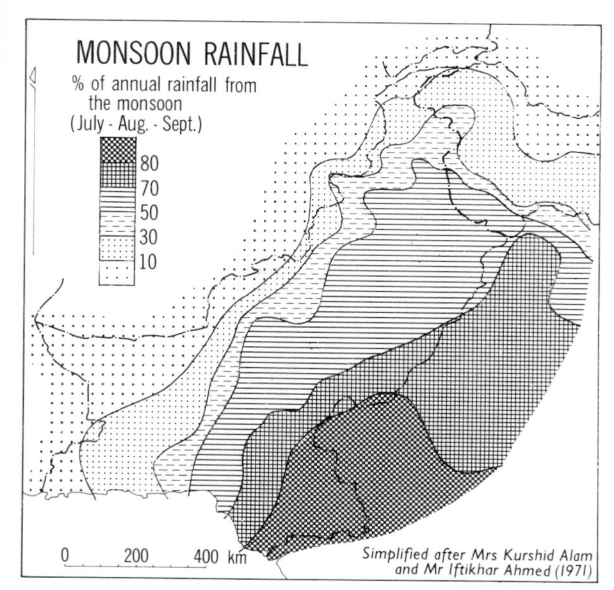

FIG. 4.7

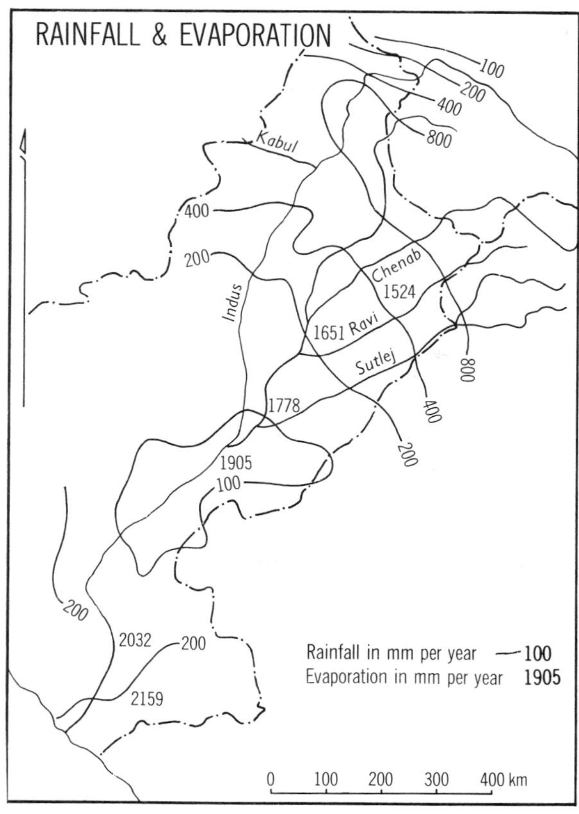

FIG. 4.8

fall. Such conditions promote the capillary movement upwards of moisture in the soil, a process responsible for concentrating in the surface layer salts from the subsoil, to the serious detriment of plant life.

Based on runs of up to 30 years of monthly rainfall data, Fig. 4.9 plots the dispersion of rainfall for 9 meteorological stations. For each station the diagram shows for each month the absolute maximum rainfall ever recorded, and the absolute minimum (which in many cases is nil.). The shaded portion of each column is the interquartile range, denoting that half the individual recordings for that month lie in the range of the shading, this indicating a probability that as likely as not in any future year rainfall will occur within those limits. The bar within the shaded section marks the median or central value in the series analysed, while the dot shows the mean or average value, almost invariably lying *above* the median because of the affect of the very occasional extremely high value. The median rainfall value is a more useful measure than the mean for most purposes.

The relative importance of rainfall outside the summer monsoon months is readily seen in the diagrams. Apart from Peshawar where the summer and winter rainfall is more or less in balance, and Quetta and Pasni where winter is the wetter season, all three being western stations, the remainder show a strong dominance of monsoon precipitation. Murree, a hill station at 2126 m, receives the highest total, 1618 mm on average, but as Fig. 4.6 shows, the restricted areas with totals over 1000 mm are limited to the outer, southern margins of the northern mountains. Inwards, the mountains, although rising very high, become increasing dry and more so the valleys between the ranges. At Murree, winter precipitation is quite substantial, and every month has a better than even chance of at least 25 mm. Zero rainfall is an occasional phenomenon in any month from October to January. The very high rainfall of July and August demonstrates the impact of the monsoon air flow arriving from the southeast as a stream of considerable thickness and high absolute humidity. Caused to rise against the Murree Hills, copious precipitation is triggered off from vast masses of towering cumulus. An August night in these hills can be a continuous spectacle of brilliant lightning, with

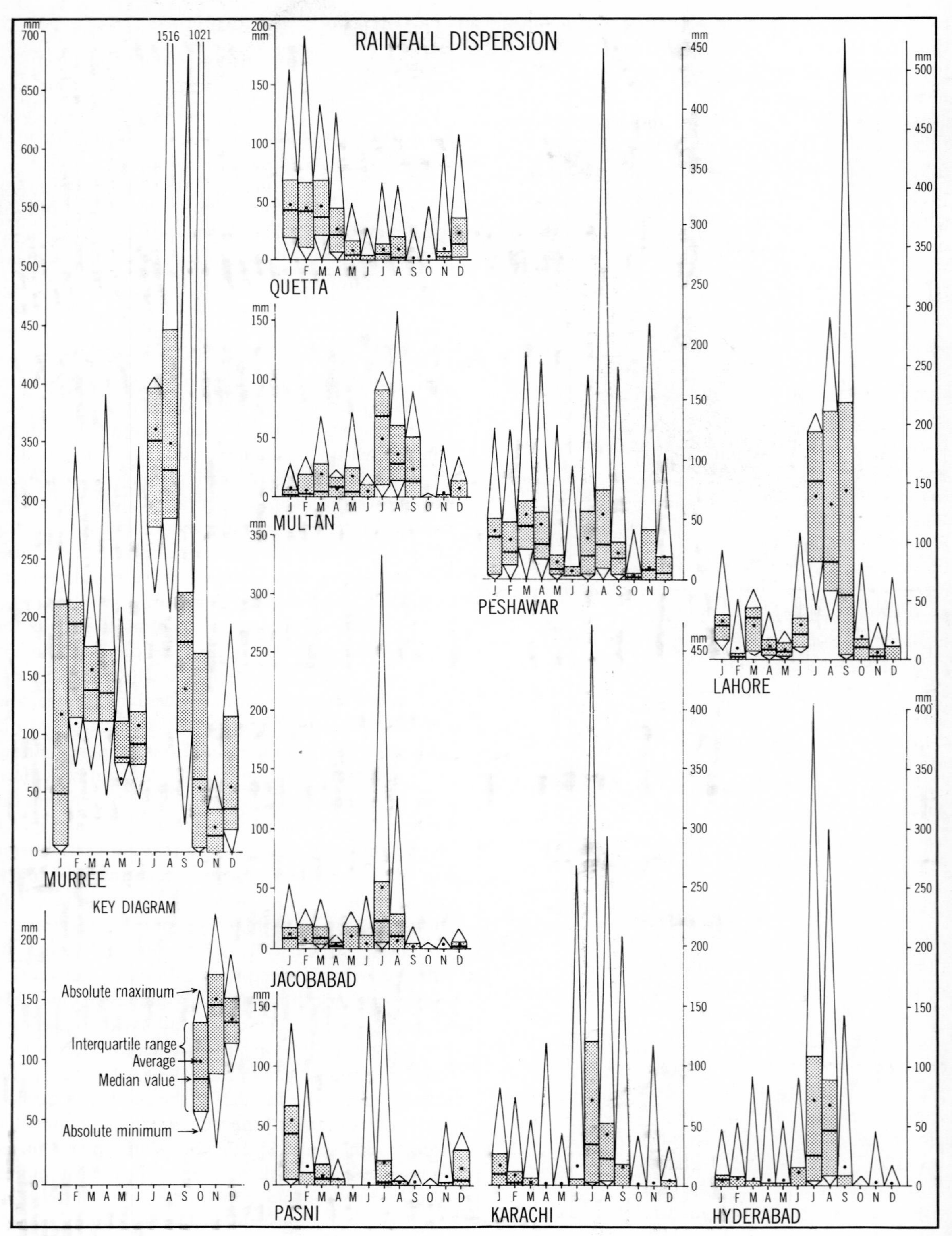
RAINFALL DISPERSION
MURREE
QUETTA
MULTAN
JACOBABAD
PASNI
PESHAWAR
KARACHI
LAHORE
HYDERABAD
KEY DIAGRAM
Absolute maximum
Interquartile range
Average
Median value
Absolute minimum
mm
J F M A M J J A S O N D
1516
1021

FIG. 4.9

thunder reverberating in the valleys, the whole performance perhaps culminating in a deluge of more than 25 mm in an hour.

It is the thickness of the monsoon air mass that accounts for the occasionally very high rainfalls at other stations. What is required to precipitate an exceptional downfall of rain is a stimulus to instability in an excessive thickness of moist air. Generally there is present over Pakistan in summer an inhibiting canopy of stable dry air. This represents the eastern extension of the mid-latitude high pressure cell that stagnates over Saharan Africa and Arabia in summer. It limits the scope for upcurrents in the surface monsoon air stream to rise very high and so to generate as much energy and rainfall as they otherwise might. Occasionally, as will be described in a later chapter in the case of a specific flood situation, this lid is removed and is replaced by an inflow of moist air from the Arabian Sea. Added to the general monsoon flow advancing up the Ganga valley, an extreme thickness of humid air arrives over the northern plains and hills, which once triggered into ascent is liable to deliver extremely heavy rainfall.

Winter air streams arriving over Pakistan from the west are shallower, cooler and so contain far less moisture than those just described. Frontal rainfall occurs, accentuated by the hills, as at Murree and Quetta, but is generally gentle. However, this winter precipitation, where it occurs, is of the greatest value to farmers, since it arrives when evaporation is at a minimum and so when a little rainfall goes a long way.

Wet and Dry Seasons

Away from the northern hills and their narrow piedmont belt, from Rawalpindi to Sialkot, north of Lahore, where annual precipitation averages between 500 and 1000 mm, the plains become increasingly arid and the rainfall more variable and less well distributed through the year. For the farmer looking to rainfall to enable him to raise a crop, or at least to complement his irrigation, the effectiveness of rainfall is more important than the amount. Granted the quality of the meteorological data available from the more remote stations, it is impracticable to apply the most sophisticated methods to analysing precipitation effectiveness.

In Fig. 4.10 of wet and dry seasons, and in the appendix of climatic statistics a simpler technique has been applied which has been found satisfactory in practice when applied to the climates of South Asia. In this system a dry month is one in which the average rainfall in millimetres is less than twice the temperature in degrees Celsius. Fig. 4.10 has been constructed using this formula, and it shows well the extreme brevity of the wet season throughout the plains. Effectively no month is wet in southern Punjab and northern Sind, an area with an average of less than 125 mm rainfall in many parts; similarly in the western interior of Baluchistan. The longer monsoon in the Himalayan piedmont brings 4–5 and even 6–7 wet months in a narrow belt which continues into the western hills on account of winter rainfall, which though light

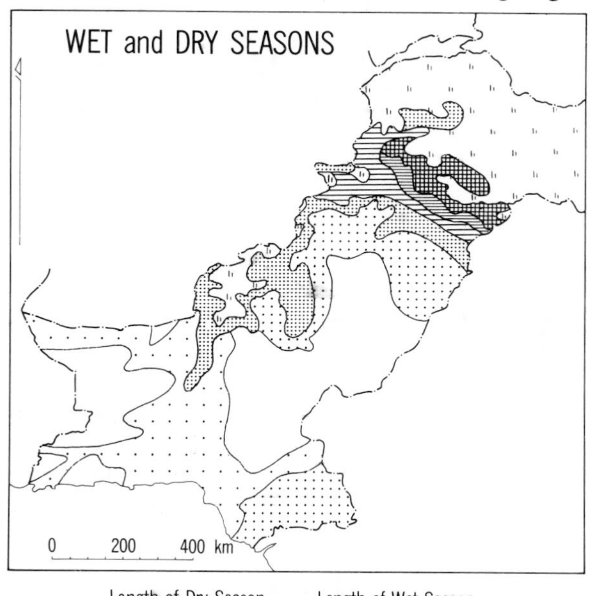

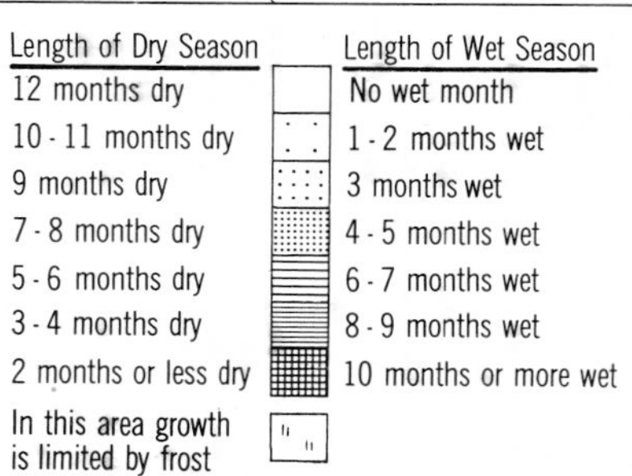

Fig. 4.10

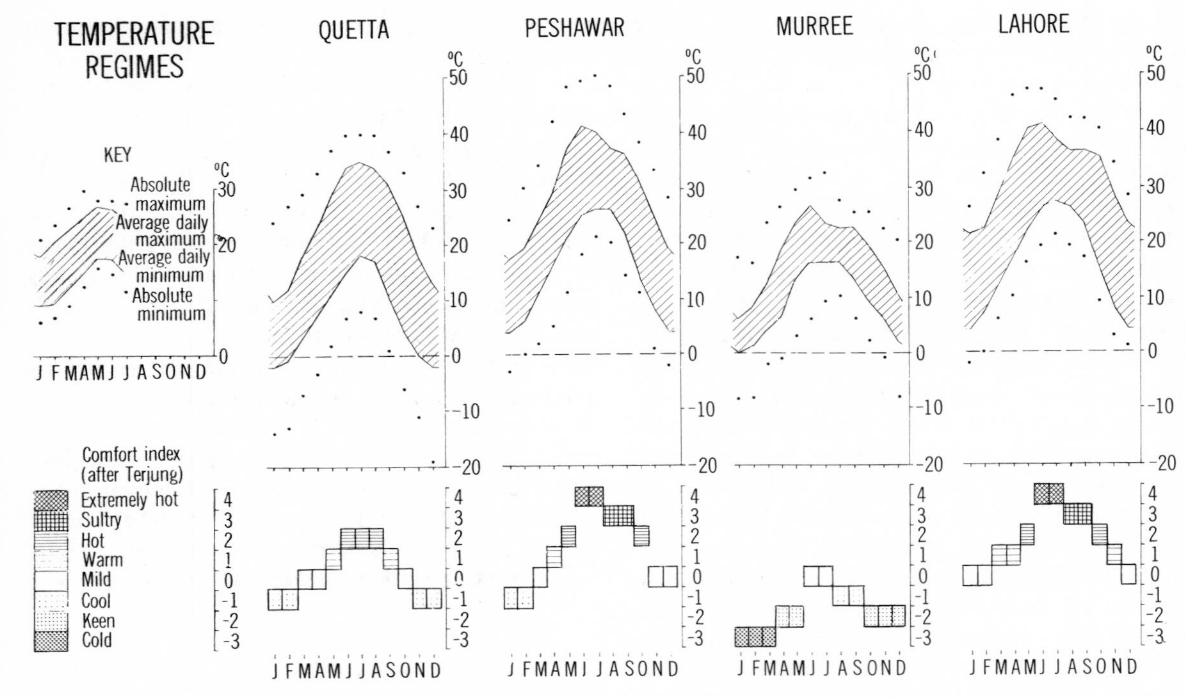

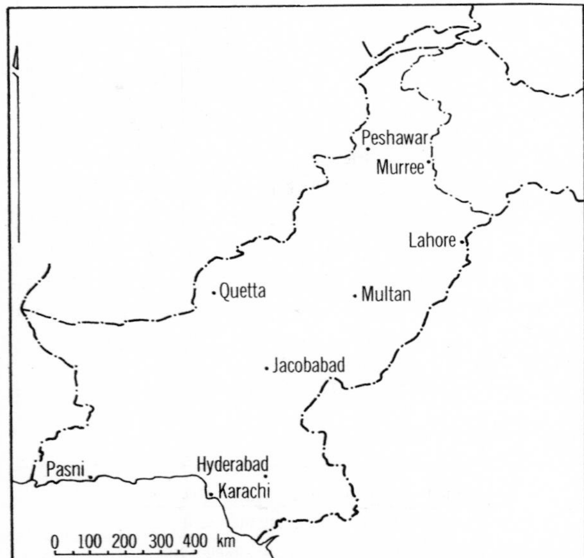

FIG. 4.11

is shown as more effective by the formula. The high hills around Murree have 10 or more wet months. The highest areas, irrespective of rainfall, have been excluded from this analysis since agriculture is constrained here by winter frost rather than by moisture deficiency.

The formula may usefully be applied to a time series of data to arrive at a measure of the probability of a month being dry. This has been done in the appendix to this chapter. In Murree no month is ever dry. In Lahore, April, May and November are always dry and only in July, August and January is there a better than even probability of the month being wet, with the lowest probability of dryness being 10 per cent in July. Peshawar shows much greater uncertainty of having a wet month in the monsoon, with probabilities of being a dry month standing at 90, 80 and 80 per cent for July, August and September respectively. The winter rains are rarely heavy, and at best reduce to 20 per cent the probability of a dry month in March.

The record for Multan is characteristic of the central plains. On average *no* month is a wet one, though occasionally this may happen as shown by its lowest probability of a dry month, 55 per cent in

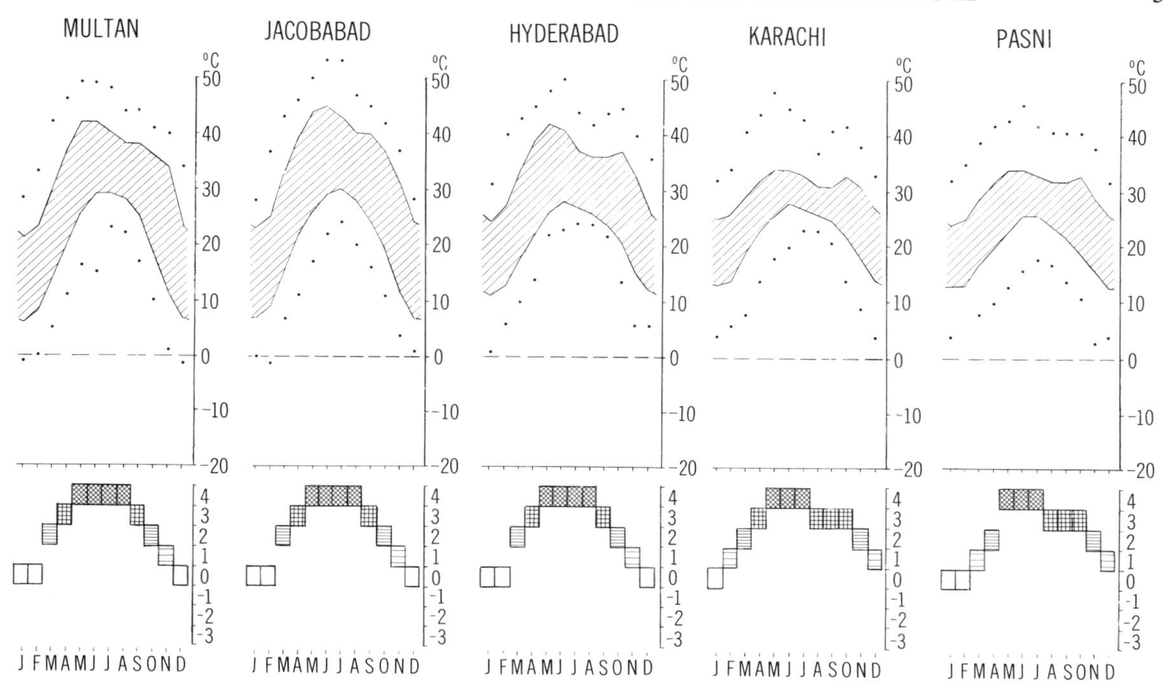

July. For all other months the probability is 85 per cent or more. Hyderabad is similar but more extremely seasonal in that only July (54 per cent) August (71 per cent) and September (86 per cent) diverge from 100 per cent probability of a dry month. Karachi, however, from June to December has a chance of being wet, extremely slight except in June when the chance of dryness is 54 per cent. These conditions reflect its position on the coast closest to the source of the monsoon. By contrast Pasni is at the extreme edge of monsoon influence, and June–July are dry months 92 per cent of the time. January's 69 per cent, February's 85 per cent and March's 92 per cent indicate the better prospect of winter rain.

For many of the stations of the central and southern plains and arid Baluchistan, these discussions of the probability of dry or wet months are academic when it is seen how variable rainfall is from year to year. Dera Ghazi Khan in southwest Punjab serves as a random example but typical of much of Pakistan. In the nine year period 1964 to 1972 its rainfall averaged 159 mm but in two years there was no rain at all. The yearly totals (in millimetres) were thus: 164, 239, 0, 595, 102, 26, 253, 0, 56. Pasni on the Arabian Sea coast west of Karachi has a long term average annual rainfall of 128 mm, a poor enough prospect, but made worse by its extreme variability. In the five years 1969 to 1973 it received 14, 152, nil, 48, and 3 millimetres of rainfall. For the same period Sibi had nil, 11, nil, 21 and again nil millimetres, against an average of about 17 mm. Clearly in these places rainfall incidence can have no place in the farmer's planning strategy. The problems of such extremes of variability in areas where irrigation cannot be provided from exotic sources of water are obviously acute. They are discussed further in Chapter 7 below.

Temperature and Comfort

One way of resolving the sum of climatic factors is through an analysis of comfort. This is possible for each station for which monthly data of rainfall and temperature are provided: Terjung's 'comfort index' which takes into account heat and cold, and indirectly, humidity, is plotted in Fig. 4.11; the annual extremes of temperature, January average daily minima and maxima, and similarly for June, are mapped in Figs. 4.12 and 4.13 respectively.

Although occasional ground frost is quite widespread in the northern plains, sustained cold nights below freezing are experienced only in the mountain rim as far south as Kalat in Baluchistan and along that province's border with Afghanistan in the Chagai Hills. The plains as a rule have winter minima averaging between 5° and 10°C. The moderating influence of the Arabian sea is clearly shown by the pattern of the 10° and 12°C isotherms parallel to the coast. If winter nights can be chilly, especially when as frequently happens, clear skies permit rapid radiation, the same cloudlessness brings sunny days when temperatures are quite comfortable. The mountains understandably are coolest with maxima below 10°C on average, but the plains range from 15°C in the north to 25°C on the coast. Temperatures thus may slow plant growth in the northern plains but certainly do not constrain the wheat crop seriously.

For the summer extremes of temperature June figures are mapped rather than July's (which are conventionally used in world climatological maps) because June usually shows the highest maxima of the year, July temperatures being reduced by the appearance of monsoon clouds. In terms of diurnal comfort it may be said that even the hottest days are tolerable if the nights bring relief and if humidity is low. Fig. 4.13 shows average minima for June and offers little respite in these terms. The hills alone give relief with minimum temperatures below 25° or 20°C in the highest regions. The plains remain very warm at night: Sibi averages 32°C, and across the middle Indus plains 29°C is common. Southwards the reduction to 28°C in Karachi is illusory comfort indeed, as oceanic humidity ensures that one is constantly bathed in sweat, night and day. Conditions are worse there in July and August but *may* be made bearable now and then in a downpour of rain.

Some idea of the extremes of discomfort are suggested by temperature and humidity figures for Las Bela, a little inland from Karachi. Las Bela

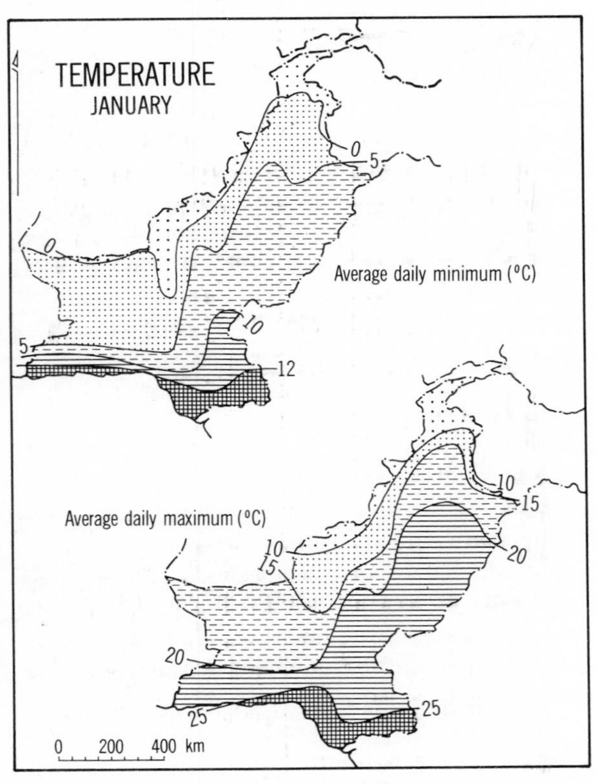

Fig. 4.12

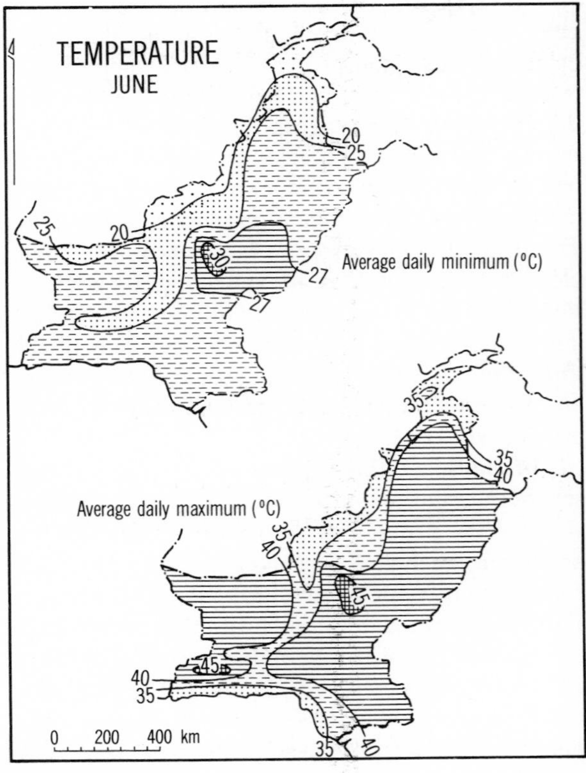

Fig. 4.13

had no rain at all from 1970 to 1973 inclusive, yet its summer humidity was high. May–June–July–August 1971 registered average daily maxima of temperature of 41°, 40°, 39° and 38°C accompanied by average humidities of 85, 84, 83 and 83 per cent respectively. For the average Pakistani without air conditioning or even a fan, these figures spell a most unpleasant season, worse perhaps than that of Sibi, where in 1973, May–June–July, had average temperature maxima of 44, 47 and 43°C with humidities of 65, 68 and 65 per cent. Throughout Lower Sind, houses are constructed to keep out heat and daylight, and are surmounted by fixed ventilators of mud brick or timber, facing southwestwards to catch any breeze that may bring relief. Flat roofs and mud-walled compounds to the houses throughout Pakistan allow the family to sleep outside in hot dry weather.

The record of the average daily maxima for June has parallels in few parts of the world: in southern Sahara, perhaps, and northwestern Australia. Two extreme hot spots, one in the Turbat valley in Makran, Baluchistan, the other in the Sibi-Jacobabad reentrant, have *average* maxima exceeding 45°C; Jacobabad's absolute maximum is 53°C. These are extremes, but the greater part of Pakistan swelters at averages over 40°C. In such conditions whatever work has to be done is carried out in the dawn hours, and by midday life has gone dormant, and man lies prone indoors in the dark, to re-emerge in the late afternoon. The hills give refuge for the fortunate, the higher settlements enjoying maxima of less than 35°C. Similarly the coast is cooler, though as mentioned already the 'moderate' 34° of Karachi, Pasni and Jiwani obscures their high humidity. Sea breezes may be some compensation, but most people would prefer to summer at 41°C in Lahore rather than in Karachi; at least Lahore offers 14°C diurnal range and a lower humidity before the rains, against only 6°C range in Karachi.

In the lower half of Fig. 4.11 are shown comfort indices derived from Terjung's maps. Eight levels of comfort are used, ranging from cold to extremely hot.* The hill station Murree with its mild summer

18 A homestead at Chhor, Tharparkar, Sind, close to the edge of the Thar Desert. This area is particularly hot in summer, and houses are often built with fixed ventilators on the roof to introduce any breeze that blows from the southwest into the window-less interior. The walls are of sun-dried bricks plastered over with mud. The roof is of thick straw or reed thatch supported by beams and mud-plastered.

months provides the possibility of relief from the extremely hot plains for those who can afford to travel. Quetta's climate is moderated to a lesser extent by altitude, but is still appreciably more comfortable than the adjacent plains stations of Multan and Jacobabad, which with Hyderabad suffers four extremely hot months. On the coast the moderating influence of the sea reduces the extremely hot season to three months at Karachi and Pasni but delays the cool season, holding August to October at the 'sultry' level. The winter months, December to February, are generally mild and most comfortable. Murree becomes uncomfortably cold from January to March, but Quetta's winter is more tolerable with four cool months.

Conclusion

The map of climates, Fig. 4.14, brings together the main factors discussed above. Rainfall amount is used as the primary criterion for classification, with cold temperature to separate the northern mountains and some western patches of high country. Within the semi-arid belt (250–500 mm) it is useful to distinguish the region with mainly summer rainfall from that most influenced by winter depressions, the 'Mediterranean' climatic type. The scheme might be further elaborated to show at

*Terjung, W. H., 'World Patterns of the Distribution of the Monthly Comfort Index', *International Journal of Biometeorology*, 12, 1968, pp. 119–51.

what season the scanty rainfall of the arid zone was received, but in view of the high variability of rainfall the exercise seemed somewhat unrealistic. Reference may be made to Fig. 4.7 to note that the greater part of the plateau and coast of Baluchistan has its scanty rainfall in winter. Two features in the north merit mention. The deep valley of the Indus enjoys a Mediterranean climate, but its more open upper valley and the inner mountains generally are subhumid to arid depending on aspect. Most of the summer monsoon air mass is blocked by the outer ranges and the Karakoram receive much snow from the winter depressions travelling beneath the southern arm of the jet stream along the edge of the high ranges.

These, then, are the environments to which man has had to adapt in Pakistan. The extent to which he has succeeded without recourse to modern technology or to capital resources beyond his ability to repay, is the subject of Chapter 7 below. It is a story of endurance in the face of adversity that helps explain the character of the peoples of Baluchistan and the Frontier hills. In the chapters that immediately follow are studied the more prosperous and sophisticated developments of the Indus plains using canal waters in defiance of climate, and providing the mainstay of Pakistan's economy.

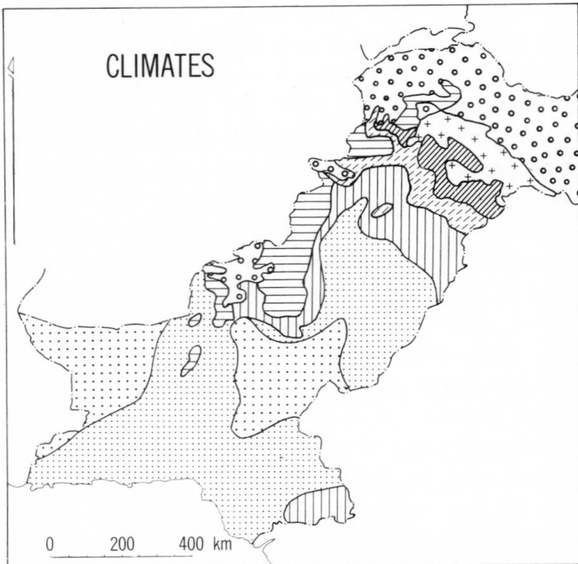

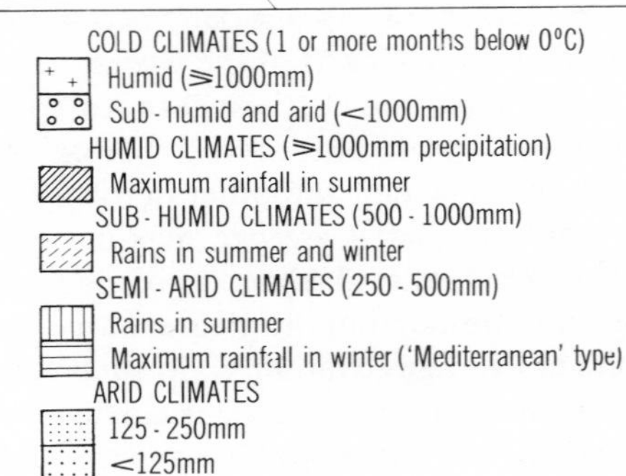

FIG. 4.14

CLIMATIC DATA APPENDIX

Station	*No. of years*	*Jan.*	*Feb.*	*March*	*April*	*May*	*June*	*July*	*Aug.*	*Sept.*	*Oct.*	*Nov.*	*Dec.*	*Year*
QUETTA		30° 15′N 66° 53′*E* *Height above sea level 1587 m*												
Temperature														
Av. daily max	30	10	12	18	23	29	34	35	34	31	25	18	13	23
min		−2	−1	3	7	11	15	18	17	10	4	0	−2	7
Absolute max	30	24	27	29	33	37	40	40	40	37	33	27	23	40
min		−14	−13	−7	−3	2	7	8	7	1	−6	−11	−19	−19
Av. temp.	Clino	4	6	1	16	21	24	27	25	21	14	9	5	15
Av. rainfall		37	43	42	12	7	1	18	4	1	1	6	23	195
P > 2T		√	√	√									√	
Probability month will be dry %	13	15	23	38	85	100	100	92	100	100	100	86	38	
Rainfall Dispersion														
Absolute min		1	1	0	0	0	0	0	0	0	0	0	0	99
1st quartile	43	17	12	21	6	0	0	0	0	0	0	0	2	167
Median		40	40	37	21	3	0	4	1	0	0	1	13	223
3rd quartile		68	65	68	41	14	3	13	19	0	3	6	36	274
Absolute max		162	191	135	129	50	29	68	64	29	47	94	108	548
Mean		48	46	48	26	9	4	9	11	2	3	8	24	237

CLIMATIC DATA APPENDIX continued

Station	*No. of years*	*Jan.*	*Feb.*	*March*	*April*	*May*	*June*	*July*	*Aug.*	*Sept.*	*Oct.*	*Nov.*	*Dec.*	*Year*
PESHAWAR				34° 01′N	71° 34′E		Height above sea level 359 m							
Temperature														
Av. daily max	33	17	19	24	29	37	41	40	37	36	31	25	19	29
min		4	6	11	16	21	25	26	26	22	14	8	4	15
Absolute max	30	24	30	34	42	48	49	50	48	43	38	33	28	50
min		−3	0	2	5	11	18	21	20	14	11	1	−2	−3
Av. temp.	Clino	11	13	17	23	29	33	33	31	29	24	18	13	23
Av. rainfall		39	41	65	42	40	7	39	41	14	10	10	15	363
P > 2T		√	√	√										
Probability month will be dry %	10	70	60	20	70	100	100	90	80	80	90	80	70	
Rainfall Dispersion														
Absolute min		0	0	0	1	0	0	0	0	0	0	0	0	102
1st quartile		3	12	25	15	3	0	5	10	3	0	0	0	279
Median	67	36	23	46	29	9	3	20	29	16	1	9	5	330
3rd quartile		56	49	68	56	20	11	61	76	36	5	41	19	406
Absolute max		129	127	191	187	131	98	175	451	178	42	216	110	711
Mean		39	35	53	46	18	8	36	55	21	5	11	16	342
LAHORE				31° 35′N	74° 20′E		Height above sea level 214 m							
Temperature														
Av. daily max	23	21	22	28	35	40	41	38	36	36	35	28	23	32
min		4	7	12	17	22	26	27	26	23	15	8	4	16
Absolute max	10	26	32	38	46	47	47	45	42	42	40	34	28	47
min		−2	0	6	10	16	19	21	19	17	9	3	1	−2
Av. temp.	Clino	12	15	21	27	32	34	32	31	30	25	19	14	24
Av. rainfall		31	23	24	16	12	38	122	123	80	9	3	11	492
P > 2T		√						√	√	√				
Probability month will be dry %		40	90	70	100	100	90	10	30	60	80	100	90	
Rainfall Dispersion														
Absolute min		9	0	0	0	0	6	42	30	0	0	0	0	278
1st quartile		15	2	8	4	2	11	83	60	2	0	0	0	395
Median	10	27	3	36	9	7	20	151	83	54	11	1	0	660
3rd quartile		37	6	44	16	13	34	191	212	218	15	9	11	684
Absolute max		95	52	61	40	22	109	209	292	526	80	33	71	870
Mean		33	9	27	12	8	30	138	134	143	18	17	14	572
PASNI				25° 16′N	63° 29′E		Height above sea level 9 m							
Temperature														
Av. daily max	20	24	25	29	32	34	34	33	32	32	33	29	26	31
min		13	13	17	20	23	26	26	24	22	19	16	13	19
Absolute max	20	32	35	39	42	43	46	42	41	41	41	38	32	46
min		4	0	8	10	13	16	18	17	14	11	3	4	0
Av. temp.	Clino	19	20	23	27	30	31	30	28	28	27	24	20	25
Av. rainfall		43	32	8	7	2	6	12	13	1	0	2	12	128
P > 2T		√												
Probability month will be dry %	13–14	69	85	92	100	100	92	92	100	100	100	93	93	
Rainfall Dispersion														
Absolute min		0	0	0	0	0	0	0	0	0	0	0	0	24
1st quartile		1	0	0	0	0	0	0	0	0	0	0	0	66
Median	14	41	0	3	0	0	0	1	0	0	0	0	3	105
3rd quartile		64	10	17	3	0	0	21	3	0	0	1	30	201
Absolute max		137	91	44	23	0	146	160	7	13	4	55	45	258
Mean		53	16	7	3	–	0.4	21	2	2	0.4	6	15	126

CLIMATIC DATA APPENDIX continued

Station	*No. of years*	*Jan.*	*Feb.*	*March*	*April*	*May*	*June*	*July*	*Aug.*	*Sept.*	*Oct.*	*Nov.*	*Dec.*	*Year*
MURREE		33° 54′N 73° 24′E Height above sea level 2168 m												
Temperature														
Av. daily max	10	6	8	12	18	23	26	23	22	22	19	15	10	17
min		0	1	4	9	13	16	16	16	13	9	6	2	9
Absolute max		17	16	23	26	29	31	32	27	25	25	22	20	32
min		−8	−8	−2	−1	3	6	9	10	6	2	−1	−8	−8
Av. temp.		3	5	8	13	18	21	20	19	18	14	10	6	13
Av. rainfall		116	108	155	103	61	106	360	348	138	53	21	54	1618
P > 2T		√	√	√	√	√	√	√	√	√	√	√	√	
Probability month will be dry %	10	10	0	0	0	0	0	0	0	30	40	50	10	
Rainfall Dispersion														
Absolute min		0	71	69	48	56	43	221	203	23	0	0	0	
1st quartile		5	114	112	112	76	74	227	284	102	3	0	18	
Median		48	193	137	135	79	91	351	325	178	61	13	36	
3rd quartile		211	213	175	173	112	119	396	445	221	168	36	114	
Absolute max		267	345	236	389	208	345	404	1516	681	1021	66	193	
Mean		(as average above)												
MULTAN		30° 12′N 71° 26′E Height above sea level 126 m												
Temperature														
Av. daily max	33	21	23	30	37	42	42	40	38	38	36	29	23	33
min		6	8	14	20	26	29	29	28	25	18	12	7	19
Absolute max	33	28	33	42	46	49	49	48	44	44	41	40	34	49
min		−1	0	5	11	16	15	23	22	17	10	1	−1	−1
Av. temp.	Clino	13	17	22	28	33	36	35	33	32	27	20	15	26
Av. rainfall		7	10	13	6	8	8	45	33	20	1	2	5	158
P > 2T														
Probability month will be dry %	12–14	92	92	85	100	85	100	55	100	92	100	93	92	
Rainfall Dispersion														
Absolute Min		0	0	0	0	0	0	0	0	0	0	0	0	55
1st quartile		1	0	0	0	0	0	11	14	0	0	0	0	94
Median	14	1	1	5	9	5	0	69	30	13	0	0	1	176
3rd quartile		5	18	28	15	25	12	91	61	53	0	1	13	223
Absolute Max		28	33	68	22	75	19	107	162	89	4	46	37	358
Mean		7	6	19	8	17	4	50	38	23	0.3	5	9	184.5
KARACHI		24° 48′N 66° 59′E Height above sea level 4 m												
Temperature														
Av. daily max	43	25	26	29	32	34	34	33	31	31	33	31	27	29
min		13	14	19	23	26	28	27	26	25	22	18	14	22
Absolute max	43	32	34	41	44	48	45	43	37	41	42	38	33	48
min		4	6	8	14	18	20	23	23	21	14	9	4	4
Av. temp.		19	21	24	27	29	30	29	28	27	27	25	21	26
Av. rainfall	Clino	7	11	6	2	0	7	96	50	15	2	2	6	204
P > 2T								√						
Probability month will be dry %	13–14	100	100	100	100	100	92	54	86	79	93	93	93	
Rainfall Dispersion														
Absolute min		0	0	0	0	0	0	0	0	0	0	0	0	12
1st quartile		0	0	0	0	0	0	1	2	0	0	0	0	85
Median		9	1	0	0	0	0	31	22	0	0	0	0	155
3rd quartile	55	26	11	4	0	0	5	122	52	14	0	0	5	242
Absolute max		86	75	57	121	47	269	473	295	210	40	118	36	605
Mean		15	9	7	3	2	16	72	43	14	5	3	4	188

CLIMATIC DATA APPENDIX continued

Station	No. of years	Jan.	Feb.	March	April	May	June	July	Aug.	Sept.	Oct.	Nov.	Dec.	Year
JACOBABAD		28° 18′N		68° 28′E			Height above sea level 56 m							
Temperature														
Av. daily max	30	23	25	33	39	44	45	43	40	40	37	31	24	35
min		7	9	16	22	26	29	30	28	24	19	12	7	19
Absolute max	10	28	37	43	46	50	53	53	47	45	42	37	28	53
min		0	−1	7	11	17	22	24	20	16	11	4	1	−1
Av. temp.	Clino	15	18	24	30	35	37	35	34	32	28	22	17	27
Av. rainfall		8	8	7	2	4	6	37	22	1	0	1	3	99
P > 2T														
Probability month will be dry %	13–14	92	100	100	100	100	100	92	93	100	100	100	100	
Rainfall Dispersion														
Absolute min		0	0	0	0	0	0	0	0	0	0	0	0	17
1st quartile		0	0	2	1	0	0	6	0	0	0	0	0	45
Median	14	10	2	9	1	0	0	22	9	0	0	0	1	96
3rd quartile		17	18	17	4	18	10	53	27	2	0	0	5	153
Absolute max		54	34	43	10	30	43	333	129	16	4	10	14	408
Mean		11	5	9	2	8	1	51	7	1	0.4	2	3	99.8
HYDERABAD		25° 23′N		68° 25′E			Height above sea level 30 m							
Temperature														
Av. daily max	20	24	27	34	39	42	41	37	36	36	37	32	26	34
min		11	13	18	22	26	28	27	26	24	21	15	12	20
Absolute max	10	31	40	43	45	48	50	44	42	44	45	40	36	50
min		1	6	10	14	22	23	24	24	22	14	6	6	1
Av. temp.	Clino	17	21	26	31	34	34	33	31	31	29	24	19	28
Av. rainfall		4	5	1	2	4	6	69	44	15	8	1	3	157
P > 2T														
Probability month will be dry %	13–14	100	100	100	100	100	100	54	71	86	100	100	100	
Rainfall Dispersion														
Absolute min		0	0	0	0	0	0	0	0	0	0	0	0	26
1st quartile		0	0	0	0	0	0	2	10	0	0	0	0	78
Median	54	7	0	0	0	0	0	24	47	0	0	0	0	150
3rd quartile	58	9	8	4	1	1	13	109	93	10	0	0	0	275
Absolute max		49	55	92	86	56	91	402	300	145	10	48	17	537
Mean		6	7	5	3	4	12	71	67	16	0.3	2	1	193

Sources: US Department of Commerce, *World Weather Records, 1951–60*, Vol. 4 Asia, Washington D.C. 1967.
UK Meteorological Office, *Tables of Temperature, Relative Humidity and Precipitation for the World*, Part V Asia. London 1966.
Wernstedt, F. L., *World Climate Data*, Climatic Data Press, Pennsylvania 1972.

Note: Clino stands for climatological normal, or long-term mean value.

CHAPTER FIVE

MAN'S TRANSFORMATION OF THE HYDROLOGICAL ENVIRONMENT

SUMMARY*

The irrigation system developed in the Indus plains over the past hundred years represents the most complex and extensive example in man's experience of the transformation of a natural hydrological system to serve his agricultural needs. The flow characteristics of the rivers that constitute the Indus basin drainage system are analysed as to their relative volume, seasonal fluctuation and annual variability. The use of river and ground water to supplement rainfall or even to enable man to cultivate crops where rainfall is inadequate is of great antiquity in the Indus basin. Progressively man has evolved engineering systems better to control the rivers and distribute their waters. His mastery of water management has increased greatly since the scientific and industrial revolution began during British rule to have a direct impact in the areas now forming Pakistan. Partition of the Punjab in 1947 was a serious set back, but from the responses to the challenge it provided, a vast, intricate and integrated system of water use has been devised and continues to be developed. It is not without problems, however, some of which are dealt with in more detail in Chapter 6.

*Technical and other background for this chapter comes mainly from:

Michel, A. A., *The Indus Rivers*, 1967, Yale University Press. UNECAFE Bureau of Flood Control – *Flood damage and flood control activities in Asia and the Far East*, 1950.

Raikes, R. L., 'The Ancient Gabarbands of Baluchistan', *East and West*, 15, No. 1–2, 1964–65., pp. 26–35.

Lieftinck, P., Sadove, A. R. and Creyke, T. C., *Water and Power Resources of West Pakistan* IBRD/John Hopkins Press 1968.

WAPDA, *Annual Reports*, 1974–75, et seq. and *WAPDA Yesterday, Today, Tomorrow*, 1978.

Irrigation may be taken to mean a system of cultivation in which water is brought to a farmer's fields as a substitute for, or as a supplement to, an inadequate rainfall. In essence use is made of precipitation which fell elsewhere. Such water may be obtained from the surface flow in rivers or may be recovered by a variety of means from groundwater below the surface.

For reliable agriculture in the Indus plains the major sources of irrigation water have been the large exotic rivers of the Indus system: the Indus with its major right-bank tributary the River Kabul, and the five rivers of the Punjab, Jhelum, Chenab, Ravi, Beas and Sutlej. All these rivers obtain most of their flow from their catchments within the mountains. In each case their discharge is marked by great seasonal variation. Table 5.1 below shows the relative importance of the several catchments measured by run-off within the river basins above a point close to where they enter the plains.

The average regimes of the three major rivers are illustrated in Fig. 5.1. Those of the smaller eastern rivers, now utilised by India under the Indus Waters Agreement, are closely similar to that of the Chenab. The diagrams of average monthly flow and the mean annual total mask considerable variation from year to year. Thus the average discharge of 115,000 million m^3 in the Indus at Attock lies within a range of between 75–118 per cent of normal. On the Jhelum the variation from its mean runoff of 28,000 million m^3 is greater, between 65–135 per cent of normal. The Chenab's higher monsoon flow compared with that of the Jhelum is due to its more easterly catchment closer to the monsoon airflow from the southeast.

TABLE 5.1
Runoff and Discharge in the Indus Rivers

Catchment	*Average annual runoff (thousand million m³)*	*Discharge ('000 m³ per second)* Maximum	Minimum
Indus (including Kabul)	115	26	0.48
Jhelum	28	22	0.11
Chenab	32	20	0.11
Ravi	9	7	0.03
Beas	16	10	0.06
Sutlej	17	14	0.08

A better impression of variability of flow may be had from Fig. 5.2 in which the actual flows in the kharif (April–September) and rabi (October–March) seasons is shown for the three rivers over a period of eleven years. Rabi flows follow kharif but at a much lower level, since winter precipitation is relatively small and is in any case in the form of snow in the mountains, thus not becoming available as melt-water until the spring time, when the rivers start to rise from March onwards.

With the onset of the monsoon rains in late June and July the rivers approach their peak flood levels of July–August at the 'rim stations' where they issue onto the plains, and the flood wave passes downstream to reach Sind a month or so later.

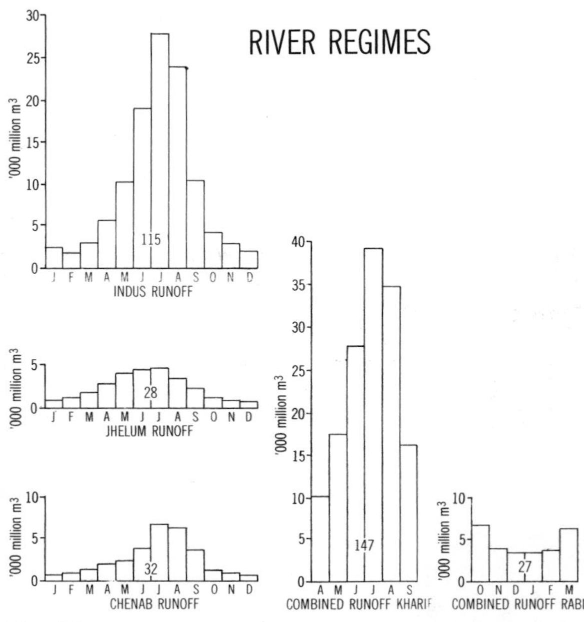

FIG. 5.1

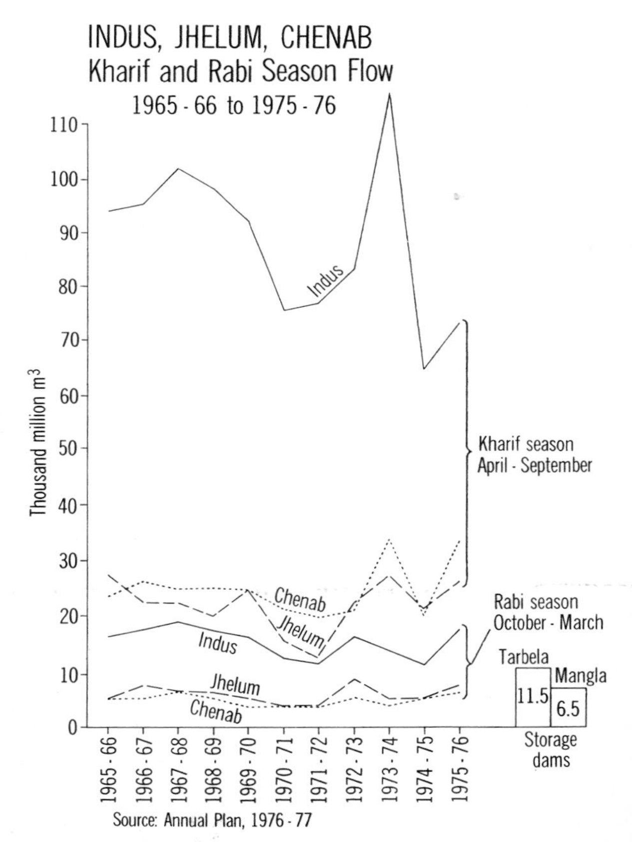

FIG. 5.2

TRADITIONAL IRRIGATION

Modern water engineering brought about such a demographic and economic revolution in the twentieth century in the region that is now Pakistan that one tends to forget the very considerable impact of irrigation even before the period of British rule.

Before the Indus rivers were brought under control by the weirs and barrages constructed to divert their water into canals, thus reducing the flood levels downstream of such works, they behaved in their natural fashion, overspilling their banks during the high water period of summer, and occasionally changing their courses. Such natural flooding was no doubt the essential basis for early agriculture in the drier parts of the plains, though it may well have been supplemented by water drawn from shallow wells. The Indus rivers, flowing as they do in a plain of deep alluvial sediments, are bordered by a zone within which the ground water table is relatively close to the surface, due to seepage from the river, and in which it is relatively easy to obtain water by digging wells.

The Indus civilization which supported the cities of Mohenjo-Daro and Harappa between 2500–1500 BC contended with a climate not significantly different from that of today. Clear evidence is lacking as to whether or not the farmers of that time diverted flood waters out of the active flood plain by inundation channels, but in order to subsist in so arid a region they must surely have cultivated the flood plains themselves by minor diversion works when the rivers were not in spate. Lift irrigation by simple 'shaduf' pole and basket systems are suggested by one of the seals found in Indus valley excavations, and in the *Rig Veda*, one of the earliest epics in the Hindu tradition, predating written records and thought to be approximately contemporary to the later stages of the Indus civilisation, reference appears to be made to well digging and 'circles of pots' which might describe the Persian Wheel common today for raising water. Such systems are likely to have been in use in the Indus plains for many centuries.

During the first millenium BC irrigation was being actively developed by Persians and Greeks in Iran and Afghanistan. In Pakistan's area, Greek influence was strong in the Himalayan valleys and piedmont from the Kabul valley and Swat through Taxila to Sialkot, regions where today it is customary to supplement the moderate rainfall by well irrigation and minor channel diversions. The Greek writer Strabo quoted from Megasthenes, an Ambassador of the court of Chandragupta Mauryam *c.* 300 BC, concerning the appointment of officials 'to superintend the rivers, measure the land ... and inspect the sluices by which water is let out from the main canals into their branches, so that everyone may have an equal supply of it.' This could well refer to a system of inundation canals.

Islamic influence particularly from the fourteenth century brought further stimulus from Iran where the Muslims had developed a reputation for water management, horticulture and the design of pleasure gardens. Firoz Shah Tughlak was perhaps the first ruler to use canal irrigation as an instrument of colonisation, in the interfluve between the Sutlej and Yamuna, east of Pakistan's border with India. The continuing fame of the Shalimar Gardens watered from the Ravi just east of Lahore dates from the efforts of Ali Mardan Khan, from Afghanistan, in 1639; in tapping the Ravi close to its exit from the hills, he came near to the basic design of modern perennial canal systems.

19 A Persian Wheel of more or less traditional design, powered by a blindfolded bullock. It turns a horizontal wooden wheel geared to a vertical wheel at the distant end of the shaft. This carries the vertical metal wheel in the left foreground, to which is attached a chain of earthernware pots. The pots raise water from the well and spill their contents into the channel that leads to the fields.

In the dry heart of the Indus plains, the Moghul governors of Multan and the Emirs in Sind, and following them in the nineteenth century, the Sikhs and later the British, cut inundation canals to divert the summer floods. These were a great improvement on natural flood irrigation and on the more limited flood plain channel diversions possible only at low water. The British military engineers brought to the challenge of the Indus rivers knowledge and skills from the centre of the world's scientific, industrial and economic revolution. Many pre-existing inundation canal systems were renovated and extended and new projects were undertaken.

As a general rule inundation canals were highly seasonal in operation. They were fed from cuts in river bank and would normally fill only when the river level was high. Where the level of the intake allowed the river to be tapped at a moderately low stage irrigation might be extended to rabi crops, but the principal consideration was always the watering of the kharif crops with the provision of a pre-watering for planting a rabi crop as an occasional bonus. Some inundation canals utilised abandoned anabranches of the river, and might provide perennial irrigation if lift mechanisms could be arranged. Naturally gravity flow was preferred, but frequently Persian Wheels or shadufs had to be used to raise water to fill the distributary channels.

Irrigation by inundation canal was limited to the areas fairly close to and parallelling the river, the active and meander flood plain. A sketch map of Sind in about 1860 shows the pattern of distribution channels developed at that stage (Fig. 5.3) while the more formal surveys later in the century depict the situation in the lower Punjab and Sind. In such belts wells dug to the relatively shallow water table near the river provided a valuable complement to the canals. Together with river bank lift systems, wells extended, as they do still, the period of irrigation somewhat, at least for a part of the cultivated area. Multan district had 14,975 wells in 1884–85, 5.9 per 100 ha, and Jhang 12,132, 8.6 per 100 ha.

Fig. 5.4 shows the pattern of irrigation canals in the Indus plains in 1891–1901. Morphology was largely responsible for the fact that inundation canals had greater application to Sind and the southern Punjab than further upstream (see also Fig. 4.4, above). In the submontane Punjab as far south as Multan, there are old alluvial tracts, the 'scalloped interfluves' or bar lands, which stand well above the active flood plains. Inundation canals could not command such tracts. At a lower level the height of the meander and cover flood plains relative to the rivers reduces progressively downstream. It was the comparative ease with which the flood flow of the Indus, lower Chenab and lower Sutlej in the southern Punjab and of the Indus in Sind could be diverted that mainly explain the pattern of inundation canals that developed.

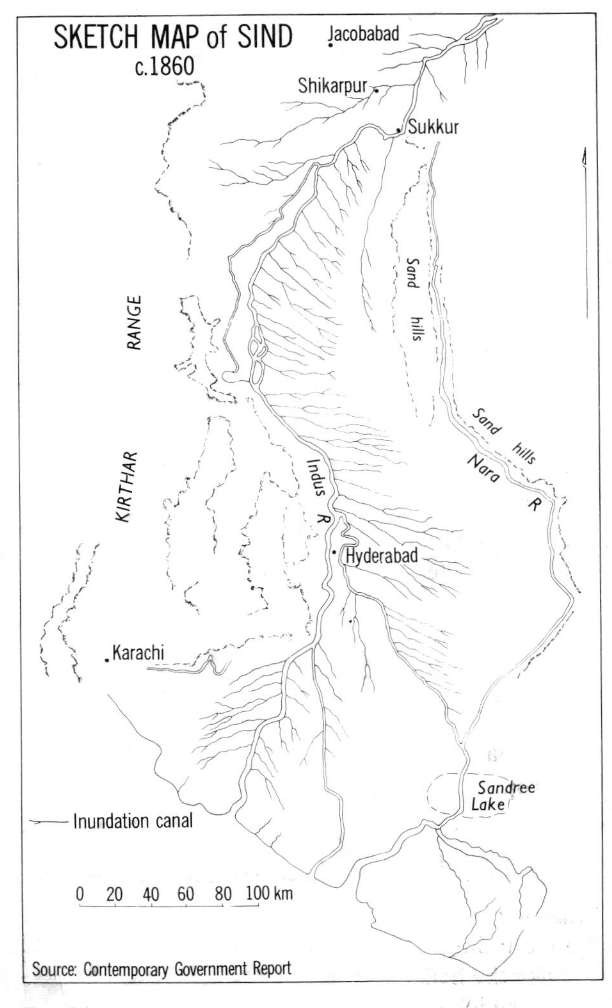

FIG. 5.3

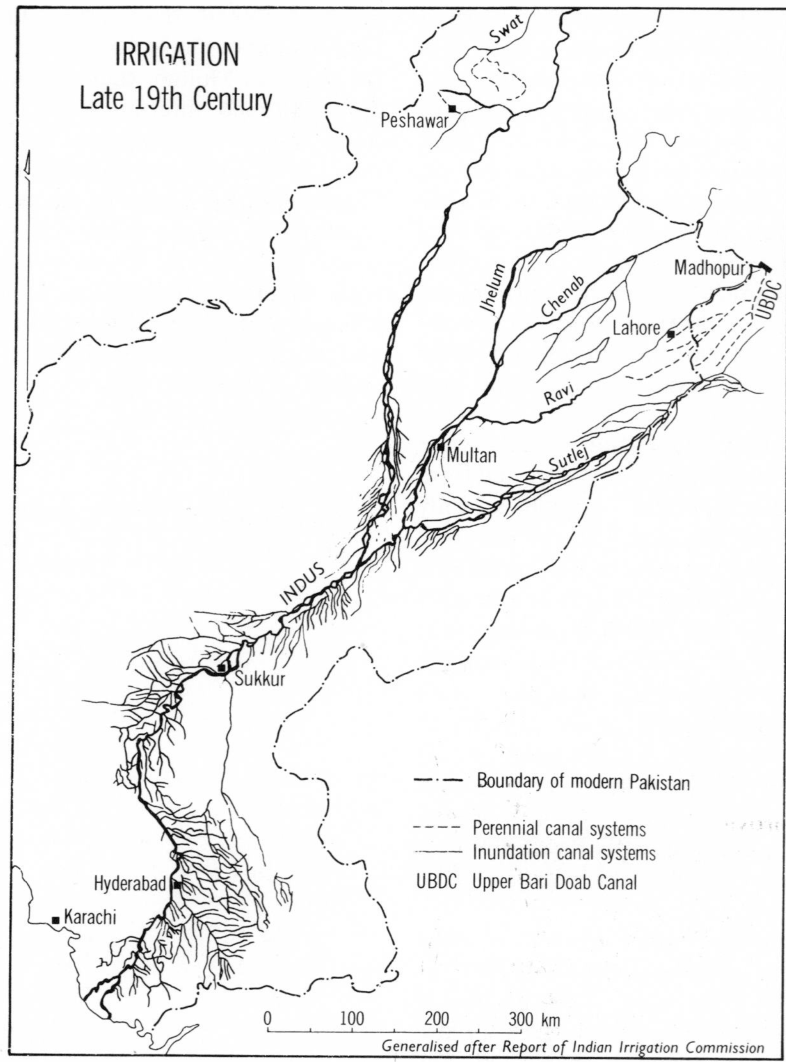

FIG. 5.4

As today, wells were an important factor in the agriculture of the submontane Punjab where they augmented the moderate rainfall in areas where canals watered at best 20 per cent of the crop. Table 5.2 illustrates this point.

PERENNIAL CANAL IRRIGATION

Experience gave the British engineers confidence: from building simple weirs across the river channel to raise the head of water and so extend the season during which water flowed in the canal, they moved to the construction of gated barrages which gave them a much greater measure of control of the water and its transported load of silt. The problem of the heavy silt load of the Indus rivers as they slacken their gradients on entering the plains has been a perpetual headache to irrigation engineers. Silt deposition has sometimes threatened to render canals useless.

The British made their first 'breakthrough' to perennial canal construction with the Ganges canal in India in 1854, to be followed in 1859–61 by the Upper Bari Doab Canal in the Punjab. While the Ganges canal was constructed as a protection against famine, the UBDC was dug for political reasons, to provide areas in which to settle Sikhs who lacked livelihood at the end of the Sikh Wars

TABLE 5.2
Irrigation in select districts of the Punjab, 1884–85

District	Percentage of cultivated area				
	Canal watered	Flood watered (in channel)	Rainfed	Wells per 100 ha	No. of wells
Sub-montane belt					
Gujrat	–	8.5	67.0	2.7	8,235
Sialkot	1.1	12.0	39.0	4.9	17,539
Gujranwala	–	7.0	21.0	4.0	10,902
Lahore	22.0	2.4	53.0	3.2	10,083
Southern Punjab					
Multan	74.0	19.0	1.3	5.9	14,975
Jhang	68.0	30.0	1.1	8.6	12,132
Muzaffargarh	35.0	26.0	0.0	6.9	11,793

by which the British annexed the Punjab in 1849.

The new found ability to control the flow of rivers by barrages yielded two important results. Water could now be diverted at all times of the year independent of the level of flow. This gave a measure of security to the cool season rabi crop which it had never previously enjoyed, while extending the support available to the hot season, kharif crop. More important, however, was the facility which barrages constructed near the point of entry of the river onto the plains gave for canals to be constructed along the higher parts of the interfluves. Such canals were able to command areas, like the *bet* lands, which because of their relative elevation above the river flood plain could not be reached by inundation canals and which, because of the generally excessive depth of the watertable, could not economically be irrigated from wells. The UBDC took its water from a weir on the Ravi at Madhopur where the river enters the plains. The weir raised the level of water in the Ravi so as to maintain a flow into the canal at all seasons. As Fig. 5.5 shows, the canal traces a course along the height of the Ravi-Beas interfluve (hence the acronym 'Bari' Doab) enabling it to command an extensive area quite beyond the reach of inundation canals.

A small scheme, the Sidhnai canal project on the lower Ravi followed the UBDC in 1886. Intended to be a perennial canal, the Ravi failed to provide enough flow in the rabi season. Very much more important was the Chenab canal system opened in 1892 to command the Chenab Canal Colony which attracted some 444,000 migrants mainly from more eastern districts of the British Punjab. The impact of these and later canals on the density of population in the districts of the Punjab is seen in Figs. 12.1–12.5 discussed below in Chapter 12.

20 Marala barrage on the Chenab is the headworks of link canals that cross the Rechna Doab to the River Ravi. The machines for raising the gates of the barrage can be seen to the right of the road.

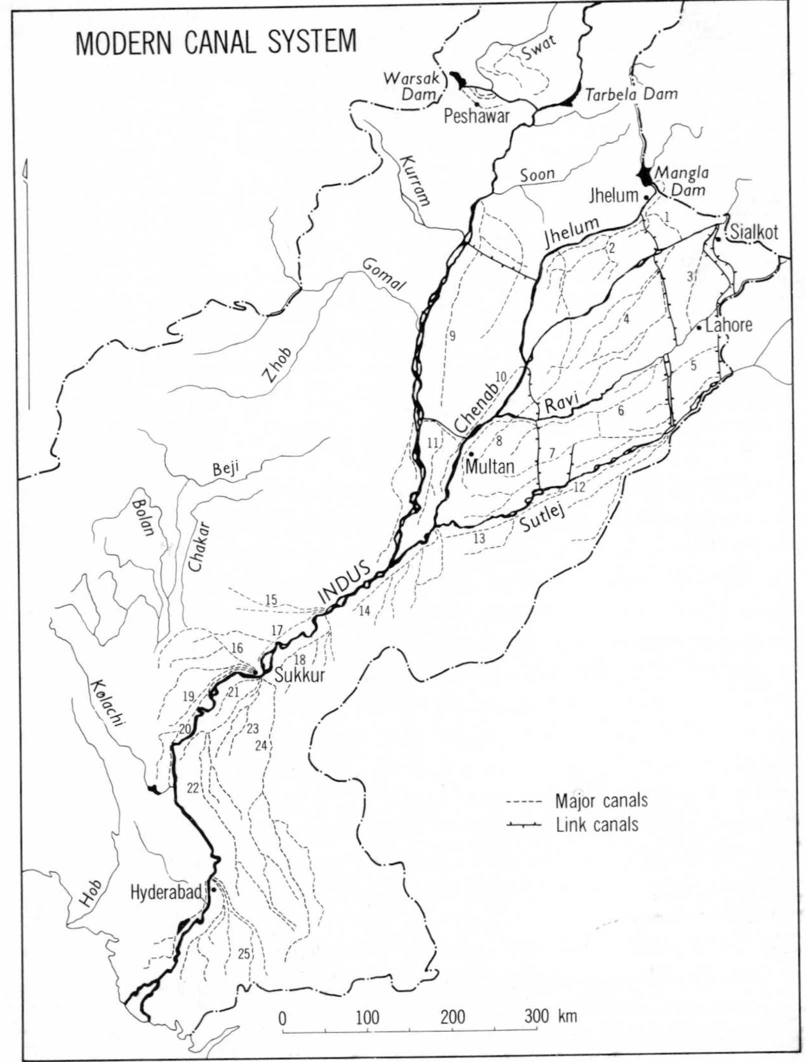

1. Upper Jhelum Canal; 2. Lower Jhelum C.; 3. Upper Chenab C.; 4. Lower Chenab C; 5. Central Bari Doab C.; 6. Lower Bari Doab C.; 7. Pakpattan C.; 8. Sidhnai C.; 9. Thal C.; 10. Rangpur C; 11. Muzaffargarh C.; 12. Fordwan C.; 13. Bahawal C.; 14. Panjnad C.; 15. Pat Feeder C.; 16. Northwest C.; 17. Begari Sind C.; 18. Ghotki C.; 19. Rice C.; 20. Dadu C.; 21. Khairpur C. West; 22. Rohri C.; 23. Khairpur C. East; 24. Eastern Nara C.; 25. Fueli C.
FIG. 5.5

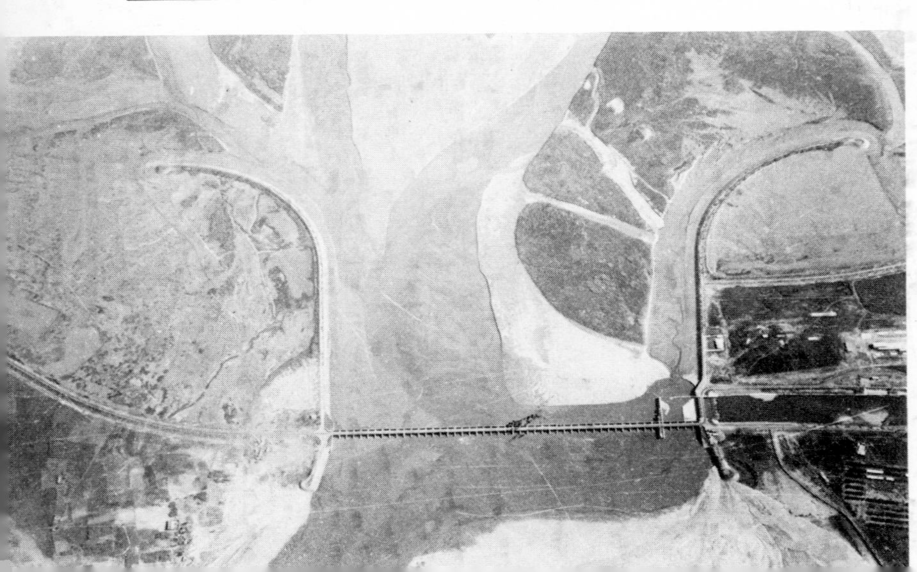

21 Aerial view of a typical barrage in the Punjab. Water is diverted into the canal on the right. Protective embankments upstream prevent flood waters from outflanking the barrage, which also serves as a road bridge.

The Jhelum canal followed in 1901 and the Punjab plains east of the Jhelum were set on the path to prosperity. West of the Jhelum the Sind Sagar Doab, or Thal region, separated that river from the Indus. This area of sand hills and sandy plains deterred the British engineers who while they saw the possibility of commanding the doab by constructing a barrage on the Indus at Kalabagh, realised the unsuitability of much of the rolling sand dune tract for irrigation. Development of irrigation here had to await the application after Pakistan's independence of new earth-moving and levelling technology epitomised in the bulldozer.

Downstream of the Indus – Panjnad confluence the Indus flows in a broad 'active' flood plain with relatively low plains on either side. For long the Province of Sind had to rely exclusively on inundation canals which flowed in the summer months from May to September watering the kharif crops and moistening the soil for planting rabi crops where well irrigation could subsequently bring them to maturity. The very scale of the River Indus and its wayward behaviour within its active flood plain, made its control by a barrage a formidable proposition, but until this could be achieved reliable perennial irrigation was out of the question. Lack of urgency and the availability of more obviously profitable fields for investment in the Punjab kept the Sukkur (Rohri) project on the shelf from its inception in 1847 till its eventual approval in 1923. Irrigation started in 1932 in the world's largest single scheme ever undertaken at that time.

The markedly different cropping systems under perennial and inundation canal commands in 1870–71 are shown in Fig. 5.6. The Lower Sutlej and Chenab canals and the Indus canals of the southern Punjab supported farming systems in which kharif crops dominated, to the extent of 66 per cent and 72 per cent of the cultivated area respectively. These were typical inundation canal systems. By contrast in the UBDC command area the ratio was reversed, kharif crops accounting for only 32 per cent, and rabi crops for 68 per cent.

The overall situation for as much of modern Pakistan as was covered by district agricultural statistics at the end of the nineteenth century is depicted in Fig. 5.7. Because of the separate administration of Punjab (including the present NWFP), Sind and the British-administered area of Baluchistan, the data on which this figure is based are not precisely contemporaneous: the Punjab statistics are for 1884–85, those for Sind 1890–91 and for the Baluchistan districts, 1900–01.

Apart from Hazara District in the Himalayan foothills enjoying the highest rainfalls of anywhere shown on this map, and able to support a rainfed kharif crop, dominance of kharif over rabi does not occur southwards until one reaches Dera Ghazi Khan in the Indus canals (inundation) commands. Here, and in the districts west of the Indus in Sind, crops were taken occasionally (as now) when torrents in spate could be diverted to water the fields for whatever appropriate crop could subsequently be sown. In Sukkur and Larkana, the canals although technically inundation canals, could in a year when moderately high river levels were sustained provide enough pre-watering to establish a rabi crop which would then take its chance on the basis of remanent soil moisture.

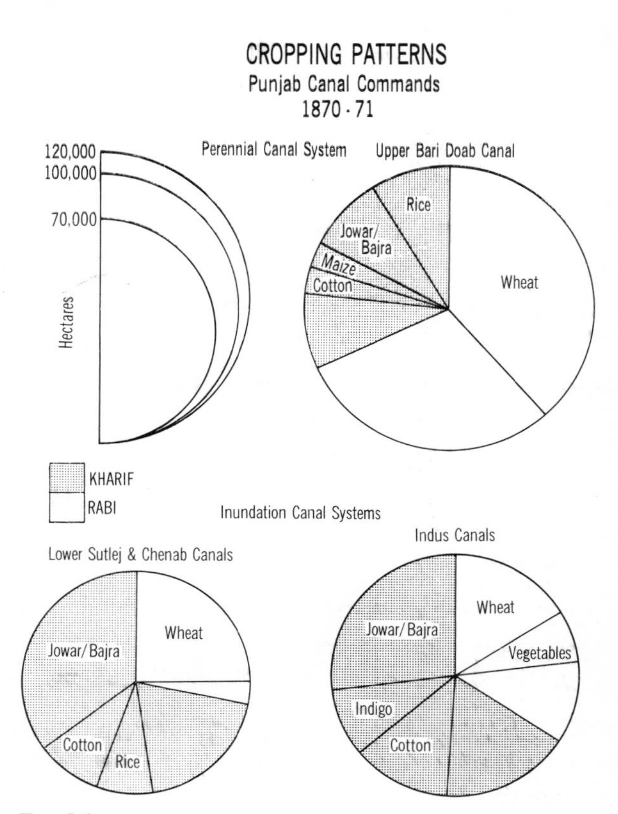

Fig. 5.6

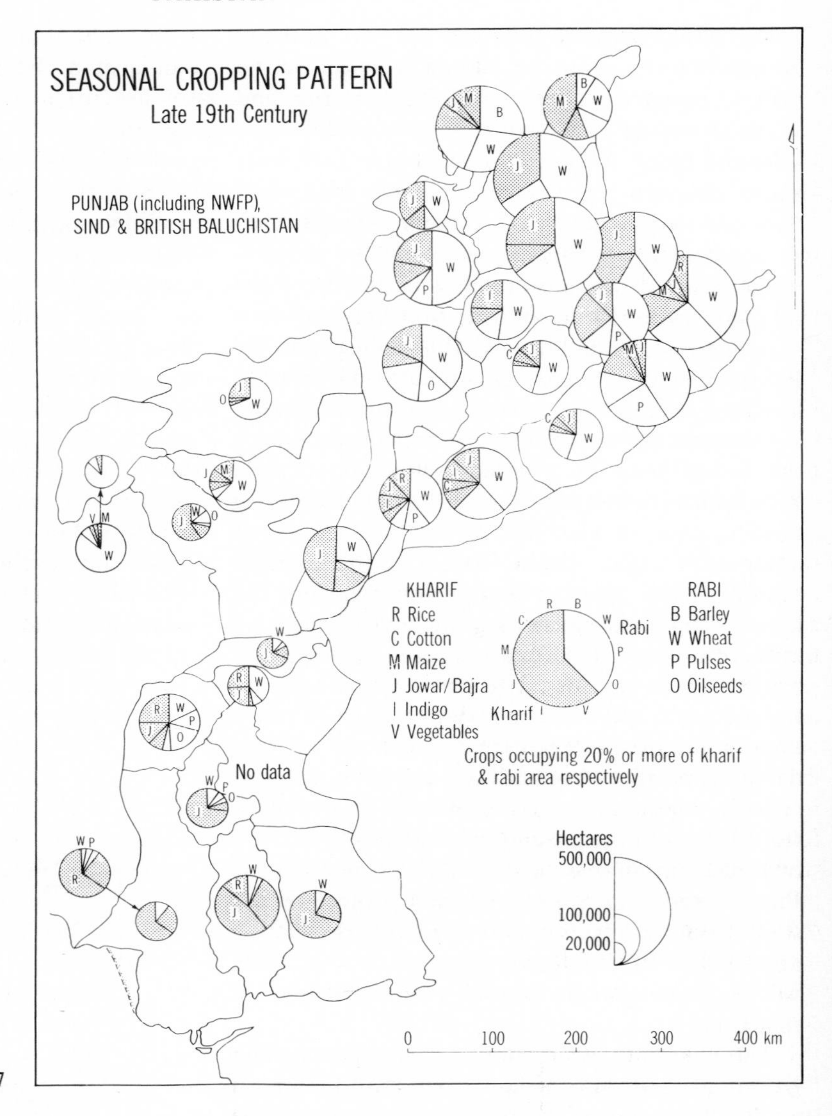

FIG. 5.7

Elsewhere in Sind, kharif crops were strongly dominant, reaching 84 per cent in Hyderabad in 1890–91 (92 per cent in 1880–81), 78 per cent in Karachi and 90 per cent in Thar Parkar. To the extent that there could be some unreliable kharif rainfall in these districts nearest the Arabian Sea (cf. Figs. 4.6 and 4.7), the contribution of inundation canal irrigation to the level of kharif dominance may be a little exaggerated, but was an unquestionable fact. The change that took place when perennial irrigation became available is seen in Fig. 8.24 discussed below.

ELABORATION IN THE PUNJAB, 1905–15

Within the Punjab it was realised that an integrated approach to river development was necessary when the water of the three eastern rivers, Ravi, Beas and Sutlej proved insufficient to meet expectations. This was partly due to overriding obligations to assure water to inundation irrigators in southern Punjab and Bahawalpur State. The western rivers Jhelum and Chenab had surplus water available. Their rates of run off and discharge are substantially greater than those of the rivers to the east (see Table 5.1 above) but the doabs they command are less in area than those of the eastern rivers. Engineers hit upon the ingenious expedient of transferring water from the western rivers to supplement those further east. In the Triple Canals Project, carried out between 1905 and 1915, Jhelum water from Mangla was transferred by the Upper Jhelum canal to the Chenab above the Khanki barrage. Chenab water was extracted further upstream at Marala to feed the Upper Chenab canal leading ultimately to the Ravi, which it crossed at Balloki to feed the lower Bari Doab canal flowing past Harappa. The Triple Canals Project was to provide in later years a blueprint for solving the post-partition problems of water distribution between Pakistan and India.

In the inter-war period it was realised that ultimately efficient irrigation in the Punjab would depend on the creation of storage reservoirs to provide a carry-over of surplus water from the monsoon flows into the rabi season and the early part of the subsequent kharif season before the next monsoon rains recharged the rivers. A high dam on the Sutlej at Bhakra was contemplated as early as 1919, but along with other major schemes in the Punjab it was viewed with apprehension by Sind. Inter-provincial rivalry for the waters led to the setting up in 1935 of a Committee on Distribution of the Waters of the Indus from whose decisions the provinces could appeal to the government of India in Delhi.

The problems with which this Committee had to contend were small, however, compared with those that were brought about by the partition of Pakistan and India in 1947. To say that partition took place in an atmosphere of mutual suspicion and ill-will is an understatement. Eventually the boundary between the new independent dominions in the Punjab had to be arbitrated by Radcliffe mainly on the basis of communal population distribution and sensitivities about holy places. The outcome ignored the existing facts of water engineering and inevitably cut across the arteries and veins of the Punjab's irrigation system, separating headworks from areas they irrigated. The most serious problem concerned the Upper Bari Doab canal supplied from the Madhopur barrage on the Ravi, now in the new east Punjab state of India. The UBDC watered Lahore, capital of Pakistan's west Punjab. The Ferozepore barrage, given to India, watered districts in Pakistan on the right bank of the Sutlej. It was obvious that cooperation of the kind necessary to the successful maintenance and future development of the five Punjab rivers as an integrated system could not be achieved, and radical alteration of the existing pattern of works would be necessary on either side of the new frontier. India holding the Ferozepore and Madhopur headworks could (and did for a while) cut vital supplies to Pakistan. India assumed the right to do what it liked with the waters of the three eastern rivers whose headworks it controlled (thereby abrogating in 1956 the Barcelona convention of 1921–24). This posed a long-term threat to Pakistan, which was forced to arrange to transfer more water from the western rivers to protect Lahore and the Bari Doab.

By the Indus Waters Treaty of 1960, made practicable by grants and loans from friendly governments (notably USA) and the World Bank, the impasse was finally overcome. Pakistan was allowed full use of the Indus, Jhelum and Chenab, while India ultimately gained sole use of the Ravi, Beas and Sutlej. Pakistan now needed additonal link canals to replace by transfers from the west the lost water in the lower doabs of the eastern rivers. Not only water transfer was needed, but water storage also, for although the total mean flow of the Indus, Jhelum and Chenab was in theory adequate for Pakistan's needs, its distribution over time did not match the seasonal demand. Difficulties were experienced in the low water period that affected two critical agricultural stages, first the maturing of the kharif and the sowing of the rabi crop as the monsoon peak flows subsided in September–October, and second the maturing of the rabi and the sowing of the kharif in March–April. Storage of surplus flow from the monsoon peak for use for sowing kharif in the following year could solve this problem.

The link canals constructed from the initiation of the Triple Canals Project up to the present time are located in Fig. 5.8 which also shows the areas of Pakistan formerly irrigated from the eastern rivers whose water was allocated to India under the Indus Waters Treaty. Fig. 5.9 shows the capacity of the canal system to carry water, measured in cubic metres per second. It should be noted that

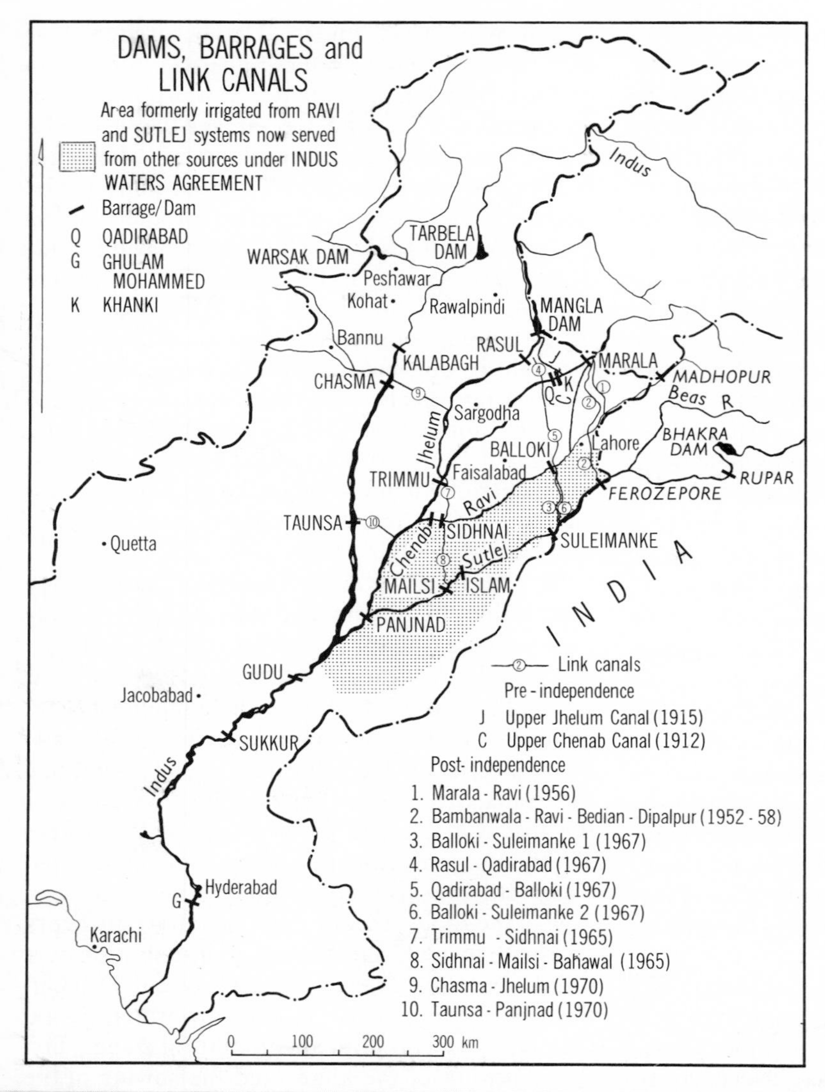

FIG. 5.8

22 Mangla dam, on the River Jhelum, with the old Mangla Fort beyond. This picture shows only a small part of the long stone-faced earth dam system. The hills of the Salt Range in the background.

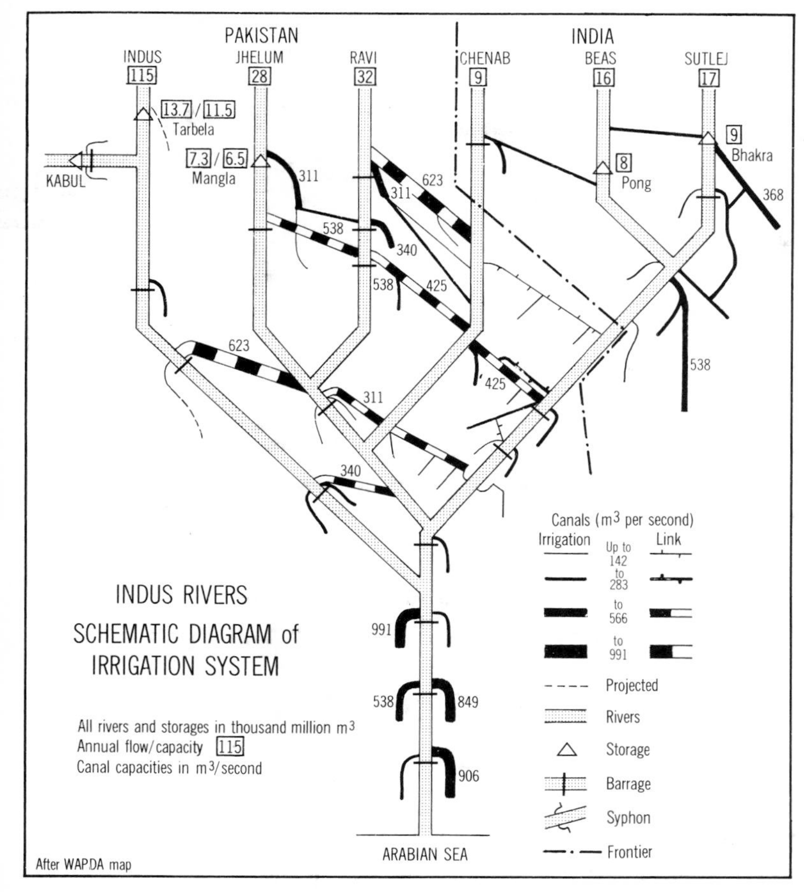

FIG. 5.9

canals do not flow all the time, and these figures indicate capacity to deliver water *when available* and do not imply continuous performance throughout a year or season. The most radical departure from pre-independence days has been the transfer of Indus water across the Thal. New barrages were necessary on the Indus to make such transfer possible. From the Chasma barrage a link canal feeds into the Jhelum while one from the Taunsa barrage feeds to the Chenab below its confluence with the Jhelum and Ravi rivers. These links replace water lost from the three rivers to links taking off water further upstream. The Thal itself was provided with perennial irrigation water from the Jinnah barrage at Kalabagh, the point where the Indus escapes from the gorge section that extends from Attock. The barrage was built in 1942 but the scheme did not advance till the end of World War Two. The Main Line Upper Canal bifurcates at the northwestern corner of the Thal, one branch running east parallel to the Salt Range piedmont as far as the Jhelum, the other parallelling the left bank of the Indus to command distributaries 200 km to the south.

Under the Indus Waters Treaty, Pakistan was assisted with finance and technical assistance to construct two major storage reservoirs. The Mangla dam on the Jhelum completed in 1968 with a live storage of 6.5×10^9 m^3 is designed to be enlarged to give capacity of 11.8×10^9 m^3, and to be further raised later to counter the inevitable loss due to sedimentation. Currently live storage represents 23 per cent of the mean annual flow. It was estimated that the Mangla dam will lose 30 per cent of its initial capacity within 50 years. Serious as this may appear to the uninitiated in these matters it pales into insignificance beside the 90 per cent loss anticipated for the Tarbela dam on the Indus within 50 to 55 years. Tarbela has a live storage capacity of 11.5×10^9 m^3, 14.5 per cent of the average annual flow at the site. Plans are already in preparation to make good the anticipated loss by building a dam of comparable capacity on the Indus above Kalabagh just downstream of its confluence with the Soan. Gross storage of 11.5×10^9 million m^3 will be ponded in the gorge of the Indus, which extends upstream as far as Attock. Since the Tarbela dam will effectively trap Indus sediment for some time, the useful life of this new storage should be considerably longer. As part of the Indus Waters Agreement before construction of Tarbela was approved, the feasibility of several alternative storage systems had to be investigated of which the Kalabagh dam project was one.

The flow characteristics of the Indus and of the combined rivers available to Pakistan are shown in Table 5.3 with the actual water used in 1957 and in prospect for 1985. It is expected that by 1985 both Mangla and Tarbela dams will be fully operational. Tarbela is capable of storing an amount equivalent to its total rabi season flow. With Mangla added, the total storage represents 61 per cent of the rabi flow of all three rivers, and the expectation is that water utilization in the rabi season will expand. Before Mangla dam was built, on average (1961–66) 105×10^9 m^3 was used in a year, 33×10^9 m^3 in rabi, 72×10^9 m^3 in Kharif. By adding 6.3×10^9 m^3 to the combined rabi flow of 27.6×10^9 m^3, Mangla restored Pakistan to its position prior to the Indus Waters Treaty. Water from Tarbela dam has enabled real expansion to take place though technical difficulties have delayed full realisation of the potential. By 1985 rabi use will have increased 22 per cent and kharif use 17 per cent.

The water budget of the canal commanded areas of Pakistan in 1975 is shown diagrammatically in Fig. 5.10 based on a World Bank projection. It will be noted how little rainfall contributes to crop growth, a mere 8.6×10^9 m^3 in a total of 80.2×10^9 m^3 actually used by crops. Evaporation takes its toll of water on the surface, but a substantial amount infiltrates to groundwater from canals and fields, some of it to be recycled through the irrigation system by pumping.

Where groundwater is not saline it can be regarded as a valuable resource which can be tapped for irrigation. Useable groundwater underlies much of the Indus plain particularly in the vicinity of the major riverflood plains and in the Himalayan piedmont. Where it is close to the surface traditional wells have long been used to exploit it. At greater depths and for maximum efficiency, tubewells equipped with power pumps are now

TABLE 5.3

River Flow Characteristics and Use (All figures in thousand million m^3)

	Indus at Tarbela	*Indus (Attock) + Jhelum + Chenab 1968*	*Actual use 1957*	*Projected use 1985*
Oct.	3.3	6.8	9.3	9.9
Nov.	2.0	3.9	4.6	4.9
Dec.	1.6	3.5	4.3	4.9
Jan.	1.4	3.5	4.9	6.2
Feb.	1.4	3.7	7.3	7.4
Mar.	1.9	6.3	6.9	7.4
Rabi season	11.5	27.6	37.3	40.7
Apr.	2.6	10.1	6.3	8.6
May	5.4	17.5	8.8	9.9
June	12.7	28.0	13.0	16.0
July	20.7	39.5	13.6	17.3
Aug.	19.7	35.0	13.8	17.3
Sept.	8.4	16.3	12.2	14.8
Kharif season	69.6	146.5	67.6	83.9
Year	81.0	174.2	104.8	124.6

Note: Totals may not tally exactly owing to rounding off in conversion.

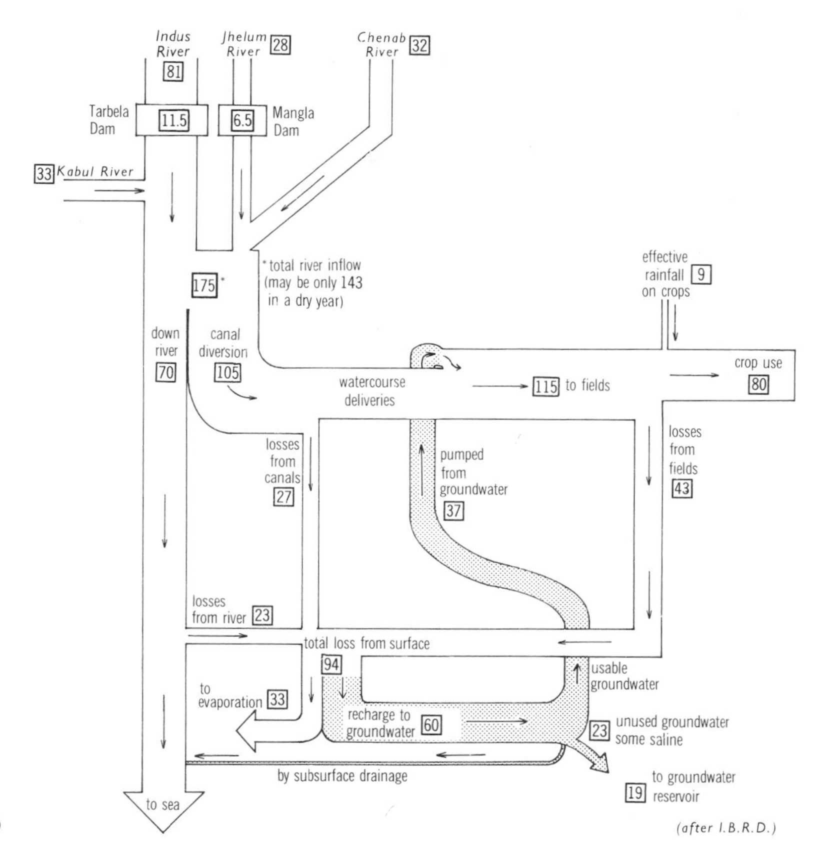

FIG. 5.10

23 The Tarbela dam on the Indus holds back a vast lake. The service and auxiliary spillways are seen here.

often used. The role of groundwater resources in the total water budget is increasing. The more groundwater is exploited in the Punjab, the less water will ultimately flow to Sind since every cycle of use in irrigating crops involves some loss through evaporation, transpiration and fixation in the crop itself.

The budget illustrated supported an average cropping intensity of 110 per cent, i.e. 10 per cent of the fields irrigated carried a second crop in the year. Projections forward to 1985 and 2000 increase the intensity to 126 and 142 per cent respectively, levels achieved by a higher factor of utilization of water on its way through the system, particularly by the reuse of groundwater.

THE PARTICULAR PROBLEMS OF SIND

Until 1932 when the Sukkur barrage was completed agriculture in Sind had been very dependent on inundation canal irrigation. With the completion of the Sukkur barrage, Sind became acutely interested as Michel puts it 'in every drop of water flowing in the Indus rivers between the melting of the snows on the Himalayas and Karakorams and the arrival of the monsoon.' Such interest pointed to the need to establish more effective planning for the development and use of the waters of the Indus rivers as a physically integrated and interdependent system. Sind needed most of all early kharif water from April to August when a period of uncertainty came between snow-melt and the augmentation of the flow in July with the beginning of the monsoon, after which the supply was reliable until the immediate pre-snow-melt period of February–March, the end of the rabi season. The intensification of water use in the Punjab inevitably reduced the flow reaching Sind as much water is lost by evaporation and transpiration by cultivated and other vegetation in the irrigated tracts. Sind's anxieties in this connection have in no way diminished over time as Punjab's irrigated cultivation intensifies under the stimuli of population and prosperity.

Two additional barrages were built on the Indus in Sind after independence, the Ghulam Mohammed barrage at Kotri close to Hyderabad in 1955 and the Gudu barrage in northern Sind in 1962. While these have given greater security to the areas formerly watered from inundation canals the long term prospects for Sind cannot be regarded as good and are tied to overall planning for the whole Indus basin. Increasing difficulties for Sind are likely to arise from the rising level of salinity in the waters entering the province from the Punjab.

Potentially at any rate, the Punjab seems capable of reclamation, but a large problem remains for Sind, well downstream in the Indus system. Eventually whatever salts are removed from the Punjab less whatever quantity may be removed into expensive and complex salt collecting basins, flow into Sind. It may be necessary for Sind to restrict its irrigated area in order to ensure that whatever water is applied is given in sufficiently copious quantities to avoid salinification. One recently completed irrigation scheme in Sind, the Ghulam Mohammed barrage at Kotri, has its own problems of aggravated salinification, due to the excessively low slope in the irrigated area, which makes drainage sluggish at best and contributes to waterlogging of the flat clay terrain.

It is clear, therefore, that irrigation in Pakistan while it has brought greater and greater benefits to more and more people as technology had developed, has been accompanied by problems increasingly difficult to overcome. These are examined in the chapter which follows. It is at any rate fortunate that with the completion of the works contemplated under the Indus Waters Treaty, Pakistan had become a sole possessor and operator of the world's largest irrigation complex. Success will depend in the long run not only on the engineer and the cultivator applying scientific knowledge, but to a significant degree on sound political administration to resolve the conflicting claims of the Provincial authorities, themselves responsible for a multitude of districts whose interests are not always mutually compatible.

CHAPTER SIX

HAZARDS – MAN-MADE AND NATURAL

SUMMARY

The transformation of the natural hydrological environment of the Indus rivers into a controlled irrigation system was a remarkable achievement. The early engineers did not forsee that the system they considered entirely beneficial carried with it a capacity under certain circumstances to ruin the land it had been designed to render productive. Waterlogging and salinity are widespread problems resulting from irrigation and were inadvertently inherent in the designs of early systems. Their control to prevent further deterioration and the reclamation of affected areas are now being vigorously pursued.

The physical engineering works constituting an irrigation system are expensive capital investments. They are designed to withstand the reasonable expectation of climatic hazards in the form of floods to the best of the knowledge available, but within financial constraints. They are, however, very complex systems and may be endangered by combinations of natural events impossible to forecast fully. As such time the engineers on watch throughout the system must take preventative action to protect key installations such as the barrages from damage that might endanger the livelihood of thousands of farming families for the seasons that follow. The closer settlement of the country increases the chance of hazard. Road embankments, railways and buildings tend to grow up in the landscape in a fashion uncoordinated with the needs of the canal system and may themselves add to the natural hazard of overland floodings.

Even in the most arid part of the country there is a risk of occasional flooding by torrents in spate. Local villagers have enough experience to avoid settling where such floods might threaten. The floods in Karachi in 1977 caught unawares thousands of refugees who had settled in river beds which they, with only a decade or two of experience in the region, had presumably assumed to be abandoned.

WATERLOGGING AND SALINITY

It must not be thought that the engineering marvels of modern irrigation have brought nothing but good. As early as 1859 the problem of waterlogging due to bad alignment of the West Jumna canal was remarked upon. Towards the end of the century the Australian statesman Alfred Deakin made a study of Indian irrigation in which he said 'The lessons of the danger of over-watering, of the necessity for adequate drainage, of the value of unremitting cultivation, fallowing, fertilizing and rotation of crop . . . will bear much repetition . . . . In some older districts, patches of snow white *reh* (salt), deserted stretches of sodden and marshy land, growing only coarse reedy grass or semi-aquatic plants, tell their own tale of neglected drainage.'

Many others have referred to this man-induced problem of waterlogging and associated salinification. In simple terms waterlogging arises from the introduction of irrigation water into a region without adequate provision being made for drainage (Fig. 6.1). Water not used by crops, evaporated or transpired or drained away on or below the surface, accumulates as ground-water, until such time as its level rises to the surface to produce a swamp. In many areas ground-water has not

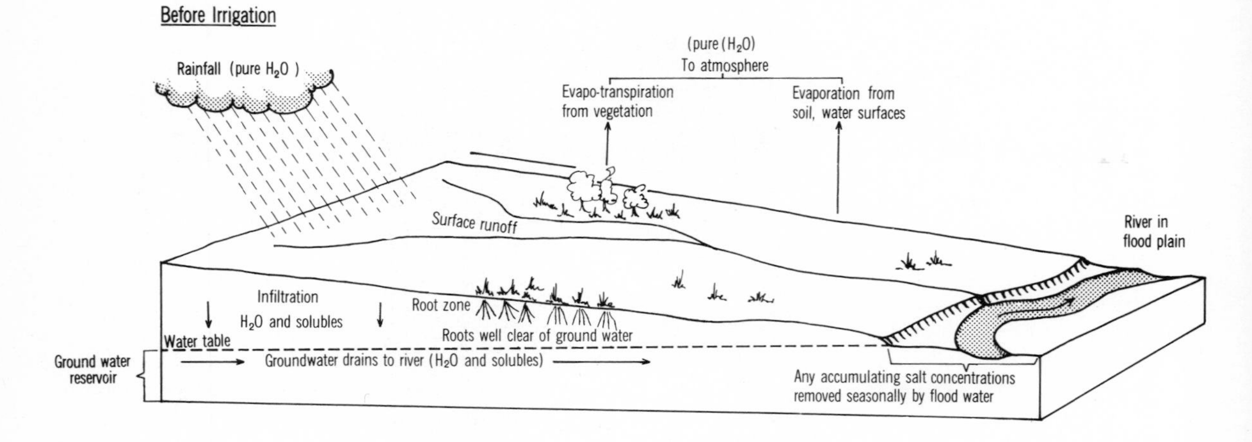

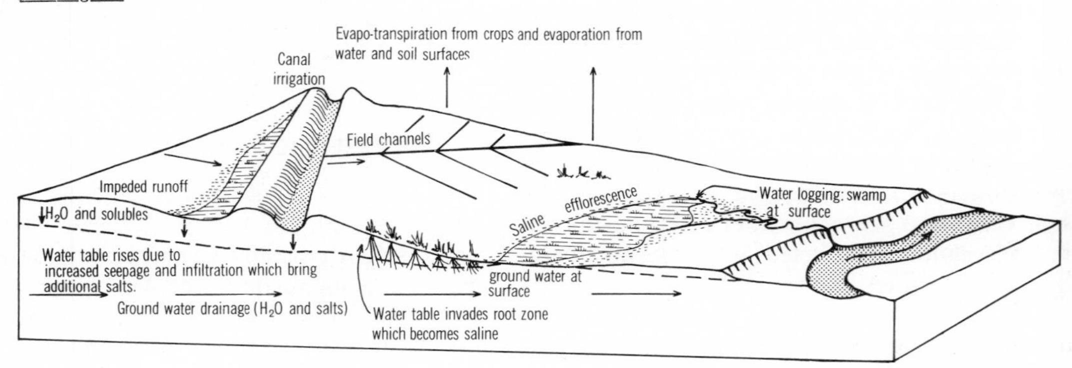

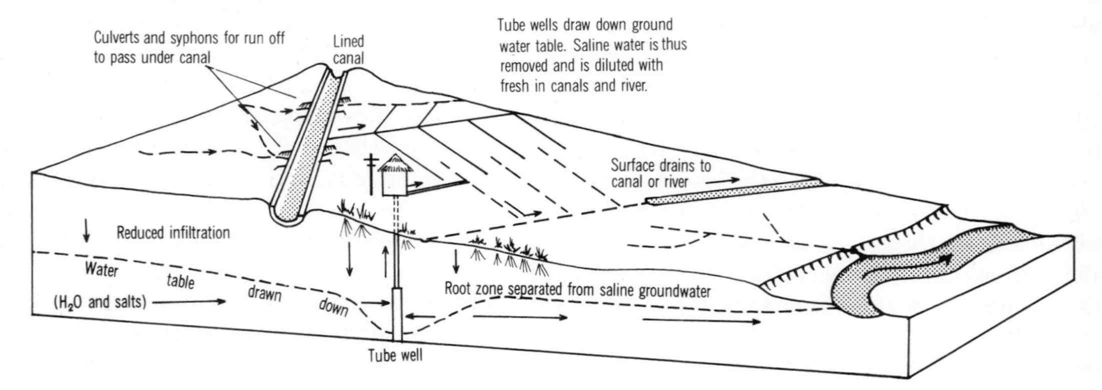

FIG. 6.1 Salinity: development and reclamation.

24 Aerial view of a typical irrigated landscape in the Punjab plains. A canal runs from the top of the picture, a river on the left. A drainage channel runs across the picture from the top right to discharge into the river. The whitened area to the left of the main canal and parallelling the drainage channel in places is caused by salinity. Note the closely nucleated villages.

25 Saline efflorescence whitens the surface of this fallow field near Kohat, NWFP. A meagre crop of wheat flanks the fallow. The trees on the left shade the main road.

accumulated as subsurface drainage removes it elsewhere, but where the subsurface structure inhibits from drainage, as in the case of the sub-alluvial rock ridge running southeast across the Punjab from Shahpur, or where reduced permeability occurs, extensive waterlogging may result at the surface. Impediments to surface drainage, such as road, railway, or canal embankments running across the slope of the land, are a major cause of ponding back surface run off, which then has time to percolate to groundwater. Seepage from unlined canals and distributary channels further adds to the groundwater reservoir and often induces a rise in the watertable.

A greater menace even than waterlogging is salt in the soil. All river water contains salts in solution, and so the irrigator applies to his fields a normally very dilute saline solution. When crops transpire, and when water evaporates, it is pure water vapour that goes into the air, leaving the saline content behind to accumulate. Often, therefore, the surplus water from irrigation which joins the groundwater body is of increasing salinity. Salts in the soil itself may also be taken into solution. Where saline groundwater lies close to the surface, many crop plants will be unable to tolerate the condition and land may have to be abandoned unless remedial action is taken. This has been happening on a large scale in Pakistan as a result of several generations of irrigation. A major defect in the design of early schemes, only now fully appreciated, was to provide only enough water for a crop to be taken. Under-watering prevented the leaching out of accumulated salts, under-draining encouraged their retention in the soil. Up till now it has been no exaggeration for an engineer to say 'Heroic measures are essential if the Punjab is not to be destroyed'. The form the rescue operation is taking, is first to separate the saline watertable from the surface zone where irrigation is supporting crops. This is achieved by closely spaced tube wells which pump the saline water into canals where it can be diluted to safe levels, or into river courses. Surface drains may have to be dug to prevent surplus irrigation water sinking to add to the groundwater and so bringing the watertable closer to the surface.

Fig. 6.2 shows the extent of severe salinity and waterlogging of the farm area; it does not reveal the whole picture as much land has been rendered unuseable and has consequently been abandoned. The major areas in which groundwater is saline are shown on Figs. 6.3 and 6.4. It will be noted that these area are generally in the middle zone of interfluves and not close to the active flood plains as a rule. In lower Sind the problem is more extensive. In not all of these areas is the watertable close enough to the surface to create a toxic environment for crops, but nonethless the groundwater cannot be regarded as a useable resource and needs to be prevented from invading the root zone. The first major reclamation programme or Salinity Control and Reclamation Project (SCARP I) was undertaken in the Rechna Doab between the Rivers Chenab and Ravi, in 1959. Other SCARPS followed, SCARP II in the Chaj Doab (Chenab-Jhelum interfluve) in 1962, III in the Lower Thal Doab in 1965, IV in the Upper Rechna Doab. By 1974, 6844 tubewells and 460 km of surface drains had been installed to reclaim 1.6 million ha of canal command in the Punjab alone.

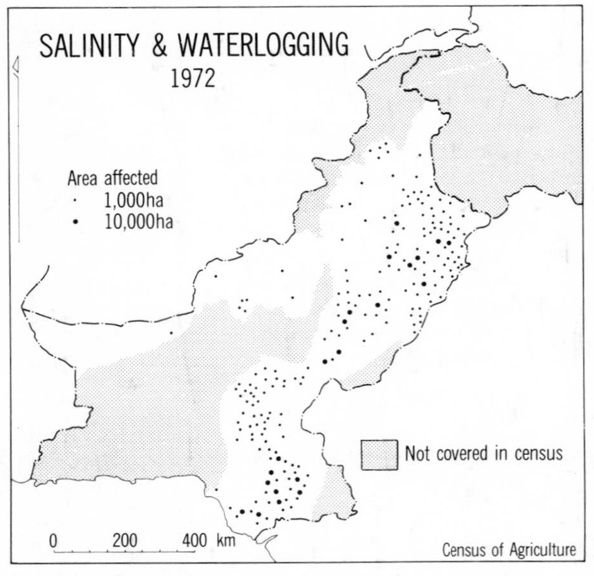

FIG. 6.3

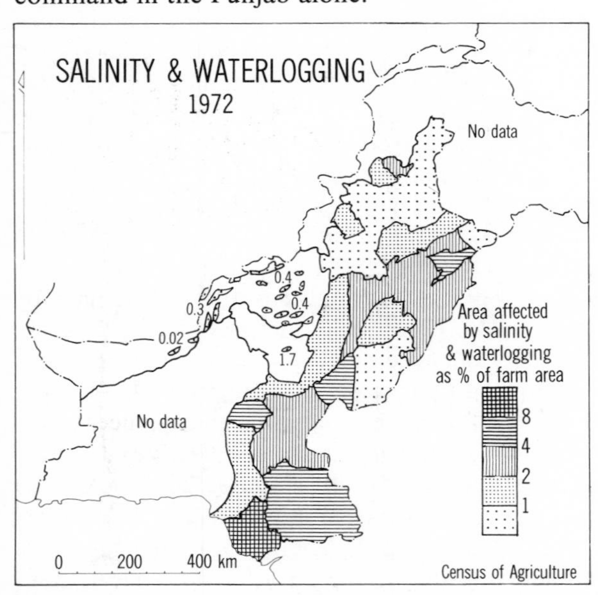

FIG. 6.2

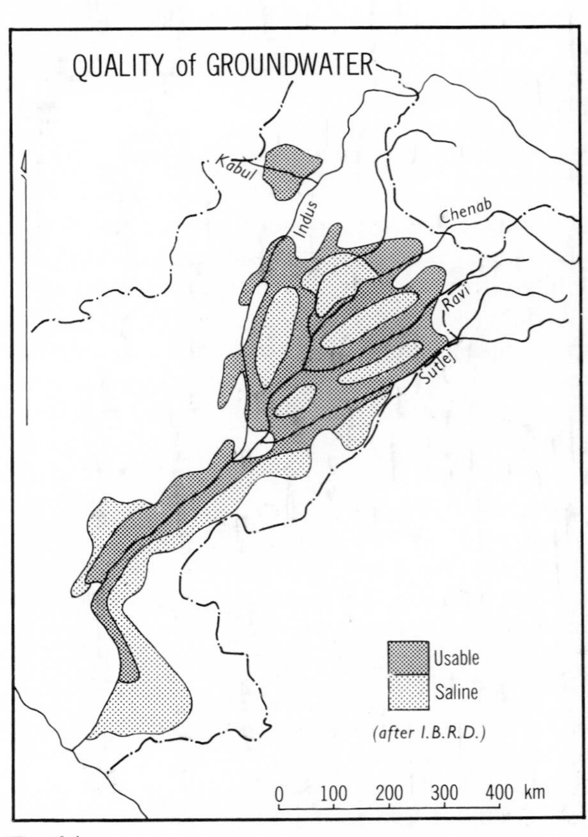

FIG. 6.4

TABLE 6.1
Extent of waterlogging (million ha)

	Gross canal command area (C.C.A.)	Area affected: Severely (watertable 0–1.5 m)	% of C.C.A.	Moderately (1.5–3 m)	% of C.C.A.	Total	% of C.C.A.
Punjab	8.2	0.6	7.0	2.7	33	3.3	41.0
Sind	5.1	1.2	23.0	1.8	33	3.0	56.0
NWFP	0.4	0.04	10.0	0.01	1	0.11	11.0
Baluchistan	0.24	–	–	–	–	0.002	0.3
Total	13.94	1.84	13.2	4.5	32	6.41	46.0

Note: Waterlogging is measured as in April when at a minimum. By September–October it rises to 1.5 to 2 times the April level.

TABLE 6.2
Extent of soil salinity 1973–74 (million ha)

	Area surveyed	Area affected: Moderately affected (0.2–0.5%)	%	Severe (>0.5%)	%	Alkali-affected	%	Total	%
Punjab	8.2	0.4	5	1.2	15	2.1	26	3.7	45
Sind	5.1	3.2	63	1.9	37	–	–	5.1	100
NWFP	0.4							0.04	9
Baluchistan	0.24							0.04	18
Total	13.94	3.6	26	3.1	22	2.1	15	8.9	64

Source: Annual Plan, 1976–77.

Table 6.1 and 6.2 summarize the extent of the problems of waterlogging and salinity respectively.

Where it is fresh, groundwater is of course a valuable and accessible reservoir of irrigation water and the pumping of tube wells to reclaim waterlogged areas can have its compensations. Under the SCARPs public tube wells are delivering almost 10×10^9 m^3 of water annually while, private wells yield a further 24.67×10^9 m^3 together making a very substantial contribution to the water circulating in the Indus basin irrigation system (cf. Fig. 5.10). Separate maps illustrate the programme of projects being undertaken to control waterlogging and salinity in the two operational zones (Figs. 6.5 and 6.6). The problems and their solution differ somewhat between the northern zone which extends as far south as Gudu barrage in the extreme north of Sind, and the southern zone. The major SCARP projects completed in the northern zone and their salient features are tabulated in Table 6.3 and mapped in Fig. 6.5.

The tube wells of SCARPs I–IV yield 7.8×10^9 m^3 to help water 2.02 million ha. In addition to the SCARP areas tabulated totalling 2,247,000 ha, a further 272,238 ha are the subject of schemes under active development in the Punjab, involving 3330 public tube wells and 2300 private wells, and plans to deal with 3.9 million ha are being formulated. NWFP has a smaller problem in the Vale of Peshawar, Bannu and Dera Ismail Khan, where 173,948 ha are being reclaimed and protected from further deterioration. The relative permeability of the subsoil in this northwest corner, the abundance of water and a less arid climate are helpful factors rarely present together elsewhere.

In the southern zone (Fig. 6.6), the problems of salinity loom larger and groundwater is seldom a welcome resource. When extracted to reduce the

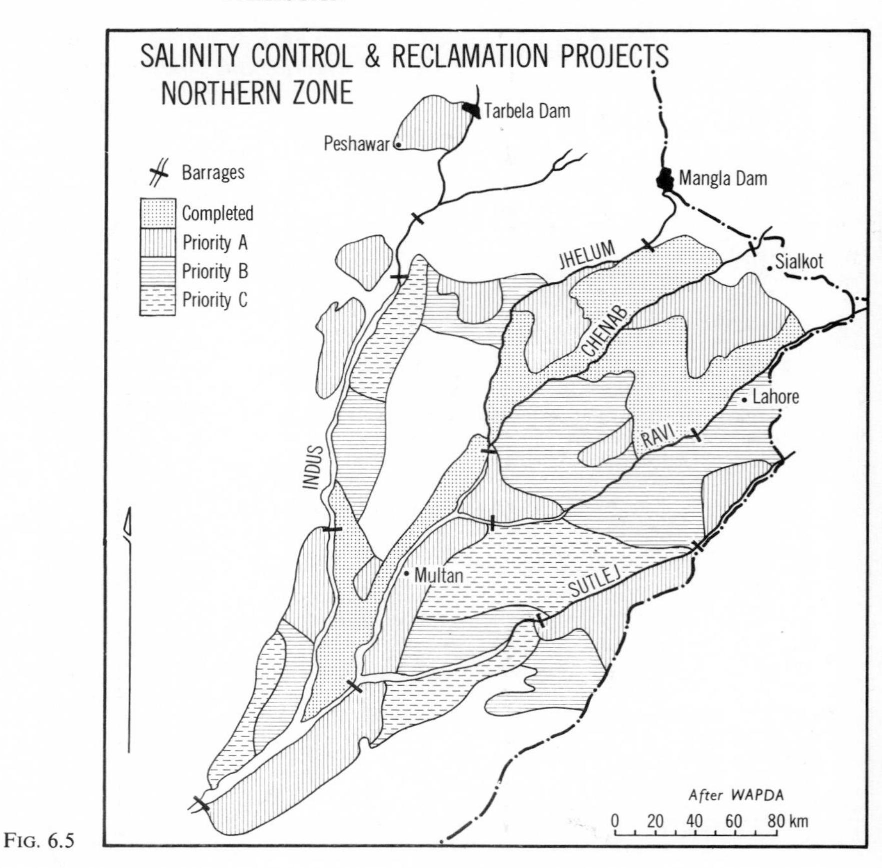

FIG. 6.5

TABLE 6.3

Salinity Control and Reclamation Projects: Punjab

Project	*Location*	*Area ('000 ha)*	*Fresh groundwater tube wells*	*Capacity m^3/sec*	*Saline ground water tube wells*	*Surface drains (km)*
SCARP						
I	Central Rechna Doab (Chenab–Ravi Interfluve)	494	2,069	164		
II	Chaj Doab (Jhelum–Chenab Interfluve)	984	2,205	240	823	603
III	Lower Thal Doab	518	1,635	177	61	185
IV	Upper Rechna Doab	251	935	105		

level of toxic groundwater, the effluent has usually to be removed to the river or to canals with sufficient capacity to dilute its saline content. Where the terrain becomes increasingly flat as the Indus passes through Sind to the sea, the removal of saline drainage water becomes difficult. Furthermore as the alluvium becomes finer down stream, it is often impracticable to drain affected areas by pumping from wells. Surface drains and subsurface tile drainage beneath the fields are being used increasingly. Fortunately, in the short run, the level of salinity while excessive for wheat, can still be tolerated by appropriate rice varieties in many areas. Table 6.4 sets out the major projects completed or in hand in Sind and the adjacent area of Baluchistan.

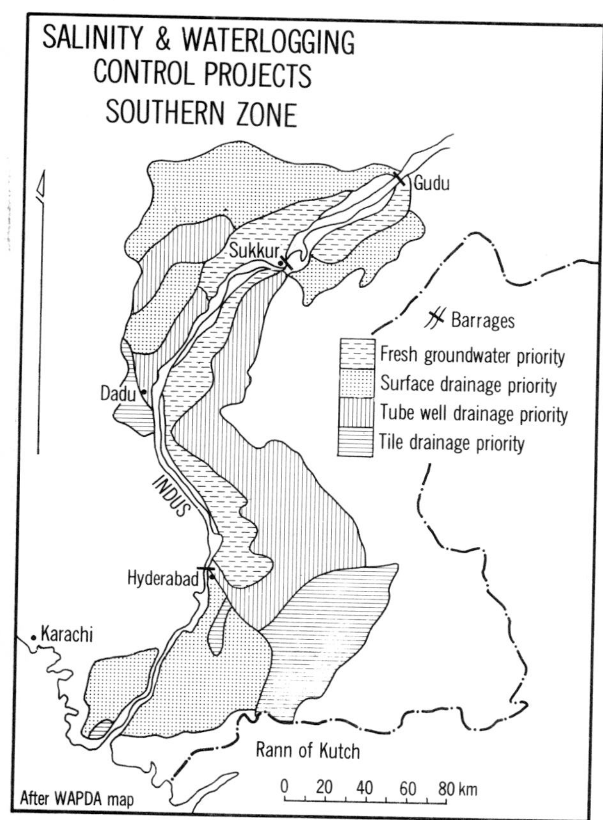

FIG. 6.6

Important among the projects now being planned to affect 2.95 million ha in Sind is the Right Bank Outfall Drain (RBOD) to run for 161 km along the Nara valley west of the Indus and to discharge into Manchar lake, a scheme which alone will aid 1.4 million ha.

FLOOD HAZARD IN THE INDUS BASIN

Introduction

But for the work of man spreading water on the soil the problems of salinity would be of small importance and that of waterlogging non-existent away from the river channels and backswamps. To some extent it might also be said that high flood levels in one or more of the rivers of the Indus basin would have had very limited impact on man before he himself began to manipulate the system with engineering structures.

In more humid climates than Pakistan's the occasional severe drought is a major hazard. In Pakistan, drought is chronic and man has learned to practice adaptive techniques in order to survive. The occasional flood, however, is of much more serious consequence. One which caused great damage in Karachi's housing areas in 1977 came as a shock to a population that rarely sees rainfall, and so is reluctant to plan for the eventuality of flood.

Another natural hazard to which Pakistan occasionally falls victim is earthquake, occurring in the unstable structures of the surrounding mountain areas. Quetta was destroyed by an earthquake in 1935, and in 1975 a severe shock struck the northermost areas of NWFP.

Floods in the Punjab

The accounts of two floods in 1955 and 1973 which follow are based on a Punjab University thesis by M. Saleem Nasir, 'Floods on the River Ravi' (1965) and on papers presented at a seminar organised by the West Pakistan Engineering Congress in Lahore in 1973, respectively.

The Punjab plains are in effect the low angle alluvial fans of the rivers issuing from the mountain rim. These rivers have incised their courses slightly into the plain in recent geological time, but the surface of the interfluves is so nearly level that flood waters and overland flow generally can spread very widely. The asymmetry of the regimes of the rivers has been noted above (Figs. 5.1 and 5.2) and their variability from year to year. Very much greater fluctuations in discharge can occur over brief periods when exceptionally heavy rainfall is concentrated over time and space in the catchment of one or more of the rivers. Nature accommodates such events by the rivers overtopping their banks and spreading the surplus water over their flood plains.

In the Punjab man has over the past century been diligently constructing devices to manipulate the flow of the rivers to his advantage in irrigating wide tracts of farmland that he has been able to colonize. Complicated engineering work, particularly barrages equipped with gates to control the flow of water into canals, represent a huge investment of capital. Since the agricultural production of the nation depends on irrigation such works have to be

TABLE 6.4
Salinity Control and Reclamation Projects: Sind and Baluchistan

Project	*Location*	*Area ('000 ha)*	*Installation fresh ground water tube wells*	*Saline ground water tube wells*	*Saline effluent disposal channels (km)*
SCARP Khaipur	Flanking Rohri Canal	142	161	379	566 + 5 outfall pumping stations
SCARP North Rohri	Sukkur Barrage Left bank command	324	1,392		
Larkana-Shikarpur Drainage Scheme I	Indus right bank west of Sukkur	235			214 + 5 outfall pumping stations
Scheme II					1097 + 2 new and 4 expanded pumping stations
Left Bank Outfall Drain Project·I	Left Bank of Sukkur command	733			282 main drain 1239 km subsurface drains
Sukkur Right Bank Fresh groundwater Project	Sukkur-Larkana Districts		500		
Hair Din Surface Drainage Project	In Baluchistan between Pat Feeder and Desert Canal fed by Gudu Barrage to take surface run off from rice fields				227 carrier drains + 512 main and branch drains + 1 pumping-station

26 The floods of 1973 caused much havoc in the Punjab and also downstream in Sind where the Indus overflowed its banks extensively as here, near Sukkur.

protected from damage. From the barrages, canals have been excavated following close to the contours of the land in order to be able to command as large an area as possible. Roads and railways also, raised on low embankments, run across the slope of the country adding their interference to the free flow of water in the event of flood. Thus man, seeking to control nature for his own advantage inevitably places himself in jeopardy. Not every risk can adequately be foreseen, and even those that may be calculated to a nicety by the irrigation engineers may be compounded by human factors beyond their control, when for example, roads and railways are constructed with insufficient provision for the drainage of water.

Floods in the Punjab represent a hazard to a most complicated man-designed environment in a partly artificial hydrological system over which man has not absolute control. Barrages and canal-head works are designed to pass maximum anticipated floods, but in an emergency may have to suffer unforeseen pressures. Silt accumulation up-river of a barrage may alter its capacity to allow flood water to pass. The braided channels of the rivers are notoriously fickle and may change their point of impact upon protective embankments from year to year. Thus the engineers must constantly review the system they strive to control. During a severe flood their activities take on the urgency of a military engagement: a barrage here to be relieved from the imminent flood wave, an embankment there to be breached to reduce the pressure on a more critical point, despite the fact that many thousands of hectares of crops and homes of the people will be inundated. Difficult decisions have to be taken in a hurry on the basis of incomplete information to save vital installations for the cropping seasons that must follow the flood.

The Flood Symposium of 1973 was unanimous in recognising the need for better information and its faster dissemination, and subsequently efforts are being made to programme the factors in a major flood so that a computer can help speedily calculate where dangers are most likely to occur. Sind is particularly vulnerable to floods coming down the system, and needs to know when and in how concentrated a fashion a flood wave will arrive. The Sukkur barrage is at a very vulnerable site chosen because of some topographic advantages and secure foundations. It is however at a point where the Indus could, and has in the geological past, followed a course to one side of the present main channel. Were it to change course now because of rupturing of the training embankments it might never be persuaded to resume its present channel, and the vast system dependent on Sukkur could be left without water. Fortunately the completion of the Gudu barrage provides some protection for Sukkur, and if breaches in banks have to be made to spill and spread the floods and so reduce the size of the peak wave, it can be done above Gudu.

The Floods of 1955 and 1973

Figure 6.7 shows the isohyets for a 72-hour period 4–6 October 1955, Figure 6.8 for a similar span 6–9 August 1973. The synoptic situation for the 1973 flood storm is shown in Figure 6.9 and closely resembled that described for the 1955 event. The progression of monsoon lows up the Ganges valley from the Bay of Bengal is normal. What was abnormal on both occasions was the reinforcing of the deep monsoon low by moist air flowing in from the Arabian Sea to pile up a great thickness of humid air over the Punjab foothills. In the 1973 storm, the monsoon depression brought a current not more than 1500 m thick over Agra by 4 August. Over the submontane region it had thickened to 6100 m as it pressed against the Himalaya. A closely following depression from the Bay of Bengal kept up the flow of moisture on the 6 August. On the 7 August a thermal low standing over

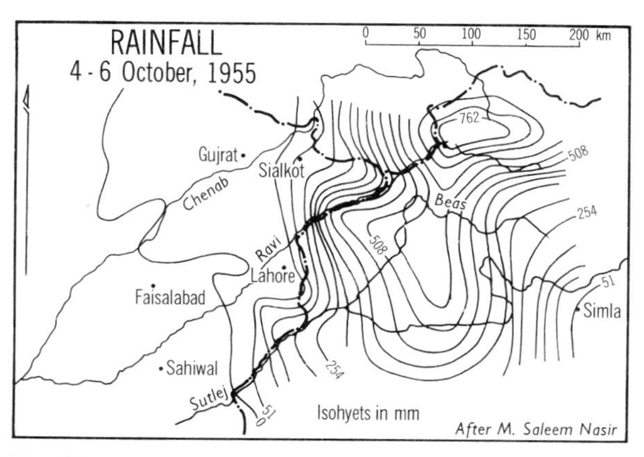

FIG. 6.7

Quetta and Sind and lower Punjab increased its activity, bringing a current of Arabian Sea air 2100 m thick to join the Bay of Bengal stream over the catchments of the Ravi and Chenab. As the two isohyetal maps show, the 1955 storm was most concentrated over the Ravi headwaters while that of 1973 straddled both Ravi and Chenab. Since these catchments lie in inaccessible areas of India and Indian occupied Jammu and Kashmir, Pakistani irrigation engineers received little immediate warning about the scale of the rainstorm being experienced.

As Table 6.5 shows, the 1955 flood produced the absolute maximum recorded flow on the Ravi, Beas and Sutlej at their rim stations. It may be noted at this point that while the Indus Waters

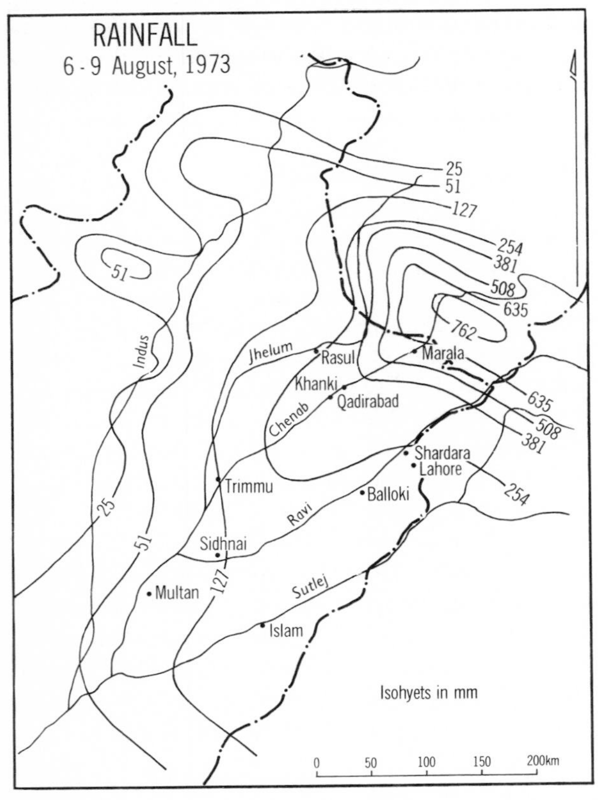

FIG. 6.8

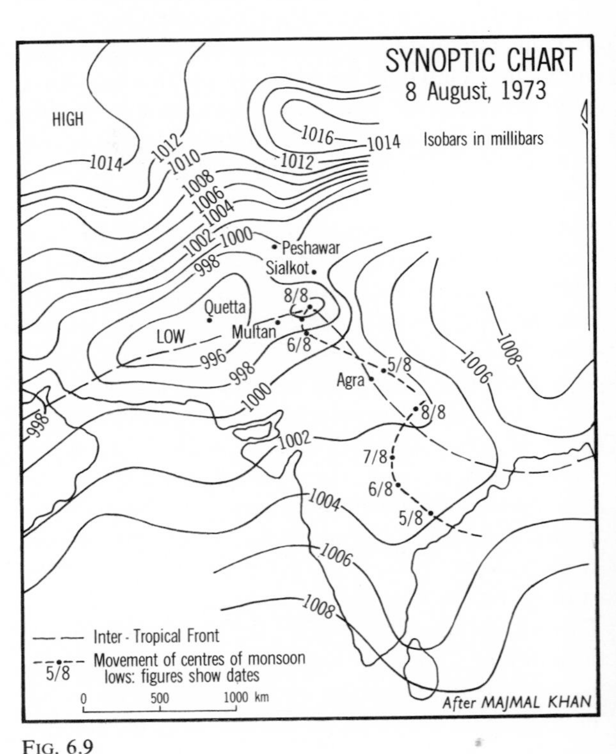

FIG. 6.9

27 This family rendered homeless by the 1973 floods takes refuge in temporary quarters on the protective embankment of Trimmu barrage, near Jhang, Punjab.

Agreement gave the waters of these three rivers to India, it did not (nor could it) insist on India exclusively accepting their flood waters. When the Ravi floods, Lahore trembles, as the waters are no respecter of frontiers. In 1955 the flood inundated 2420 villages levelling 1321 of them. Built of mud bricks, village houses do not stand up long to flooding. 400 people and 70,000 cattle lost their lives. The estimate of crops ruined was 101,911 ha. The Ravi Syphon, which carries under the Ravi the link canal taking Chenab water to the Sutlej, was destroyed. The total damage caused by the flood amounted to Rs 83 million.

It becomes clear from a study of these data that it is the combination of flood waves moving down rivers that ultimately join that can result in exceptionally high floods at a point (as happened at Panjnad in 1973) when the peaks at other stations were generally not exceptional. What the engineers pray is that peaks on several rivers will phase themselves and not arrive together in the main river.

Some account of the 1973 flood may be had from Table 6.6 which compares, with marginal comments as to damage, the flood flow with the design capacity of some of the major engineering structures on the Chenab, Ravi and lower Sutlej.

TABLE 6.5
Comparative flood data for the Indus basin

	Indus	*Jhelum*	*Chenab*	*Ravi*	*Beas*	*Sutlej*
(1) Mean annual rainfall (ins) of catchment (mm)	442	1,074	1,199	1,321	1,435	500
(2) Area of catchment ('000 km²)	102	13	11	3	7	19
(3) Mean annual runoff at rim station (Thousand million m³)	107	30	28	7	16	17
(4) Mean runoff per unit area of hilly (mm/km²) catchment	135	348	408	523	328	109
(5) (4) as % of (1)	78	84	88	102	59	57
(6) Average max. flow ('000 m³/sec)	26	22	20	7	10	14
(7) Absolute max. flow at rim station prior to 1973	27 (1942)	30 (1929)	31 (1957)	19 (1955)	8 (1955)	8 (1955)
(8) Absolute max. flow at various stations (a) Shahdara (b) Suleimanke (c) Panjnad (d) Sidhnai				(a) 15 (1955) (d) 5 (1955)	(b) 17 (1955)	
					(c) 18 (1950)	
(9) 1973 Maxima	16	8	22	(a) 5 (d) 6	(b) 5	
				(c) 23		

Note: Downstream on the Indus, at Mithankot (Panjnad–Indus confluence), Gudu and Sukkur, the 1973 flood reached record levels.

Source: mainly 'Floods of 1973', mimeo., Lahore 1973.

TABLE 6.6
Floodflow 1973 and design capacity ('000 m³/sec)

Structure	*Flood flow*	*Design*	*Comment*
Chenab			
Marala headworks	22	31	Loss of life and of live stock up stream of barrage.
Khanki headworks	28	23	Marginal bunds breached. Flooding of Lower Chenab canal.
Lower Chenab canal	453	368	Regulators damaged.
Qadirabad headworks	24	25	
Trimmu headworks	20	18	Right marginal bund breached and blasted to protect headworks Rangpur canal system damaged, villages, crops flooded. Multan threatened.
Panjnad headworks	23	20	Right marginal bund breached in two places when flow reached 20. Left bund breached in ten places. High winds drove waves to erode the bunds. Damage to Abbasia and Panjnad Canals, and to two towns.
Ravi			
Jassar bridge	6	8	In 1953 was hit by peak of 19.
Ravi Syphon	6	13	In 1953 was hit by peak of 19.
Shahdara bridge	7	7	In 1953 was hit by peak of 15. In 1973 protective bunds were breached in 9 places, Shahdara town was flooded and the Ravi shifted its course.
Balloki headworks	5	6	Breaches to left flood embankment peak of 8 in 1973.
Sidhnai headworks	6	5	Banks were cut to give relief.
Sutlej			
Islam headworks	5	8	Sutlej peak reached Panjnad 3 days after the Chenab peak.

The Karachi Floods of 1977

As can be seen from Figure 4.9 and the appendix of climatic data (chapter 4), Karachi's rainfall is extremely erratic. Years may pass without any appreciable precipitation, and then several times the annual mean fall may occur in a single month. On 1 July 1977, 229 mm of rain fell in a single day. On Figures 13.11 and 13.12 in Chapter 13, it can be seen that Karachi is laid out across the converging valleys of two major and several minor channels. The downpour in 1977 caused flash flooding in the Malir and Layari river courses, killing 280 people, rendering 18,000 homeless and destroying 5000 dwellings, many of them washed away in the bed of the Malir into which squatter settlers had overflowed from the adjacent Drigh Road Colony. Low lying areas of Karachi city were flooded to a depth of 1.5 m, railway tracks near Mauripur were washed out, telephone, power supplies and drinking water mains were cut, and the whole city was separated for several days from its sources of supply of vegetables and meat.

The flood also struck at industry: Pakistan's National Oil Refinery lost 60,000 drums of petrochemicals, scattered over 16 km of estuary.

This flood poses neatly some of the problems inherent in situations where a hazard may recur very infrequently. To the destitute and fatalistic refugee the hazard may be just one more of the 'acts of God' to which his existence has always been prone. He is forced by circumstance to settle where he can find space from which nobody will evict him. For the oil refinery the problem will be to judge the reasonable costs of protective engineering to insure against complete disaster while accepting a level of minor losses.

CHAPTER SEVEN

SUBSISTENCE BEYOND THE REACH OF CANALS: RAINFED AGRICULTURE AND THE HARSH LANDS

SUMMARY

This chapter looks at those regions of Pakistan where truly exotic water supplies are not available for irrigation. It examines how people manage to make a living in a variety of environments, most of them harsh in one way or another*. A distinction has to be made, however, between the reasonably well-watered areas in the north, where the hill country and its piedmont receives over 500 mm of rainfall annually, and the semi-arid and arid lands in the most northerly valleys in the rain shadow of the Himalaya, and much more extensively in the hills, piedmonts and plateaux lying west of the Indus and south from Kohat, as well as in the rolling sand plains of Thal and the Thar Desert fringe. For the most part the regions to be discussed fall among the lower ranks in terms of their index of development (see Fig. 2.15, Chapter 2). The exceptions are in the Punjab piedmont, the Potwar plateau and Kohat, all regions with reasonable rainfall. Fig. 7.1 showing rainfed and torrent-watered agriculture reveals the importance of the north and northwest in this connection.

Extremes of aridity, and of uncertainty of rainfall, lack of soil due to these desert conditions, extremes of slope, of sheer altitude reducing temperature, and very often dominating all, the extreme pressures of man's numbers on scant resources, all these and other factors make for a hard life in the harsh lands that extend outwards from the canal-irrigated 'garden of Eden' of Pakistan's basic ecumene.

*The term 'Harsh Lands' is borrowed from the title of a comprehensive study of such regions: Grigg, David, *Harsh Lands: a Study in Agricultural Development*, St. Martins Press, London 1972.

In most of these areas man has been able to survive, albeit in small numbers and at very low density, by his ingenuity at husbanding what little water comes into and can be wrung from his environment. Either he guides water from where he can find it, on or beneath the surface, to water the cultivable pockets of alluvium, or he lays in wait, his dams, ditches and fields prepared, for the occasional rains direct to his soil, or more usually, for the sudden torrents carrying the rapid runoff from rocky catchments. These he diverts with cunning and circumspection to

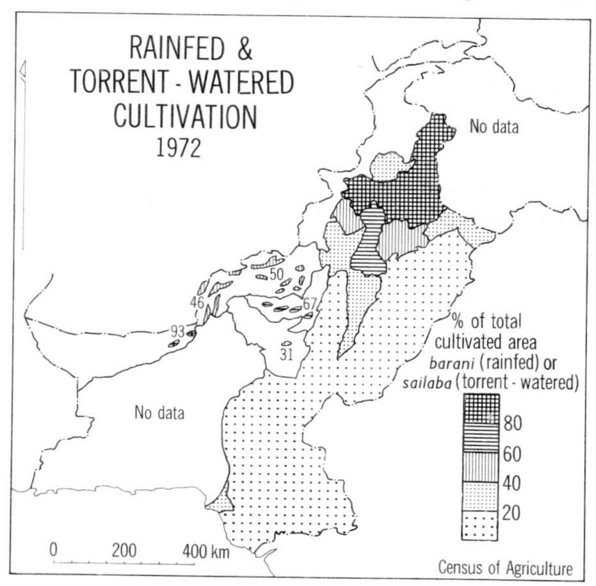

FIG. 7.1

28 A mixed flock of sheep and goats forage among the stones. Such sparse vegetation is typical of much of Baluchistan.

soak into the fields to support the crop he has waited to sow. In the most arid regions such supply of moisture is so precarious that reliance solely on crop cultivation for livelihood is too hazardous, and man has turned to exploit the other possibility of his surroundings, using livestock, goats, sheep and camels as his intermediaries with nature. Their foraging over wide expanses of sparse vegetation represent a harvest of the moisture that accumulates over time in the desert ecosystem. Their mobility enables the flocks and herds to seek out the most favoured patches of growth whether springing from recent localised rainfall or due to the nature of the terrain concentrating moisture from a wider catchment onto an alluvial fan or a pediment where it can infiltrate the weathered material in which plants take root to tap it.

Pastoralism, semi-nomadic and seasonally transhumant, is a common way of life, in Baluchistan particularly, but one poised in a delicate balance with the environment. The ecological system of a desert is a very fragile set of relationships which can easily be disturbed. Man's livestock, most notoriously his goats, have during generations of over-grazing insidiously reduced the palatable species in the plant association that evolved to survive and reproduce in the desert to the extent that its recovery to its 'climax' state is no longer possible in the short term. 'Desertification' is the term now used for such man-induced deterioration in the environment. The remedies are known in principle but difficult to implement in practice without radically altering the human ecological system which depends on these deserts, a system largely unconscious of the desertification process for which it bears the major responsibility. Alternative livelihoods can seldom be found in such regions, and the answer lies in encouraging on a permanent basis what already takes place temporarily, the outmigration of the people to find employment and a home elsewhere.

DESERTIFICATION AND RANGE MANAGEMENT

Short of removing the population from areas threatened by desertification, it is the responsibility of government administrators to try to ensure the proper management of such areas in order at least to minimize environmental deterioration in the future. It is not in the arid zone that range management is most needed (using 'range' in the sense of an area of natural or semi-natural vegetation used for grazing). The affects of over-grazing are perhaps more critically appreciated where forest resources are threatened by uncontrolled cutting and the invasion of herds that inhibit regrowth. Writing from the forester's point of view, H. G. Champion points to the damage accruing from failure to control such inroads on the forests.*

The forests and scrublands of Pakistan are subject to considerable pressure from man and his animals. Man enters the forest to find firewood and timber with which to build the roof to his house, and he brings his animals to forage in the clearings that he makes. In a climate where rainfall is markedly seasonal, and where in the mountains, winter cold inhibits growth, the natural grasses that livestock prefer are only seasonally available. Naturally dried hay is collected from the mountain sides wherever it grows, to be carried long distances as headloads to feed to cattle in their stalls.

*Champion, H. G., 'Effect of Human Population on the Forests of the Indian Subcontinent', *Yearbook of the South Asian Institute*, Heidelberg 1968–69, pp. 19–28.

So except during the flush period of early summer, the free-grazing stock subsist on scrub they can reach (and goats are adept at stretching to the lower branches) or on leafy branches cut down for them by the herdsman with his axe. In time over-grazing eats out the palatable species and the poisonous and unpalatable gain ground at their expense.

The areas first and most seriously affected by this process are those closest to watering points where stock must come regularly to drink, especially during the hot season. Unfortunately in the arid zone, man's technological skill has sometimes outrun his ability to consider the consequences of his actions. To sink tube wells where surface water is lacking may help support life in an arid area, but by enabling larger herds to congregate around the supply, it can also spell accelerated destruction of vegetation and with it the process of desertification.

Apart from the specific loss of grazing potential, over-grazing can be blamed at least in part for other side effects on the environment. The Potwar plateau mantled by a blanket of loess has suffered excessively and continues to suffer from gullying and soil erosion. The destruction of the vegetation cover by stock is partly to blame though man himself has sometimes handled this environment to its detriment, initiating the disturbances to the soil from which gullies can develop. The soil's capacity to hold moisture is generally less under grass or crops than under forest or scrub. When the latter is removed, run off can increase and so encourage gullying; the natural springs tend to dry up, water-storages become clogged with the soil wasted down stream, and the river beds become overloaded with sediment and more prone to flash flooding.

Reclamation of damaged areas and the protection of those threatened by desertification requires a measure of social discipline of which the rural community in its present state of development is quite incapable. It is not just that the rangelands are being subjected to damage irreversible in the short run. Excessive utilisation prevents them reaching their full potential to support livestock; at present they achieve between 10 and 50 per cent of what they might if properly managed. An estimate of the area of rangelands in Pakistan by province is as follows:*

Punjab	97,000 km^2
Sind	92,700 km^2
NWFP	56,700 km^2
Baluchistan	324,200 km^2

REGIONAL VARIETY

There is such variety of environment and of human adaptation in the broad area under discussion, that only a regional approach can do justice to the topic. It is convenient to take a more or less anti-clockwise approach, treating first the 'barani' (rainfed) agricultural lands of the north – the Punjab piedmont plain, Potwar plateau, the humid mountains and valleys. From here an excursion is made into the semi-arid interior 'behind' the outer mountain wall which traps most of the incoming moisture to consider the vertically integrated economy of this perhaps the most deeply dissected occupied landscape anywhere on earth.

Westwards as the trend of the ranges swings around the northern extremities of Pakistan, they become drier and their rainfall regime less and less monsoonal in the hills of the Frontier. The trans-Indus piedmont plain that fringes the mountain front from Kohat to Dadu, first at the foot of the Sulaiman range and then south of the Sibi reentrant, below the Kirthar range, is crossed by the courses of hundreds of torrents that flow intermittently, but when they do, descend in spate. This is the region of 'rod kohi' (torrent water) irrigation.

Behind and west of the Kirthar range extending for 400–500 km to the Iranian border are the arid plateaux and ranges of Baluchistan, where another set of environmental challenges produces fresh variety in man's response.

Finally, within the Indus plains there remain areas mostly of rolling sand dune topography where conditions must resemble those that obtained widely throughout the interfluves and desert margins a century or more ago.

*Personal communication, D. G. Wilcox, Western Australian Department of Agriculture.

The Punjab Piedmont Plain

At the outset may it be said that by no stretch of the imagination is this small region a 'harsh' land *in toto*. It is included in this chapter because it lies outside the canal irrigated tracts of the Indus plains. The belt of country, 40 km wide, that runs along the foot of the Pir Panjal range from Sialkot to the River Jhelum has been the most extensive region of continued close settlement from time immemorial in Pakistan. Physiographically it comprises two major elements, cover flood plain south of Sialkot and piedmont plain in Gujrat, separated by a strip of the active flood plain of the Chenab. Much of the cover flood plain is underlain by a good fresh groundwater reservoir which can be tapped by shallow wells. In the piedmont plain agriculture has to rely mainly on rainfall. The landforms and landuse associations of this zone are illustrated and discussed more fully in the next chapter (see Fig. 8.30.)

With a total rainfall of 826 mm, Sialkot is substantially wetter than Lahore, 120 km to the south, but has a similar seasonal distribution. The winter rainfall from December to March was never less than 103 mm with a medium of 146 mm in the nine years 1964 to 1972. 'Barani' or rainfed cropping is general for the dominant wheat crop, well irrigation being reserved for higher value crops and fodder. Between the winter rains and the monsoon in July the risk of drought is high, especially in April and May. This is when irrigation of seedling rice and sugar cane pays dividends. The maps of well irrigation (Fig 8.3) and of tube well irrigation (Fig. 8.4) in Chapter 8 below, show the high concentration of land so irrigated in Sialkot. Traditional wells are equipped with Persian Wheels which need draft animal energy to raise water. The tube wells are powered by diesel engine or increasingly by electricity from the grid supply. In Sialkot 65 and Gujrat 64 per cent of the cultivated area is irrigated, but a difference shows up in the fact that in the former canals provide some support, and only 7.4 per cent of the total cropped area has well water *alone*, whereas in Gujrat the proportion dependent solely on wells is doubled.

Such then is the Punjab piedmont, now agriculturally transitional in fact between barani and canal irrigated systems.

Potwar Plateau

Only a tiny fraction of Potwar has canal irrigation and that is in the gentle valley that leads away from the plateau proper from Taxila into Haripur and Hazara. For the rest it is 'barani' country par excellence because of the impracticability of irrigating such a fretted landscape rather than from preference. In Attock District the cover flood plain immediately bordering the Indus has intensive well irrigation, but as a general rule agriculture depends on rainfall which at Rawalpindi totals 960 mm, of which 259 mm occurs in the winter half year from October to March inclusive. Thus fair crops of wheat are grown on the fine loessic soils. Without irrigation the high yielding varieties are not favoured here and yields are consequently low (see Fig. 10.2 in Chapter 10). Rainfall declines away from the mountain edge, and Campbellpur averages 612 mm, but the winter fall is 229 mm. In Attock and Jhelum districts rabi cropping dominates, while in Rawalpindi it balances kharif. All three districts have

29 A cultivated field in the floor of a broad gully in the loess of the Potwar plateau, near Rawalpindi. Farm houses and their circular granaries are sited on the original plateau surface. Note the vertical cliffs characteristic of loess.

92 per cent or more of the cropped area dependent on rainfall alone, the remainder having some traditional well irrigation. Wheat occupies almost 90 per cent of the rabi land throughout, gram and pulses most of the rest. In kharif, the millets jowar and bajra occupy more than half the land then cultivated in Attock and Jhelum, while maize is a little more important than these together in Rawalpindi, being the favoured crop for the terraced hill slopes.

Fig. 7.2 shows the variation in wheat yields, both irrigated and unirrigated, for Rawalpindi and Attock Districts, 1965–66 to 1974–75 and the corresponding rabi season rainfall for Rawalpindi. Since irrigation in Potwar is mainly by wells and not from sources of exotic water, there is a fairly close relationship of yields to rainfall for both irrigated and unirrigated wheat, the main difference being in the comparatively higher yield for the irrigated crop. Whether the late kharif and early rabi rains have been good apparently influences the farmer in deciding how much land to sow. The unirrigated wheat area in Attock District shows a relationship between rabi rainfall and the area sown.

An inhibiting factor on the loess loam areas in kharif is their proneness to erosion if exposed to heavy rainfall. Much capital is being expended to protect these lands from further gullying. Check dams are built along the channels and contouring of fields has been undertaken to control runoff. Where the land has been very severely cut up into buttes and gullies, dynamiting and bulldozing have been used to level it into fields. Meanwhile the hard life of the farming families goes on, many of them living in cave houses cut into the loess bluffs, warm in winter, cool in summer, and probably no darker than the average mud house.

The environment sets constraints upon development, and many men leave Potwar to find work elsewhere, the army being a common career.

The Humid Mountains and Valleys

The belt of mountains that stretches north from the Murree Hills with first the Jhelum valley to the east and then the Kagan is the best watered part of Pakistan with the monsoon rainfall predominating, but ample precipitation in winter as

30 West of Rawalpindi efforts are being made to reclaim eroded loess lands and to prevent further deterioration by checking channel flow with structures such as this. The crop in the foreground is wheat. The scrub vegetation on the rocky hillsides is largely due to overgrazing by goats.

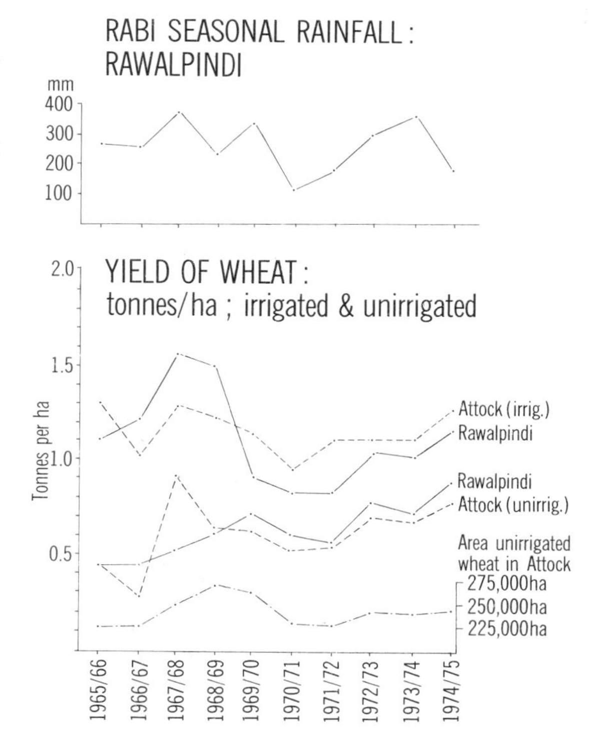

Fig. 7.2

31 The softness of the loess and its ability to maintain vertical slopes facilitates the excavation of homes in the Potwar plateau near Rawalpindi. These are warm in winter, cool in summer. A chaff cutter and water pots stand by the door; firewood is stacked against a thorn-brush corral for penning the goats in the corner.

well (see data for Murree, appendix, chapter 4). Through the Black Mountain to the west, and on into Swat and Dir the zone with an excess of 1000 mm annual rainfall persists, but with winter becoming the wetter season. These mountainous regions are carved deeply by the Jhelum, Indus, and Swat rivers and their tributaries, and present valley sides extending often from 1200 to 5000 m. With marked contrasts in altitude and aspect there is much variety in the detail of land use. The higher slopes are wooded and provide summer grazing on the way to subalpine pastures exposed above when the snow melts.

Transhumance takes place through a vertical distance from 600 m to 4300 m. Whole villages move up, locking or battening the doors of their winter homes, and staging their climb through four or five intermediate temporary huts. Some of the migratory herdsmen, the Gujars in particular, are blamed for illegal felling in the deodar pine forests. Valley sides as in the Kagan valley, central Hazara and in Swat are roughly terraced to grow maize in kharif, wheat and mustard in rabi. Landslides are a constant menace, destroying the hard won fields and homes. The broader valley floors are neatly irrigated by simple channel diversions to grow kharif rice, while rabi wheat needs no extra watering. The changing dominance from monsoonal to winter rain as one goes from Murree to Swat is reflected in the kharif/rabi cropping ratio (see Fig. 8.24) which changes in favour of the rabi crop west of the Indus.

The region contains Pakistan's best forests, the protection of which is becoming more important as population pressures increase and as water storages like Mangla and Tarbela are developed. Further deforestation would accelerate the rate of silting in these dams, so vital to power and irrigation for the plains. Timber is floated down the River Swat, eventually to be trucked to Dargai in the Vale of Peshawar.

The Inner Valleys*

North of the humid mountain belt, despite the ranges remaining high well over 5000 m, precipitation falls off rapidly. There is a mountain desert with less than 125 mm between Chilas on the Indus and Gilgit on the Hunza River, and this arid strip continues up the Indus valley through Skardu to Leh and beyond. Westward, Chitral in the Kunar valley at 167 m has 587 mm annually, and only here is rainfall enough to support barani cropping though irrigation is preferred. Elsewhere it is essential. Fortunately mountain streams coming down from the snow fields are numerous and are led often in channels several kilometres long to the areas they water. In the Indus valley these lie mainly in the valley floor, but in Chitral and Hunza the lower slopes are neatly terraced.

There is a distinct vertical zonation in land use. While conditions vary from place to place the general pattern is as follows:

*This section is based on Elizabeth Staley, *'Arid Mountain Agriculture in Northern West Pakistan'*, unpublished Ph.D. thesis, University of the Punjab 1966.

1. Up to 2000 m double cropping is possible. Wheat and maize are the staples of rabi and kharif seasons respectively. Barley, quicker growing than wheat and more reliable in difficult situations is also grown, as a rabi crop, and in the lower valleys, rice, sown in March, transplanted in July and harvested in October is an important kharif crop. Pulses are often intercropped with the grains.
2. Between 2000 and 2500 m there is a transition from double to single cropping. Cold in winter slows wheat growth to 230 days for maturation at 2150 m and to 270 days above that. At this level either wheat or maize is grown, barley replacing wheat at higher elevations. Grapes are cultivated here by the Ismailia sect who are permitted to make wine unlike orthodox Muslims.
3. Over 2500 m the choice is between a rabi crop or a kharif crop: wheat or barley as a rabi crop, maize, millets or buckwheat (the Indian *ragi*) as a kharif crop. Over 3000 m maize yields are poor; wheat extends to 3350 m, barley to 3700 m. Peaches can be grown to 3000 m, apricots a little higher, but both are important, with almonds also at lower levels. Dried fruits form a valuable part of the local diet.

Aspect exerts an influence on cropping, the south facing slopes being preferred for maize which benefits greatly from sunshine, while wheat and barley do best on north facing slopes where tillering is better and scorching less likely.

Cattle, sheep and goats are kept, their manure contributing to the high yields of crops. The smaller stock go to high pastures in the sub-alpine meadow-woodland in summer, but in winter fodder is sparse and dried lucerne from kharif cutting has to be hand-fed.

Among cash crops, charas (*Cannabis sp.*) is widely grown in Chitral, and along with opium forms a valuable export, ideal for transporting from such an inaccessible region being of small bulk but high value. Silkworms, reared on mulberry leaves, are the basis for cottage industry and for export to Swat where factory manufacture of silk fabrics is carried on. Other occupations are few: a little mining of arsenic, antimony and lead has been done in Chitral, where a road now helps link the valley to the outside world most of the year. The weaving of homespun woollen cloth for coarse clothing is a cottage industry.

It has yet to be seen how the completion of the Karakoram Highway from the China border to the former roadheads, one up the Kaghan valley and the other northeast from Saidu Sharif in Swat, will affect the economy of the region. Tourism to the 'Shangri La' of Hunza may become the most valuable 'export' once the political sensitivities of the Chinese working on the road, the Pakistan army and the local people, who have quite a reputation for independence of spirit, can be allayed.

The Frontier Hills

From Mohmand and Khyber southwards through the Tribal Areas of North and South Waziristan to the Baluchistan districts of Zhob and Loralai, the longitudinal trend of the ranges becomes increasingly pronounced. Apart from an outlier, as it were, of the humid mountain region formed by the prong of moderate rainfall in Kohat and Kurram (there is over 1000 mm precipitation at the head of the Kohat Toi valley), this is a semi-arid

32 A donkey caravan bringing tyres (maybe contraband) through the Khyber Pass from Afghanistan. Transport and trade make an important contribution to the economy of the rugged Frontier Hills.

region with between 250 and 500 mm. The higher country just east of Quetta forms an outlier of the main belt of semi-desert, separated from it by a drier corridor. The general aspect of the country, because of its barren rocky terrain, and the many centuries of overgrazing its vegetation has suffered, is more forbidding even than these figures may suggest.

Generally 70 per cent or more of the annual rainfall occurs in winter, when it is much more effective than the summer falls, which tend to come in sudden concentrated downpours and to be very variable. As a consequence, rabi cropping (of wheat almost exclusively) is usually more important than kharif (Fig. 8.24), the exception being Kurram which with its higher and more reliable rainfall can put more land under maize and even rice in summer. It will be noted in Fig. 8.24 how tiny are the areas cultivated in the Frontier Hills. Minor crops are barley in rabi, jowar and bajra in kharif.

Towards the south higher temperatures and a lesser proportion of the rainfall coming in winter, accentuate the aridity, and the comments of the Gazetter for Zhob in 1906 to the effect that most land depends on rainfall 'and from this a fair crop cannot be expected oftener than once in about three years', still hold true. Summer rains, when they occur, are generally conserved for a rabi crop.

The Trans-Indus Piedmont

West of the cover flood plain of the Indus, the land rises gently in the pediments and alluvial fans of the piedmont plain at the foot of the Sulaiman and Kirthar Ranges, the region of rod kohi. For the most part this is an arid strip of country with less than 250 mm, and in Sibi, Kachhi, Larkana and Dadu less than 125 mm of rainfall. Extremes of variability in rainfall from year to year are characteristic, and it is the more extraordinary that such elaborate earthworks are undertaken in the hope and expectation of the rivers coming into spate and so being led to water the prepared fields.

The systems used to manage torrent waters vary in detail and some are illustrated in Fig. 7.3. A major collective enterprise in Dera Ismail Khan District is shown in Figs. 7.4 and 7.5, and the overall relationship there of landuse and landforms in Fig. 8.32 (Chapter 8).

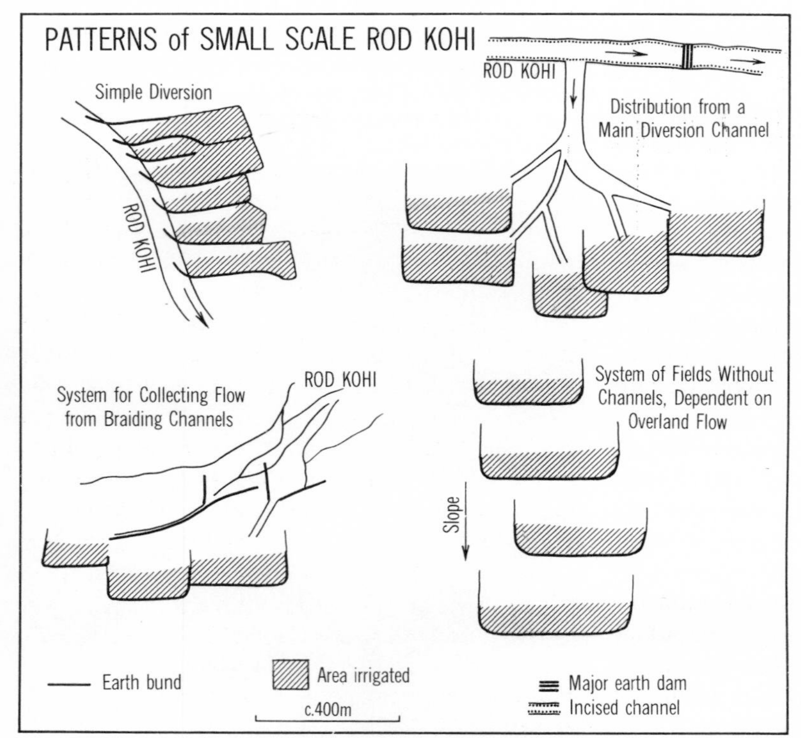

FIG. 7.3

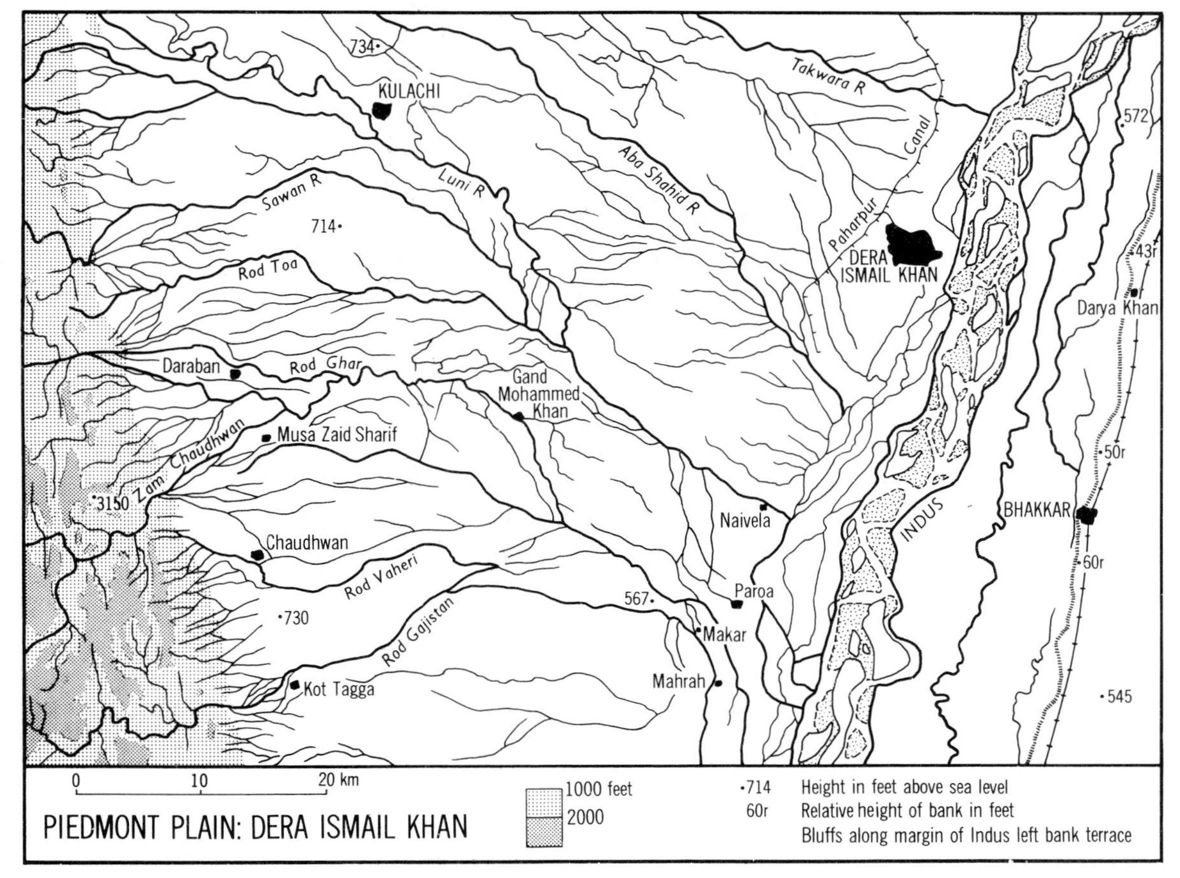

FIG. 7.4

In the north, as in D.I. Khan, the torrents may flow in late winter and in summer. The winter rainfall tends to come late, March averaging 25 mm (ranging between 6 and 32 mm in 1970–74), February 18 mm, January 12 mm and December 5 mm. Summer falls are very erratic, the July average of 64 mm represented extremes of 181 mm and 17 mm in the years 1970–74. Summer rainfall tends also to be highly concentrated over time, and a sudden rise in the torrents with a very heavy storm may wash away the earthworks built to divert them. Over a longer run of years (1964–72) at Dera Ghazi Khan, the winter rainfall was zero in four of the nine years, and there was no kharif rainfall in three years. The data for Jacobabad (see appendix, chapter 4) are characteristic of the southern end of the belt of rod kohi.

In the drier south, July and August are the most likely to produce flow in the torrents, and the expectation of winter rain is very low.

The success of the rod kohi system of irrigation, or more precisely of pre-watering the land, depends on preparedness for floods coming. In the case illustrated in Dera Ismail Khan, the system is essentially a series of earth dams, sometimes with boulder foundations, scraped up across a stream bed. The dams are intended to be breached in succession down stream as each has performed its task of diverting water into adjacent blocks of fields. Building the bunds is a communal undertaking overseen by the Deputy Commissioner's Office. Some work is done by government agencies who may provide a bulldozer. Much however is achieved by cooperation between government and the farmers, and some tasks are left to the farmers alone. The D.C. prods the farmers into repairing bunds before January. They give work with their bullock teams and scrapers in proportion to their water rights.

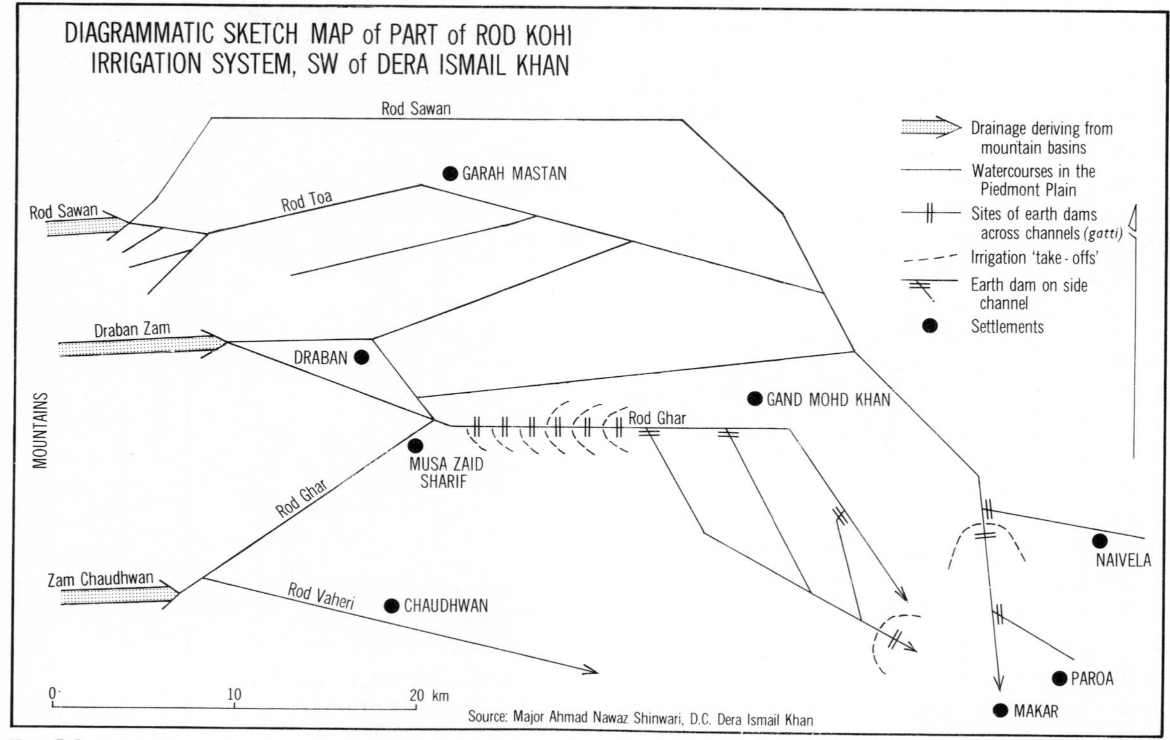

FIG. 7.5

The main crop grown is wheat, which depends on the previous summer's flood waters soaking into the soil. A kharif crop of jowar or bajra may be taken when local rainfall is adequate.

The district data mapped in Fig. 8.24 include canal irrigated lands commanded by Indus waters and so cannot be read as truly indicative of the cropping pattern supported by rod kohi.

33 The construction of a Rod Kohi dam is a co-operative enterprise. Here bullock teams with scrapers are moving soil from the river bed on the right onto the top of the dam which can be seen extending beyond the returning line of bullocks. Dera Ismail Khan, NWFP.

Baluchistan: the plateaus, ranges and basins

Extremely dry, almost half with less than 125 mm, and the rest no more than 250 mm, this region is further handicapped by extreme variability. Rain is more likely in the winter half year which is some consolation. The higher ranges, like those around Quetta, trap moisture, sometimes as snow, which percolates into the alluvial infill of the valleys from which it can be recovered by man for use in his fields. The relative accessibility of Quetta by railways gives farmers the chance of

34a The excavation and maintenance of karez tunnels is an expert and dangerous job generally undertaken by Brahui tribesmen. Here one is descending a shaft in the gravelly alluvium while his mates control the rope.

reaching the Karachi, Hyderabad and Multan markets, and indeed Lahore and the northern cities as well. This has helped set the Quetta-Pishin District apart from the rest of the Baluchistan plateau as a producer of fruits and vegetables. For much of the region, the former Baluchistan States covering Kalat, Kharan, Makran and Lasbela, comprehensive and reliable agricultural statistics have been lacking until quite recently.

The basic problem of the region is clearly water. Irrigation by karez tunnels or by damming torrents depends ultimately on rainfall to recharge the aquifer or fill the nalas. Direct rainfall is too scanty and uncertain for cultivation, and the farmer must concentrate water onto an arable patch from a wider catchment by one means or another. Of Kalat, it was said in 1907 and remains true, that 'dry crop cultivation is like hunting the wild ass', only one full crop in five years being the average expectation.

Where there is accessible ground water, as in Quetta and parts of Makran, karez are dug to lead it to the surface from the valley side fans. Farmers trace water from springs where it issues in the valley floor, digging long tunnels reached through shafts at 50 m intervals to tap the watertable and maximise the flow to their crops of wheat, gardens and orchards.

34 Cooling off in the waters of a karez as it issues from its excavated tunnel in the background. Near Quetta, Baluchistan.

35 Eventually the karez channel reaches the level of the fields to irrigate wheat crops in the floor of the valley. The hillsides are extremely barren in this semi-arid landscape near Quetta.

Fig. 7.6 shows the karez system in cross section and Fig. 7.7 the pattern they make on the landscape with their mole-like spoil heaps. The sketch map shows an area 12 by 5 km on the southern outskirts of Quetta. Karez irrigation is essential to cultivation here. The karez generally start well up the slope in the coarser gravels of the upper part of the alluvial fans that border the mountain front. Water is led out on to the valley floor to nurture fruit gardens and wheat fields. Most karez run more or less straight down the slope, but some appear to run transverse to the gradient of the fans before turning downhill to a garden. Such karez presumably tap a broader section of the groundwater flow.

Such is the variability of rainfall and of irrigation water in the Baluchistan plateau that an average situation can hardly be said ever to exist. The ratio of kharif to rabi crops shown in Fig. 8.24 is based on the Agricultural Census, itself a partial sample taken in a single year, 1972. In Table 7.1 the area under the major rabi crop, wheat and the kharif crop, jowar (tomatoes in Quetta) is traced through five years, 1969–70 to 1973–74 for Quetta, Kalat, Kharan and Makran Districts, with the relevant season's rainfall for several stations.

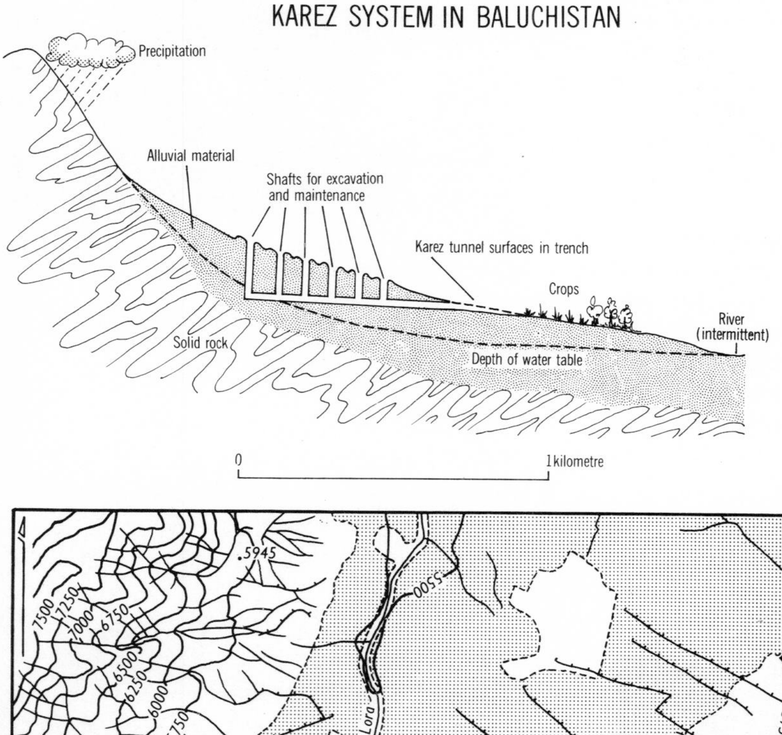

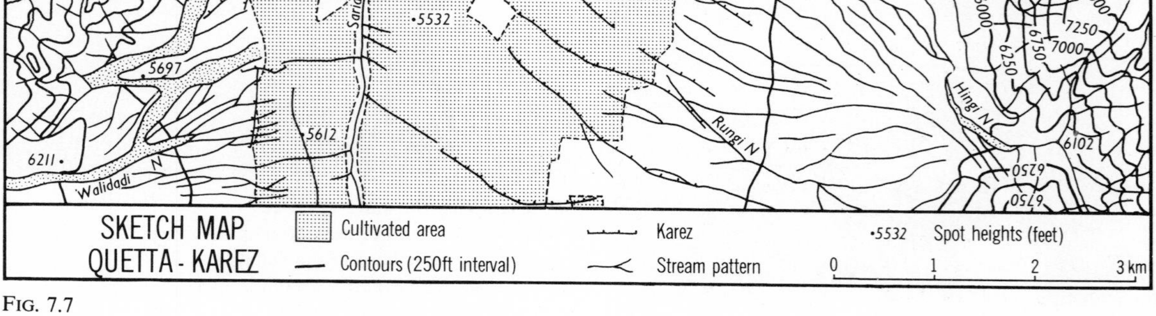

KAREZ SYSTEM IN BALUCHISTAN

SKETCH MAP QUETTA - KAREZ

FIG. 7.7

TABLE 7.1

	1969–70	*1970–71*	*1971–72*	*1972–73*	*1973–74*
QUETTA DISTRICT					
Area irrigated (ha)					
Canals	810	7,290	7,290	7,290	6,075
Tubewells	4,860	7,290	7,695	7,695	8,100
Karez	24,705	38,880	31,185	30,375	20,250
Kharif rainfall (mm)	76	27	0	51	29
Tomatoes (ha)	130	147	122	122	460
Yield (tonnes/ha)	8.02	10.17	8.02	9.98	9.35
Rabi rainfall (mm)	160	52	178	113	n.a.
Wheat irrigated (ha)	8,505	7,695	17,415	9,720	8,505
non-irrigated	4,050	19,035	45,765	35,640	20,250
total	12,555	26,730	63,585	44,955	28,755
yield (tonnes/ha)	0.74	0.28	0.20	0.22	0.50
KALAT DISTRICT					
Area irrigated (ha)					
wells	10,530	2,430	2,025	2,430	3,645
tubewells	405	4,050	6,480	6,075	6,075
Karez	15,795	20,250	18.225	15,795	10,530
Kharif rainfall (mm)	88[a]	74	0	4	0
Jowar irrigated (ha)	608	932	—	—	81
non-irrigated	3,240	—	41	41	41
total	3,848	932	41	41	122
yield tonnes/ha	0.52	0.75	—	—	0.83
Rabi rainfall (mm)	125	12	147	8	n.a.
Wheat irrigated (ha)	21,060	7,290	2,349	2,308	2,430
non-irrigated	27,094	810	3,929	2,633	3,848
total	48,155	8,100	6,278	4,941	6,278
yield tonnes/ha	0.76	0.71	0.33	0.49	0.46
KHARAN DISTRICT					
Area irrigated (ha)					
tube wells	–	405	405	405	405
Karez	3,240	3,240	4,050	4,050	2,835
Kharif rainfall (mm)[b]	7	19	0	43	13
Jowar irrigated (ha)	–	20	–	81	–
non-irrigated	1214	–	–	404	243
total	1214	20	–	485	243
yield tonnes/ha	0.33	n.a.	–	0.42	n.a.
Rabi rainfall (mm)[b]	86	18	64	23	n.a.
Wheat irrigated (ha)	4,860	567	324	2,471	–
non-irrigated	15,390	14,054	7,979	243	1,134
total	20,250	14,621	8,303	2,714	1,134
yield tonnes/ha	0.64	0.09	0.27	0.71	0.71
MAKRAN DISTRICT					
Area irrigated (ha)					
Canal	–	–	8,505	8,100	5,265
Wells	–	405	–	–	1,620
Tubewells	–	–	810	810	810
Karez	15,795	15,390	7,695	6,885	6,075
Kharif rainfall (mm)					
Panjgur	51	106	2	119	8
Jiwani	0	18	0	20	0
Pasni	1	57	0	33	18

Jowar irrigated (ha)	810	365	41	41	122
non-irrigated	–	6,683	–	–	–
total	810	7,047	41	41	122
yield tonnes/ha	0.62	0.46	n.a.	n.a.	0.83
Rabi rainfall (mm)					
Panjgur	110	5	125	5	n.a.
Jiwani	231	23	162	38	n.a.
Pasni	137	0	0	28	n.a.
Wheat irrigated (ha)	3,645	1,256	810	608	1,013
non-irrigated	2,835	770	770	405	1,256
total	6,480	2,026	1,580	1,013	2,269
yield tonnes/ha	0.89	0.29	0.44	0.50	0.44

Source: *Agricultural Statistics of Baluchistan*, 1969–70 to 1973–74, Quetta 1977.
[a] At Khuzdar. Other years at Kalat.
[b] At Dalbandin.

The table demonstrates more convincingly than mere words the variability in the environment's capacity to feed man through cropping. Karez are the main source of irrigation, though tube wells and even canals are becoming important. Non-irrigated land includes that watered in advance by torrent diversion. The kharif crop jowar is most liable to extremes of variation in area and yield in response to moisture conditions. Yields are very low for both wheat and jowar. Often a 'good' crop follows 'good' rains in the preceding half year. Apart from wheat and jowar, barley and pulses are grown as rabi crops, maize, cotton, pulses, bajra, and coriander as kharif crops.

Fruits have relatively more importance in Baluchistan than elsewhere. Melons which can grow very quickly in the intense sunshine after a shower of rain or a flood are popular on sandy soils, the flesh is used as a vegetable or fruit and the rinds are fed to stock. Tree fruits are a valuable source of food and income. Being deeper rooted than field crops, trees can better withstand the frequent droughts. At high levels such as Quetta, temperate fruits like apples grow well as well as do grapes, apricots and almonds. In Makran the humid heat near the coast favours date palms, mangoes and papaya (the last if water is available) while the higher lands inland can grow apples, peaches, plums, apricots, pomegranate and grapes if watered. The dates can be eaten green in May or harvested ripe in late summer, the best being preserved in earthen jars, baskets or skins.*

*Detail from Yahr, C. C., '*Present Economy and Potential Development of the Baluchistan States of Pakistan*', unpublished Ph.D. thesis, Univ. of Illinois 1957.

An impression of the landscapes of Makran representative of the region may be had from Figs. 7.8 and 7.9. The first is of northern Makran, a transect running through Panjgur. The east-west parallelism of the geological structure (the maps are based on geological surveys) is apparent. Alluvial basins occupy the wider valleys, with small 'hamuns' (playa lakes) in the lowest ground. In the north is the fringe of the Hamun-i-Murgho, one of several inland drainage systems in the centre of the Baluchistan plateau, with coarse gravel and sand dunes. The smaller streams lose themselves in the desert before reaching the Hamun.

Fig. 7.9 shows part of the coast of Pasni where the Shadi Khaur breaks through the ridges parallel to the sea to reach the coast. Sandy alluvium and beach sands reworked into dunes cover much of the terrain that is not bare rock.

The Rolling Sand Plains

This evocative term aptly describes much of central and southern Thal and the margins of the Thar Desert beyond the left bank flood plains of the Sutlej and the Indus. These lands were traditionally the realm of pastoralists who grazed their sheep, goats and camels on the desert flora, and raised crops spasmodically in the swales between the dunes. This still goes on though with the introduction of tube wells and power pumps, fresh ground water can sometimes be tapped to give a secure basis for cultivation, as in Thal. Barani crops are gram in rabi and bajra and jowar in kharif. Wheat has to be irrigated. Tharparkar in south-eastern Sind, beyond the command of the Eastern Nara canals, occasionally has rains from the monsoon.

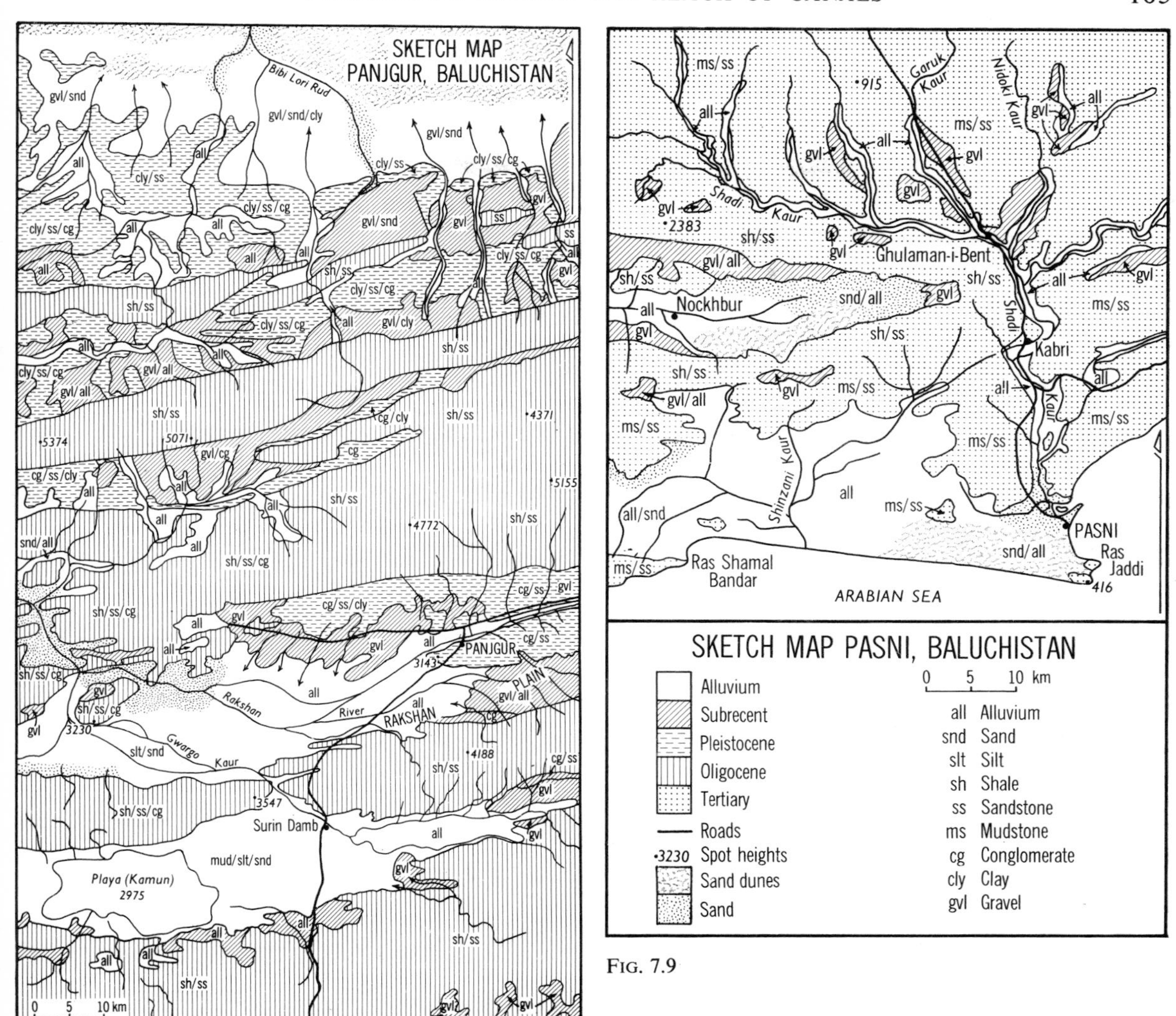

FIG. 7.8

FIG. 7.9

36 In the rolling sand plains of Thal the lower ground may sometimes be ploughed and sown with gram in the hope of getting a crop if the winter showers are adequate. Fixed sand dunes are still in evidence, but the electricity pylon carries some promise of modernization.

37 Goatherds aid and abet their herds in defoliating vegetation. All the trees in this picture in central Thal have been lopped at one time or another and only the hardiest thornbushes can withstand the grazing of goats.

38 The coming of the tube well has enabled farmers to find water in central Thal and to raise irrigated crops in the swales between the dunes. Here the crop is wheat, growing in extremely thirsty light soil.

39 A camel with a load of oil drums on the banks of the Main Line Upper Canal which feels the Thal Irrigation Project, near Daud Khel, Mianwali.

This is trapped by dams across the hollows and as it soaks into the soil the land is ploughed and rolled to seal the moisture in as far as possible, for a crop of rabi gram or barley, or kharif millets depending on the time the rain arrives. In the late nineteenth century 81 per cent of Tharparkar's cultivated area was rainfed, an area of 102,000 ha. Now 98 per cent is canal irrigated and only 7640 ha remain dependent on rainfall.

The conditions in the rolling sand plains may locally be harsh, but the relative proximity of developed irrigated settlement enables the people to link their traditional way of life to a source of income as a market for their produce or for their own and their camel's labour.

PASTORALISM

As was mentioned in the introduction, livestock browsing on the natural vegetation and able to be mobile in their search for sustenance, are a significant element in the Baluchistan economy. Although the numbers of goats, sheep and camels are not particularly high compared with the districts of the Indus plains, in relation to population they are considerably greater (see Figs. 9.1 and 9.7 in Chapter 9 below). Regular seasonal movements take place from higher pastures to lower for winter, some in fact out of Baluchistan and onto the plains.

The pastoralists' year in Baluchistan as summarised here is based on the gazeteers of the 1910s which still has relevance in this the most backward region of Pakistan:

January–February: flocks shelter from the cold, the rain and snow, in huts, caves or 'kizhdi'. Fodder collected the previous spring is fed – dried grass, and shrubs.

March: grass and shrubs begin to shoot. Fodder still used. Lambing.

April: weaning of kids and lambs to quarter milk. Risk of west winds damaging pasture.

May: abundant grass. Shi making (clarified butter) starts and goes on till July. Shearing. Cutting and storing grass.

June: grass begins to dry. Milk yield falls 'Krut' made from spare buttermilk by heating, straining the residue and making small balls to dry in the sun to eat as a relish. Male goats and rams sold off.

July: weaning and castration of wethers. Ewes go dry. Browsing on wheat and barley stubble.

August–September: grass now dry. Some green on the shrubs. Sheep shorn. Rams join the sheep in some areas. Flocks go down to the plains from Quetta.

October: rams loose generally. Sheep fattened for killing. Flocks subsisting on dried grass, leaves, shrubs.

November–December: fat sheep killed to make 'landi' dried meat. Grazing very poor.

Camels are bred, particularly in Makran, Kalat and Lasbela, for sale to the plains as draft and transport animals. From those kept, camel hair is a cash product.

FISHERIES

It is not surprising considering the sparseness of resources inland, that fishing is an important industry on the coast. Jiwani, Pasni and Ormara are the principal fishing settlements. The catch is very varied, sharks, drums, croakers, catfish, skates and rays making up more than half the total by weight. From June to mid-August when the southwest monsoon blows it is too stormy for fishing. The main season is from mid-August to mid-February.

In the overall picture, however, Baluchistan's fisheries are relatively poorly developed compared with those of Sind, centred on Karachi. More than a thousand trawlers operate on the Sind coast, together with 700 gill net boats and 4000 sail boats. Baluchistan has 60 gill net boats and 2200 sail boats. Judged by the number of fishermen engaged, inland fisheries are as important as marine: of 201,300 more than half, 104,000 are inland, 54,000 in Sind, 45,000 in Punjab and 5,000 in NWFP. The sea-going remainder is split between Sind, 72,000 and Baluchistan 25,000.

CHAPTER EIGHT

CROPS AND CROPPING PATTERNS IN THE INDUS PLAINS

SUMMARY

The seasonal availability of water is the main constraint on agriculture in the Indus plains, whether it comes from rainfall or as some form of irrigation. Canals, wells and tube wells all play a part in providing water. Canals suffer the limitations imposed by the regimes of the rivers upon which they depend. Groundwater reservoirs are less immediate in their response to changes in the rate of recharge of their aquifers, and farmers are keen to employ wells even where canal water is the major source because of the independence and assurance of supply they give. Some of the variables in water availability are noted.

The principal crops are reviewed by brief analyses of their pattern of distribution. Crop associations are studied in particular through the relationship between crops cultivated in the two cropping seasons, rabi and kharif. Seven transects characteristic of the range of terrain within the Indus plains are described in terms of their landforms and the agricultural land use associated with these, as a means of synthesizing the preliminary studies.

IRRIGATION

The map of irrigated cropping, Fig. 8.1, shows how paramount is the importance of irrigation to the Indus plains. Over a very large part of Punjab and Sind, the percentage of the cropped area irrigated in one way or another exceeds 80, and is often over 90 per cent. The barani lands of the Punjab piedmont and the southern areas of Thal are less fully provided with canal supplies but nonetheless have over 60 per cent of their crop land irrigated. The low figure of 30 per cent for Mianwali is due to the district covering a large area of the rolling sand plain in central Thal together with some rocky hill country. Some districts in Sind fall below the 80 per cent level, but only barely so in the cases of Sukkur and Dadu. Jacobabad at 52 per cent irrigated is on the fringe of canal commands and has a substantial area reliant on torrent watering or 'sailaba'. There is a problem of definition that can distort the picture somewhat. In Sind, for example, it used to be common under the earlier inundation canal system, and may still be the practice when supplies are short, to grow a rabi crop without irrigation in that season on the basis of moisture remaining from late kharif 'pre-watering' of the land. Some of the apparent deficiency in irrigation in Sind, in districts that can hardly rely on the very low and unpredictable rainfall, may be accounted for in this way.

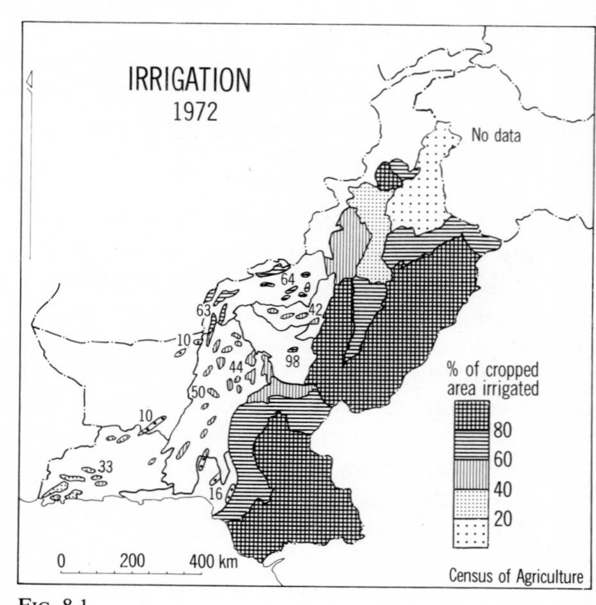

FIG. 8.1

Fig. 8.2 shows the dominance of the various irrigation modes in the Indus plains. Perennial canal supply has been the ultimate target for as much of the basin as can be reached. Seasonal canals, meaning 'inundation' canals flowing only when the rivers are running high, serve broad tracts in the Sutlej valley, now in the 'tail-end' position of the systems deriving water from rivers to the west. The old seasonal canals in Sukkur and Jacobabad have in part been converted to perennial status since the Gudu barrage was built. Wells are the dominant source in only limited areas. Those tapping the extensive groundwater reservoir in the Punjab submontane belt have been discussed already in Chapter 7. Elsewhere well irrigation is found closely paralleling the rivers where they can easily tap the shallow but fresh groundwater associated with the active flood plain. It may be said, however, that wells powered by animal teams turning Persian Wheels are found wherever there is accessible water, and as Fig. 8.3 shows, the Punjab and the Vale of Peshawar in NWFP are liberally provided with this source of water even where canals are the main supply. Sind is less fortunate due to the finer sediments there which make aquifers slow yielders of water, and also due to the high salinity of much groundwater in the lower Indus basin. Gradually, but with accelerating speed under the stimulus of rural electrification, the much more efficient tube well is displacing the traditional Persian Wheel over its dug well. Fig. 8.4 shows clearly how tube wells now dominate, particularly

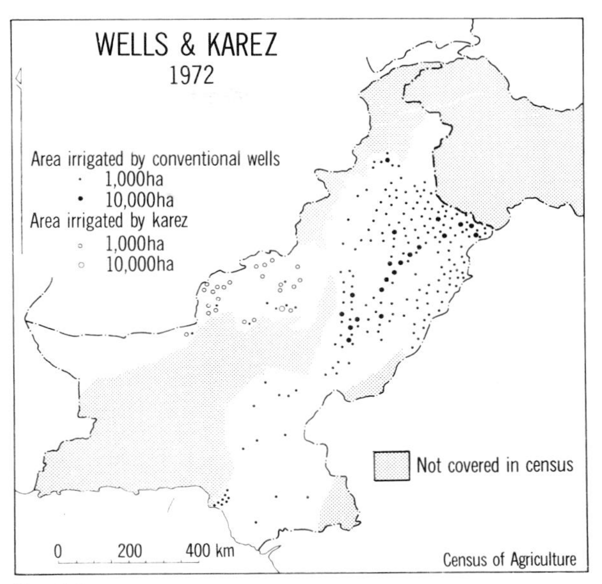

FIG. 8.3

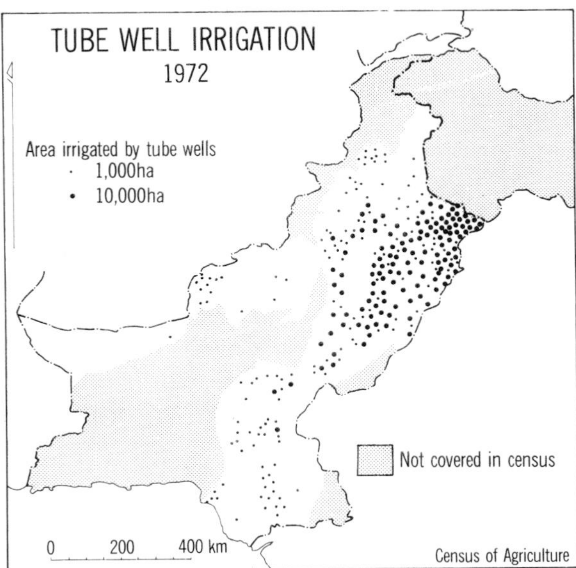

FIG. 8.4

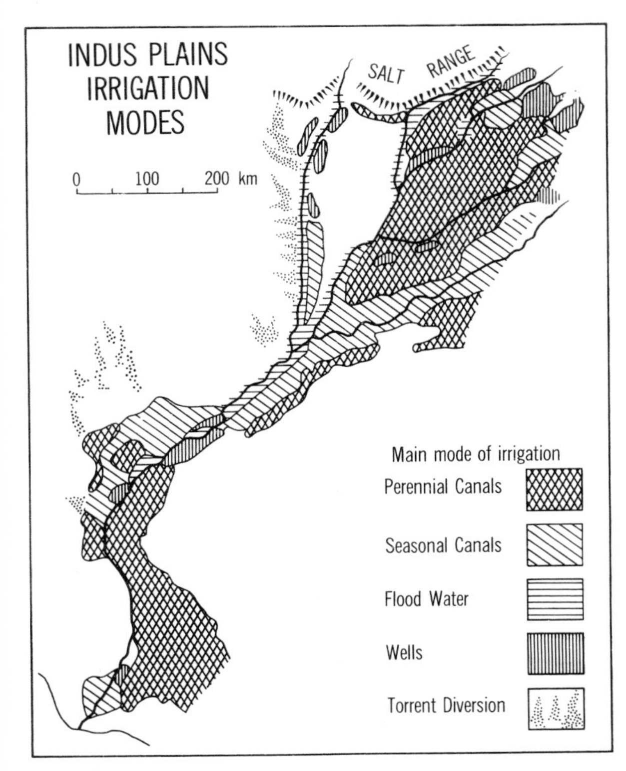

FIG. 8.2

in the Punjab. The considerable numbers in Thal are noteworthy where they may be found watering swales amid the sand dunes. Some of the stimulus to sink tube wells has come from the government in its efforts to reverse the wasting effects of salinity which continues to be rife in parts of the Punjab.

Table 8.1 shows some of the current trends in water availability over the past ten years. While the relative increase in provision by tube wells, public and private, has been very great, they contribute only 6.5 and 25.8 per cent respectively of the total delivery at the 'farm gate', a big advance nevertheless on their 3.6 and 11.8 per cent shares of the 1965–66 total. Tube wells can deliver a constant flow all year round, as the figures indicate, while the canal system is sensitive to seasonal flows. It is interesting to note, however, that the kharif share of the total from canal sources has fallen from 71 per cent in 1965–66 to 64 per cent in 1975–76, indicating the impact of Mangla storage on improving the availability in rabi.

The canal system is not simply a tap that can be turned on at will to provide the maximum flow which the canals are capable of carrying. Their flow must depend on that of the rivers they tap. As Fig. 5.2 has shown, there is considerable variability particularly in kharif flow. As the canal system is designed basically to deliver the rabi flow, knowing that the kharif will not be less, the rabi variation is the more critical to the farmer. Table 8.2 shows the monthly pattern of canal water supply as measured by canal withdrawals from 1969–70 to 1974–75 in Punjab and Sind. The months are grouped in the four irrigation periods used by the irrigation engineers, thus:

April–June: for preparing the kharif.
July–September: main kharif watering.
October–November: pre-watering for rabi.
December–March: main rabi watering.

In terms of liability to variability April is the critical month in the Punjab, the lowest supply reaching 38 per cent of the highest. Sind has always been most anxious about this pre-kharif watering when new projects for Punjab were under discussion. The temptation to plan for more use in the Punjab at this period of low supply leaving less for Sind is ever present, and suggests that a supra-provincial authority should have the responsibility for determining where the water will best be spent.

The very low withdrawals in August and September 1973 were occasioned by the disastrous floods described in Chapter 6. By and large it seems that the variation between minima and maxima are around 75 per cent in Sind with lows of 58 in January and 61 in April. In Punjab the pre-rabi and rabi flows can vary by 52 per cent in October and December, 55 per cent in January and March.

Until the 1960s canal irrigation systems were usually designed to spread water thinly in order to give some benefit to the largest area. That this policy contributed to salinity is now recognized, and gradually optimisation of supply will be the target. With storages at Mangla and Tarbela it will be possible to guarantee a greater rabi supply, thus appreciably improving the intensity of cropping possible. Fig. 8.5 shows intensity of cropping in 1972 as a percentage of the net cultivated area represented by the total cropped area including current fallow. The highest intensities are in the perennially irrigated tracts, but clearly large areas still carry single crop each year, either rabi or kharif, but not both.

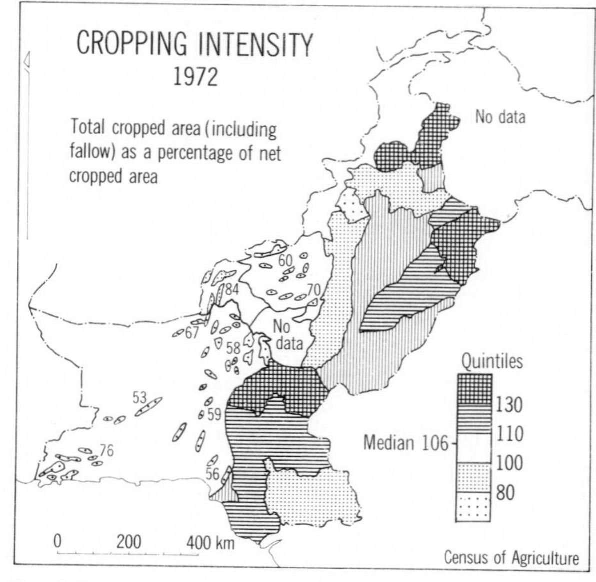

FIG. 8.5

TABLE 8.1

Changing overall availability of water (thousand million m³)

	1965–66	*1970–71*	*1975–76*	*% change 1965–66 to 1975–76*
Surface water				
at canal head				
Kharif	80.0	75.0	78.0	− 3
Rabi	32.0	32.0	44.0	+ 38
Total	112.0	107.0	122.0	+ 9
at 'Farmgate'				
Kharif	52.0	48.0	49.0	− 5
Rabi	21.0	21.0	28.0	+ 35
Total	73.0	69.0	78.0	+ 9
Public tubewells				
Kharif	1.5	2.7	3.7	+150
Rabi	1.5	2.7	3.7	+150
Total	3.0	5.3	7.4	+140
Private tubewells				
Kharif	5.1	10.1	15.0	+193
Rabi	5.1	10.1	15.0	+193
Total	10.1	20.2	30.0	+193
Grand total at 'farmgate'				
Kharif	58.0	60.0	68.0	+ 17
Rabi	27.0	33.0	47.0	+ 73
Total	85.0	95.0	115.0	+ 35

Source: *Annual Plan*, 1976–77.

TABLE 8.2

Variability in water supply: canal withdrawals (thousand million m³)

	1969–70	Total	*1970–71*	Total	*1971–72*	Total	*1972–73*	Total	*1973–74*	Total	*1974–75*	Total
Punjab												
April	4.6		3.1		1.9		3.5		4.8		4.1	
May	7.3	19.7	4.7	14.8	4.2	12.8	5.1	15.8	6.9	17.6	5.1	15.6
June	7.8		7.0		6.7		7.2		5.9		6.4	
July	8.0		7.9		8.0		8.3		6.7		7.4	
August	7.6	23.0	8.0	23.4	7.8	22.7	8.8	24.6	3.8	14.8	7.9	21.0
September	7.4		7.5		6.9		7.5		4.3		5.7	
October	5.3	9.1	4.9	8.1	3.6	6.1	6.8	10.1	6.2	10.3	3.8	6.4
November	3.8		3.2		2.5		3.3		4.1		2.6	
December	2.7		2.2		1.5		2.8		2.6		2.0	
January	2.5	10.5	1.6	7.8	1.4	6.5	1.9	11.0	1.9	10.6	1.6	9.0
February	2.1		1.7		1.6		2.7		2.6		2.3	
March	3.2		2.3		2.0		3.6		3.5		3.1	
Sind												
April	2.6		2.1		1.6		2.3		2.5		1.6	
May	3.8	13.6	3.1	11.2	2.7	11.5	4.1	14.4	4.1	15.0	3.3	11.4
June	7.2		6.0		7.2		8.0		8.4		6.5	
July	8.4		8.8		9.5		9.5		9.1		9.4	
August	7.6	22.3	7.3	22.0	8.3	24.8	8.9	26.3	7.5	23.4	9.5	26.2
September	6.3		5.9		7.0		7.9		6.8		7.3	
October	4.4	7.6	4.6	7.3	3.6	6.2	4.2	6.8	4.8	7.8	3.6	5.9
November	3.2		2.7		2.6		2.6		3.0		2.3	
December	2.3		2.0		2.1		2.0		2.7		2.0	
January	2.1	8.4	2.0	7.2	1.6	7.3	2.3	8.6	1.9	8.7	1.4	6.7
February	2.0		1.6		1.7		2.0		2.1		1.7	
March	2.0		1.6		1.9		2.3		2.0		1.6	

THE CROPS

The principal crops are listed in Table 8.3 as in 1977–78 (provisional figures)

TABLE 8.3
Principal crops ('000 ha)

Wheat	6,350	Gram	1,113
Rice	1,815	Sugar Cane	804
Bajra	648	Rape, etc.	528
Jowar	447	Cotton	1,817
Maize	657	Tobacco	124[a]
Barley	172		

Source: *Economic Survey*, 1977–78.
[a]1976–77.

A fuller table, based on the latest available volume of *Agricultural Statistics of Pakistan*, 1975 is given in Table 8.4 to show the range of crops, separated into rabi and kharif seasons.

FOOD GRAINS

Wheat

The distribution of wheat is shown in Fig. 8.6. As the country's staple food grain it is widely grown in every district. Conditions in the extreme south are least favourable because of high temperatures and salinity towards the Indus delta. Almost two-thirds of the wheat crop is now HYV Mexi-Pak variety (Fig. 8.7 and 8.8 and Table 8.5). The need to irrigate the HYV so that fertilizer can be effectively applied militates against its popularity in the barani areas like Potwar which account for almost a quarter of the wheat area. In many districts of Punjab and Sind almost 100 per cent of the wheat crop is HYV.

It is to be expected therefore that yields under irrigation will be substantially higher than those for rainfed wheat, and this is so. Under irrigation yields

TABLE 8.4
Crop areas: five-year average, 1971–72 to 1974–75 ('000 ha)

	Rabi		*Kharif*		
Food Grains					
Wheat	5,939		Rice	1,512	
(HYV)	(3,682[a])		(HYV)	(631)[a]	
Barley	172		Maize	633	
			Bajra	680	
			Jowar	520	
Pulses					
Gram	996		Mung	67	
Masoor	75		Mash	43	
Mattar	186		Guar seed[a]	230	
Oilseeds					
Rape and mustard	520		Castor seed[a]	6	Groundnut 41
					Sesamum 23
					Linseed 8
Other annual crops					
Tobacco	51		Sugar Cane	609	
Chillies	34		Cotton	1,917	
		Potato 23			
		Vegetable 135[a]			
Fodder Crops					
	1,215			1,510	
Fruits Total 1770					
Citrus	504		Pears	22	
Mangoes	582		Plums	19	
Bananas	120[a]		Grapes	25[a]	
Apples	62		Pomegranate	19[a]	
Guava	164		Dates	221[a]	
Apricots	21				
Peaches	11				

Source: *Census of Agriculture*, 1972.
[a]Denotes latest single year.

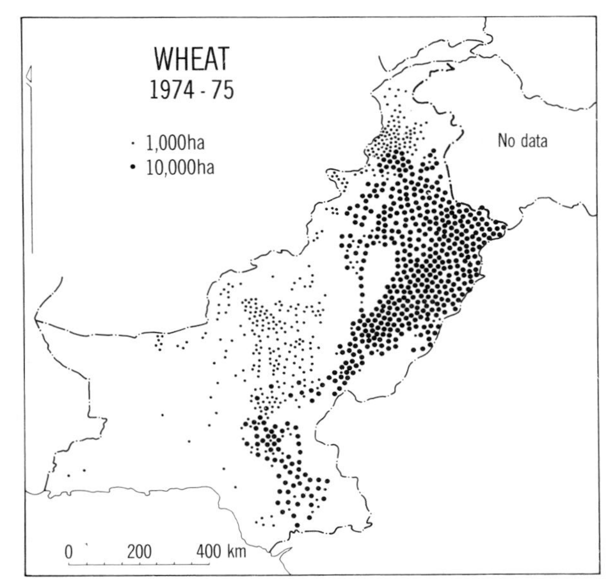

Fig. 8.6

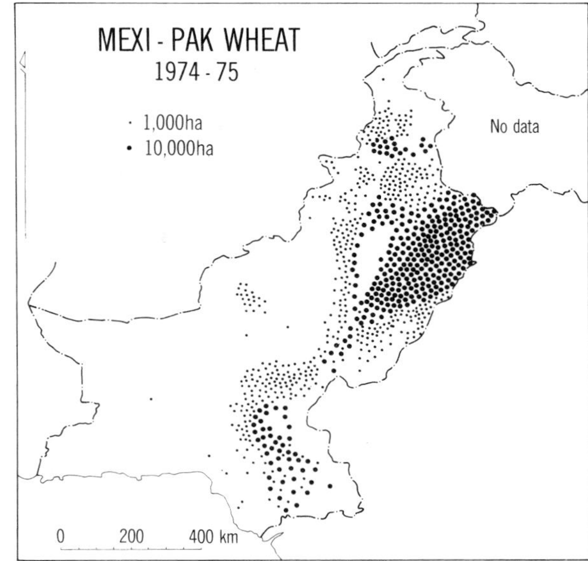

Fig. 8.8

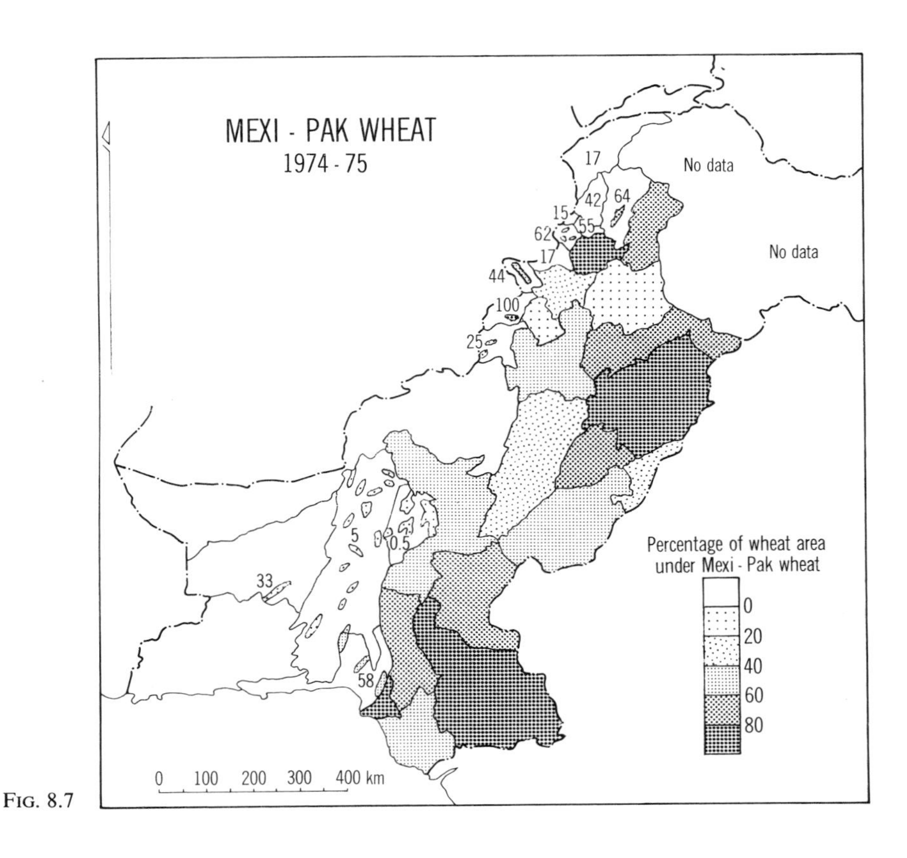

Fig. 8.7

TABLE 8.5

Wheat and rice: area (million ha), production (million tonnes) and yield (tonnes per ha), 1947–48 to 1976–77

	Wheat			HYV Wheat				Rice			HYV Rice			
	Area 1	*Prod.* 2	*Yield* 3	*Area* 4	*Prod.* 5	*Yield* 6	*4 as % 1* 7	*Area* 8	*Prod.* 9	*Yield* 10	*Area* 11	*Prod.* 12	*Yield* 13	*11 as % 8* 14
1947–48	3.95	3.35	0.84					0.79	0.69	0.87				
1948	4.29	4.04	0.94					0.84	0.75	0.88				
1949	4.18	3.92	0.94					0.93	0.80	0.86				
1950	4.37	3.99	0.90					0.97	0.86	0.89				
1951	4.11	3.01	0.73					0.88	0.73	0.83				
1952	3.82	2.40	0.62					0.91	0.83	0.90				
1953	4.22	3.64	0.86					1.02	0.92	0.90				
1954	4.26	3.19	0.74					0.96	0.84	0.86				
1955	4.52	3.37	0.74					0.97	0.84	0.86				
1956	4.69	3.64	0.77					0.97	0.84	0.86				
1957	4.61	3.56	0.77					1.07	0.88	0.81				
1958	4.83	3.91	0.81					1.15	0.99	0.85				
1959	4.88	3.91	0.80					1.20	0.99	0.83				
1960–61	4.64	3.80	0.82					1.18	1.03	0.87				
1961	4.92	4.03	0.81					1.21	1.13	0.93				
1962	5.02	4.17	0.83					1.19	1.10	0.92				
1963	5.02	4.16	0.83					1.29	1.19	0.92				
1964	5.32	4.59	0.83					1.36	1.35	0.99				
1965	5.16	3.92	0.75					1.39	1.32	0.93				
1966	5.34	4.33	0.81					1.41	1.36	0.96				
1967	5.98	6.42	1.06	0.96	2.24	1.77	16	1.42	1.50	1.05	0.004	0.01	3.02	0.3
1968	6.16	6.62	1.06	2.37	3.94	1.66	38	1.55	2.03	1.30	0.31	0.65	2.12	20.0
1969	6.23	7.29	1.17	2.68	4.76	2.34	43	1.62	2.40	1.48	0.50	1.03	2.03	31.0
1970–71	5.98	6.48	1.07	3.13	4.79	1.53	52	1.50	2.20	1.47	0.55	1.06	1.93	37.0
1971–72	5.80	6.90	1.18	3.29	5.27	1.61	57	1.46	2.26	1.55	0.73	1.41	1.94	50.0
1972–73	5.98	7.45	1.24	3.39	5.56	1.65	57	1.48	2.33	1.58	0.68	1.26	1.94	46.0
1973–74	6.12	7.64	1.24	3.48	5.73	1.65	57	1.51	2.46	1.62	0.67	1.30	1.95	44.0
1974–75	5.81	7.67	1.31	3.68	6.00	1.63	63	1.60	2.31	1.44	0.63	1.11	1.75	39.0
1975–76	6.11	8.74	1.43					1.71	2.62	1.53				
1976–77	6.32	9.15	1.45					1.70	2.63	1.55				

Source: *Agricultural Statistics of Pakistan*, 1975.
Monthly Statistical Bulletin, 1977.
Economic Survey, 1976–77.

averaged 1.546 tonnes per ha. In Potwar, where the contrast is not reinforced by the irrigated crop being also largely HYV, the difference is less marked thus:

Attock: 0.846t/ha irrigated; 0.485t/ha, unirrigated.
Rawalpindi: 0.789t/ha irrigated; 0.554t/ha, unirrigated.
Jhelum: 0.830t/ha irrigated; 0.519t/ha unirrigated.

Since independence in 1947 the wheat area has increased by 60 per cent, and fortunately, in view of the growth in population, production is 173 per cent higher. Table 8.5 shows the progress in area, production and yield per hectare. The introduction of HYV will be discussed below in Chapter 10. As Fig. 8.9 shows, imports of wheat have rarely been avoided since the early 1950s. The first impact of HYV brought imports to near zero for a year, but dislocations due to the trouble in Bangladesh allowed demand once more to outrun supply. In 1970–71 to 1974–75 total food grain production, in which wheat played the major role, averaged 11.03 million tonnes, almost double the level in 1947–48 to 1949–50, i.e. 5.64 million. Meanwhile population grew faster, from 31 to about 70 millions, the comparison being between 95.6 per cent increase in food grains to feed an increase of 126 per cent in population. Production in 1976–77 was 13.36 million tonnes.

40 Wheat is still usually harvested by hand sickle, as here in the Punjab.

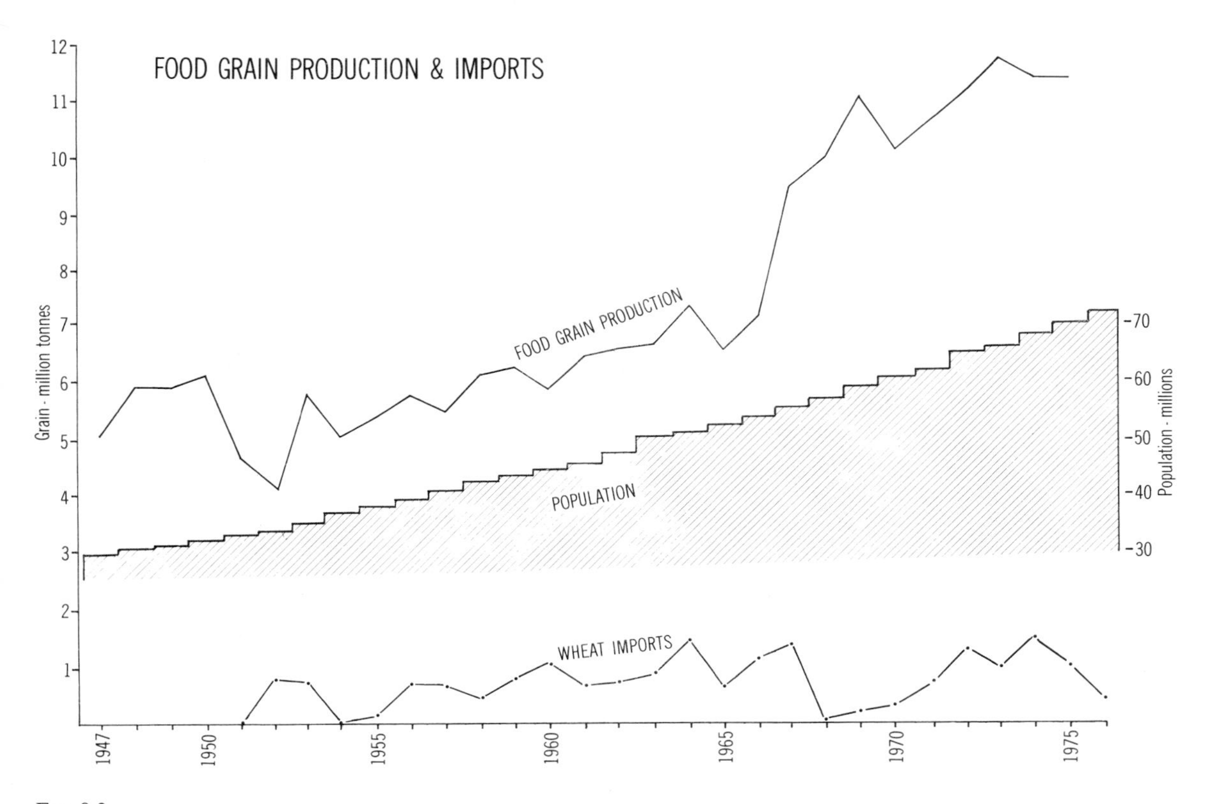

FIG. 8.9

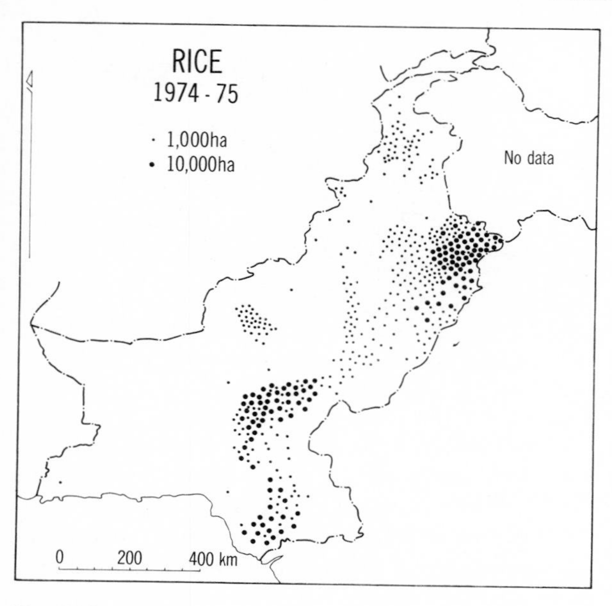

FIG. 8.10

Rice

Fig. 8.10 shows the distribution of the area under rice. It is a more clustered distribution than that of wheat because rice has to stand in water, and so needs impermeable soils and an abundance of irrigation water. Tolerance of a measure of salinity is a valuable characteristic of the crop, which is able to thrive in some areas of Sukkur, Larkana and Tharparker in Sind where other crops have difficulty. The other major concentration of rice is in the Districts of Gujranwala in Punjab and its immediate neighbours, where suitably impervious soils are found.

Although a net importer of food grains, Pakistan exports rice in increasing quantity. In 1974–75, 434,000 tonnes were exported, of which Basmati, the high quality and highly valued product of Punjab, made up 36 per cent. IRRI-Pak rice and Joshi rice contributed 25 and 37 per cent respectively. In value terms Basmati commanded Rs 7667 per tonne against Rs 4335 for IRRI-Pak and Rs 2589 for Joshi, or 57 per cent of the total earnings for rice. Basmati is grown mainly in the Lahore

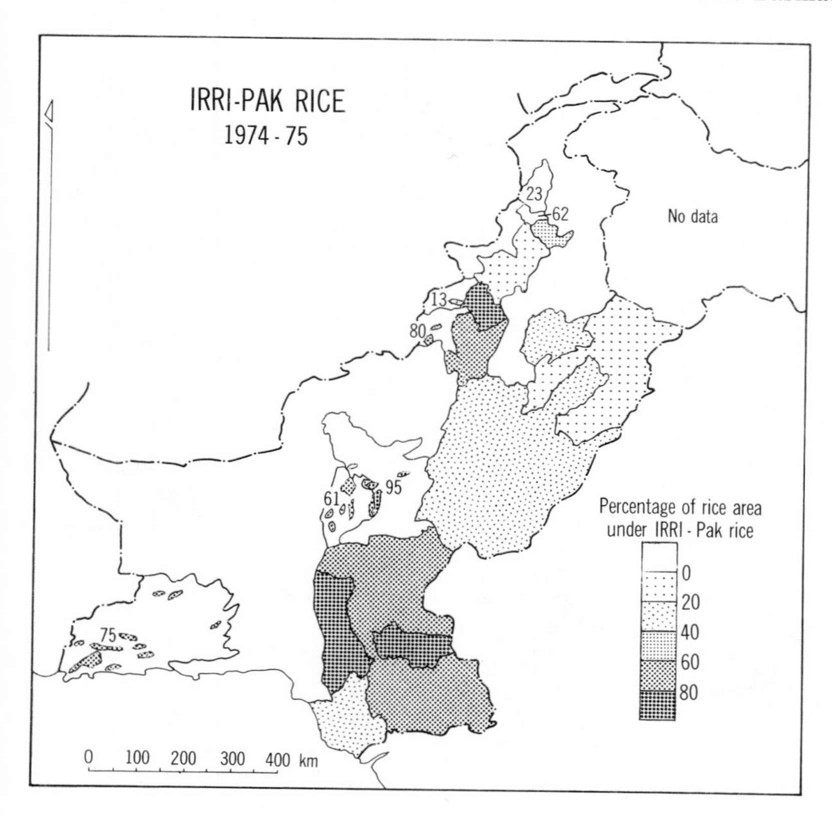

FIG. 8.11

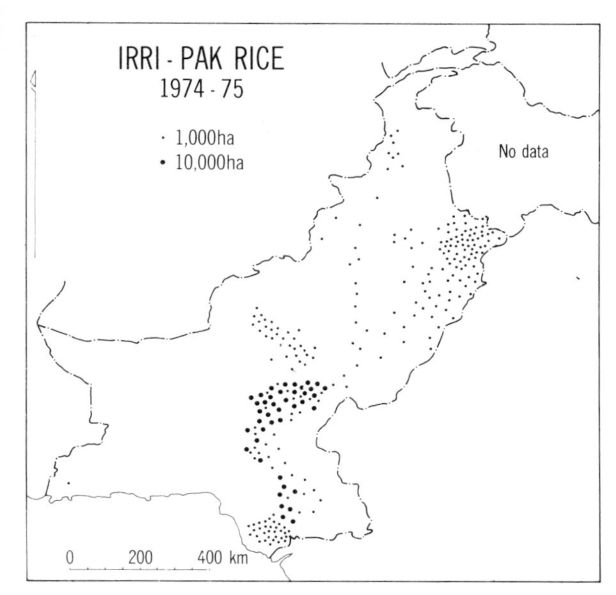

FIG. 8.12

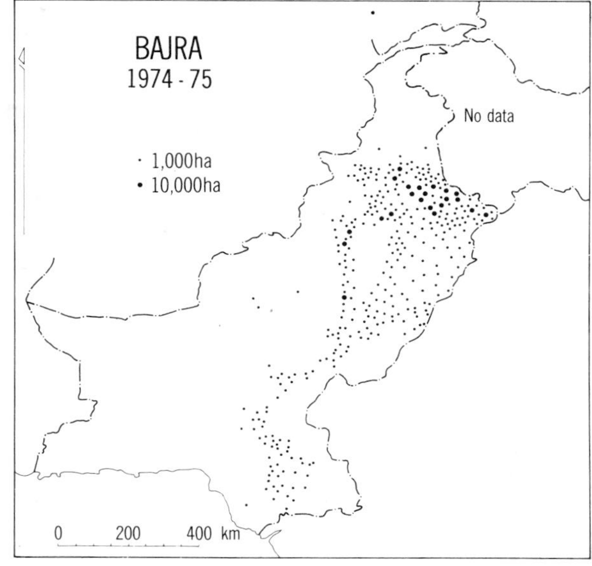

FIG. 8.13

Division of the Punjab, 27 per cent of the area being in Gujranwala alone, 21 per cent in Sialkot and Sheikhupura, 10 per cent in Lahore. Overall, Basmati occupies about 31 per cent of the rice area, while the HYV have 39 per cent. As Pakistan is less interested in coarse rice as a food for its population than with fine rice for export and the 'luxury' market, HYV is losing ground relatively to Basmati. Figs. 8.11 and 8.12 thus indicate a high acceptance of IRRI-Pak rice in Sind and Baluchistan, but only modest levels in Punjab with less than 20 per cent of the total area under rice in the prime Basmati districts.

Irrigation is essential for rice cultivation, and a major gap in its distribution (Fig. 8.10) is the Potwar plateau where the loams are too pervious and irrigation by traditional wells too laborious. The valleys of Swat and Hazara and the vale of Peshawar extend rice growing well to the north. It is everywhere a kharif crop.

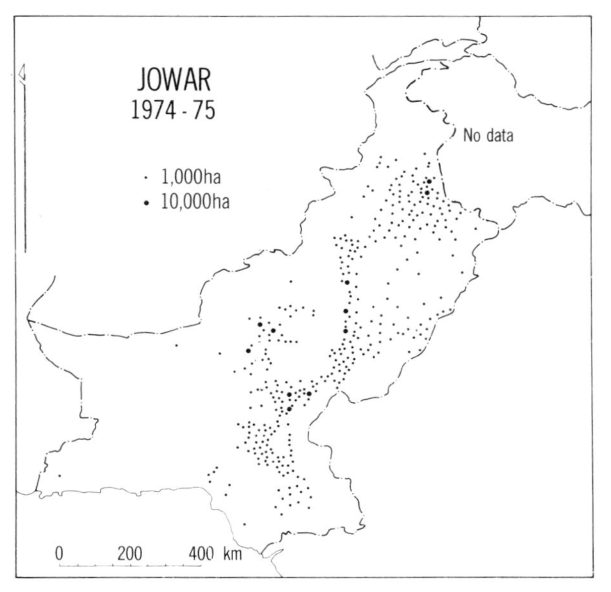

FIG. 8.14

Bajra and Jowar

Bajra (*Pennisetum typhoideum*) and jowar (*Sorghum vulgare*) may be taken together as their needs and uses are similar. They are grown as kharif crops for fodder mainly, but are also used for human food. Perhaps 80 per cent of the millets area is intended to be cut green for fodder. When grown for the grain, as in the barani tracts, the dried stalks are used for a fodder of low nutrient value. The crop grown specifically for fodder is included under that head. The chopped stalks make good fodder, green or dried. Bajra is the hardier of the two and has the potential to grow in marginal non-irrigated situations, though this does little to help explain

the minor differences in distribution of cultivation apparent in Figs. 8.13 and 8.14. What may be a factor for popularising bajra among knowledgeable farmers is that a HYV bajra has been available since about 1969. This might account for the stronger pattern of growth in the submontane Punjab.

They are a valuable crop for the poorer lands within the irrigated tracts, and as barani crops, or partially rainfed.

41 Transplanting paddy into flooded fields, NWFP. Bundles of seedlings are laid out in the field.

42 Ploughing a tiny field for a kharif crop of maize in the Kagan valley. The steep slopes are laboriously terraced into narrow strip fields supported by dry stone retaining walls.

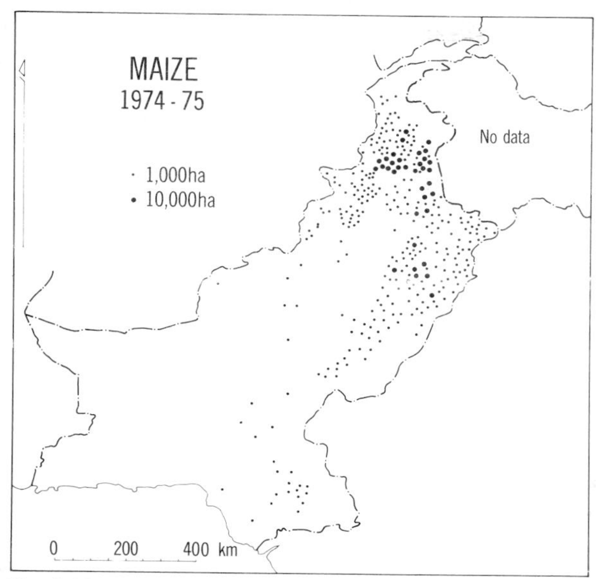

Fig. 8.15

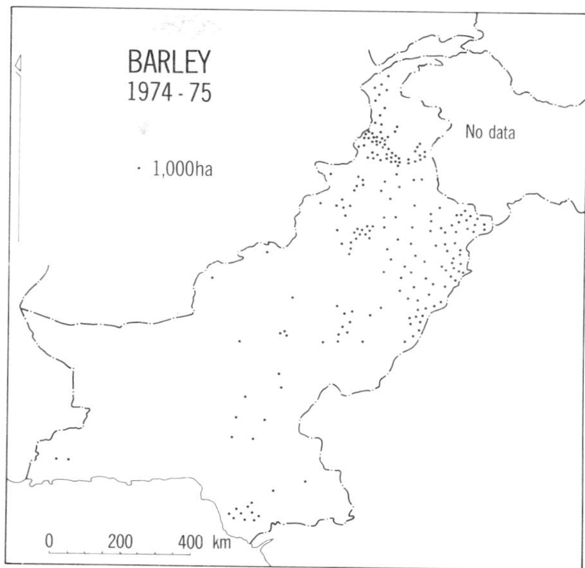

Fig. 8.16

Maize

This is the kharif food grain par excellence in the northern hill country where it stands in serried ranks of terraces marching up the slopes (Figure 8.15). It is generally a barani crop here, but is irrigated on the plains. The grain is eaten roasted on the cob, or in tasty coarse biscuits, as well as being ground into fine cornflour. As an industrial crop it is of increasing significance for extracting maize oil and starch, and its commercial value is making the crop competitive with cotton in the Punjab. It is also cut green as an irrigated fodder crop, yielding four or five cuttings in a year. In fact probably more than a fifth of the area sown with maize is for this purpose and produces no grain, in which case it is included within the fodder area.

Barley

This is the least important of the food grains, but a valuable substitute or concomitant for wheat compared with which it will stand up better to drought and to slightly saline soils. It is sometimes sown mixed with wheat as an insurance against total failure in the event of a drought. It is relatively more important in NWFP, especially in the hillier districts in marginal districts of the Punjab, such as Mianwali in the Thal tract, and Bahawalpur on the edge of the Thar Desert, and in Thatta in the Indus delta (Fig. 8.16).

Pulses

This broad group of leguminous grains plays a most important part in the Pakistani diet, as well as in the agricultural system. In the latter sense, pulses fix nitrogen in the soil and so fertilize the crop that follows. In the diet they provide a large part of the protein intake. Generally the pulses are a neglected crop in that the farmer rarely does more than scatter the seed on the land, maybe only roughly prepared to receive it. Cash returns are low, and consequently inputs are minimal. Prone to disease and to mineral deficiences in the soil, the pulses await research to re-establish their value in the cropping system.

By far the leading pulse is 'gram', chick pea or *Cicer arietinum*, the distribution of which is shown in Fig. 8.17. Two districts are outstanding: Mianwali and Sargodha together account for 53 per cent of the Pakistan total area, Sukkur and Jacobabad for a further 16 per cent. Generally unirrigated and a rabi crop, gram can give some return on sandy lands as in Thal and can root deeply in search of remanent moisture in lands that have carried an irrigated kharif crop but cannot be watered in rabi.

'Masoor' (*Lens esculenta*) or lentils, and 'mattar' (*Pisum larvers, P. sativa*) or field peas are minor rabi crops grown mainly in the Punjab. Sialkot with 19 per cent of the total and Rawalpindi (11 per cent) indicate it is not a crop mainly dependent on

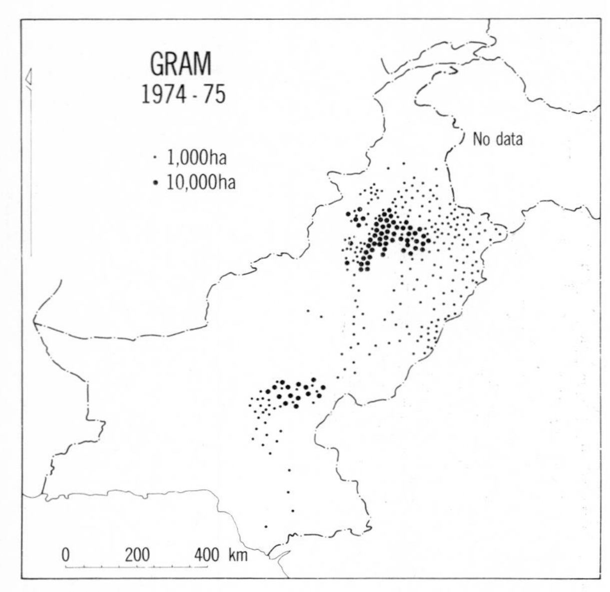

FIG. 8.17

canals, but rather a barani crop or reliant on left-over moisture. Other districts with sizeable shares of the total (over 6 per cent each) are Gujrat also in the Punjab submontane belt, and Muzzafargarh and Dera Ghazi Khan, each having plenty of poor light lands. If masoor is a Punjabi speciality, mattar belongs principally to Sind which accounts for 68 per cent of its area, mainly in the right bank districts. It seems usually to be non-irrigated but grown on land following an irrigated kharif crop, or just using some end of kharif pre-watering to get the crop started.

'Mung' (*Phaseolus mungo*), 'mash' (*Phaseolus radiatus*), both small kidney beans, are kharif pulses. Of mung, 39 per cent is grown in Rawalpindi, and the adjacent Jhelum District where it is rainfed. Sanghar only exceeds 7 per cent in Sind. Here and throughout the Punjab plains, mung is an irrigated crop. Mash similarly is irrigated in the canal areas where Bahawalpur claims 23 per cent of the crop area; Sialkot in the lead with 25 per cent, and Rawalpindi with 13 per cent both have it as a rainfed crop. 'Guar' seed, or 'gaura' (*Cyamopsis psoralioides*) is a field vetch grown as a kharif fodder in Punjab and Sind. It has become more valuable as a source of protein extract (similar to soya beans in this respect) and its derivatives constitute a minor export from a protein-poor country.

Oilseeds comprise an important group of crops providing a source of essential food and a cooking medium. For most of the population the term 'ghi' is used for the everyday cooking oil and no longer carries its original meaning which was clarified butter. Nowadays 'vanaspati', made from hydrogenated vegetable oils, has largely displaced the more expensive milk derivative.

Among the several crops grown for their oil, the rabi group, rape and mustard occupy the greatest area. Indian rape (*Brassica napus*) or 'toria', and mustard (*B. juncea*) or 'raya' together with 'colza' (*B. campestris*) or 'sarson' and 'rocket' (*Eruca sativa*) or 'taramira' ('jambho' in Sind) provide colour to the winter scene in Pakistan, their brilliant yellow and white flowers standing out against the surrounding deep green of the wheat fields, or aligned in stripes intercropped in the wheat at intervals of a metre or two. The distribution of cultivation of rape seed and mustard in Fig. 8.18 shows it to be almost universally grown, though relatively insignificant in lower Sind where salinity is a problem for non-irrigated crops in rabi. These oilseeds can be grown on remanent moisture, or dependent on winter rainfall as is usual in the northern areas or may be irrigated as is the rule in the Punjab plains and Sind. Probably only about a quarter of the crop matures to be harvested for

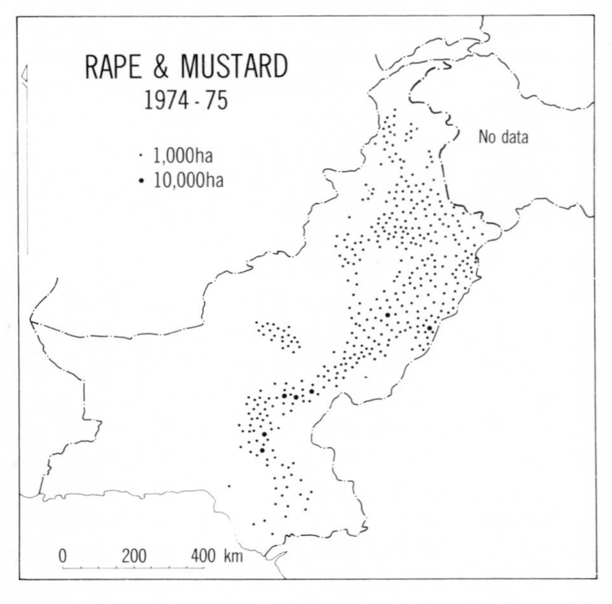

FIG. 8.18

seed. The remainder is cut as a succulent green fodder for the milking cattle. The seeds are pressed to extract oil in the village for direct use in cooking.

In kharif, sesamun (*Sesamun indicum*) or 'til', linseed (*Linum usitatissimum*), groundnuts (*Arachis hypogaea*) and castor (*Ricirus communis*) are the specific oilseeds, though not all grown to provide cooking oil. Linseed and castor have industrial and in the latter case also pharmaceutical uses. Of the sesamum, 12 per cent of the area is in one district Dadu, in Sind, and 13 per cent in Sialkot, but otherwise it is widely though somewhat eratically distributed. Linseed on the other hand has a limited distribution, 68 per cent of its area being in the contiguous districts of Sialkot, Gujrat, Gujranwala and Faisalabad. Groundnut similarly is concentrated as a barani crop on the light loams of the Potwar plateau which accounts for 86 per cent of the area.

Looked at in terms of the oils produced from the various crops, 207,000 tonnes in 1974–75, rape and mustard lead among the crops discussed above with 38 per cent, followed by groundnuts with 7 per cent. They are overshadowed, however, by cotton seed oil which accounts for 55 per cent of the total. The importance of cotton as a source of oilseeds is suggested by its inclusion with identical areas in two separate tables in the official *Agricultural Statistics of Pakistan*, 1975, first as 'cotton' and then as 'cotton-seed'. Its distribution is discussed below.

Cotton

Cotton is by far the most important cash crop because of its role in providing fibre for direct and manufactured exports. Cotton has of recent years experienced problems at home because of disease, pests and floods, and overseas because of the state of the world economy. Table 8.6 shows the area, production and yield of cotton since independence.

TABLE 8.6

Cotton: area, production and yield, 1947–48 to 1976–77

Period (averaged or actual)	*Area ('000 ha)*	*Production ('000 tonnes)*	*Yield (tonnes/ha)*
1947–48/ 1949–50	1,134	201	0.177
1950–51/ 1954–55	1,277	270	0.211
1955–56/ 1959–60	1,394	296	0.212
1960–61/ 1964–65	1,401	357	0.255
1965–66/ 1969–70	1,695	492	0.290
1970–71/ 1974–75	1,917	649	0.339
1970–71	1,735	542	0.312
1971–72	1,959	708	0.361
1972–73	2,017	702	0.348
1973–74	1,846	659	0.357
1974–75	2,032	634	0.312
1975–76	1,852	514	0.278
1976–77	1,842	422	0.229
1977–78	1,817	569	0.313

43 Punjabi women picking cotton.

Cotton is classified as Pak-Upland (long-stapled, comparable to American Upland cotton) and Desi, or local variety, short-stapled. Desi cotton accounts for 7 per cent of the area and 5 per cent of production, and is mainly grown in four districts in Punjab: Lahore, Bahawalnagar and the mutually adjacent Sind districts of Nawabshah and Khairpur, together having 64 per cent of the total. These districts stand in such marked contrast to their immediate neighbours with similar agronomic characteristics in having large areas of Desi cotton that it must be concluded that local industrial specialisation creates a particular demand.

The distribution of cotton cultivation, Fig. 8.19 shows its close tie to canal irrigation. Cotton is a thirsty kharif crop. 65 per cent of the area is concentrated in the main cotton belt of the Punjab extending from Sargodha, Jhang, Faisalabad and Sahiwal south through Multan and the three districts of Bahawalpur Division. An almost separate cotton belt in Sind comprising the left bank districts from Sukkur to Hyderabad and Tharparkar account for 23 per cent, leaving only 12 per cent for the rest of the country.

The currently declining area, production and yield of cotton are of economic concern. Economists see the crop as essential to Pakistan's export trade and to restore balance to the trading account. The problem is complex for the farmer will prefer to grow the crops that bring him the best local income, and is not directly concerned with the balance of payments situation. For him to grow more cotton and less maize or sugar cane, for example, government has somehow to persuade him through incentive subsidies to produce the desired crop.

Sugar Cane

This most valuable cash crop is grown in kharif but harvested in rabi. It requires irrigation most of the year, less than 10 per cent being unirrigated mostly in Sialkot and Sargodha. Thus it is widely grown in canal irrigated areas. Unlike cotton, there is something of a northern bias for sugar cane which is better able to stand the slight ground frosts that may occur in the Vale of Peshawar and Northern Punjab, for example, Fig. 8.20 shows its distribution. The Vale of Peshawar has 11 per cent of the crop and the Punjab 62 per cent (but not in Potwar where irrigation is inadequate).

Although some cane is crushed by the farmer and boiled down to make gur for direct sale to market or to the sugar mills for refining, this contributes less than 9 per cent of the refined sugar produced in the mills which amounted to 498,591 tonnes in 1974–75. Total gur production was 1,280,000 tonnes, two-and-a-half times more than the 452,000 made from crushing cane in the mills.

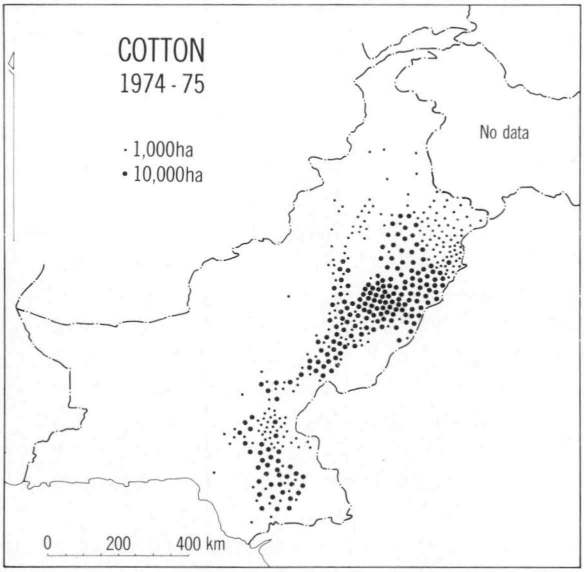

FIG. 8.19

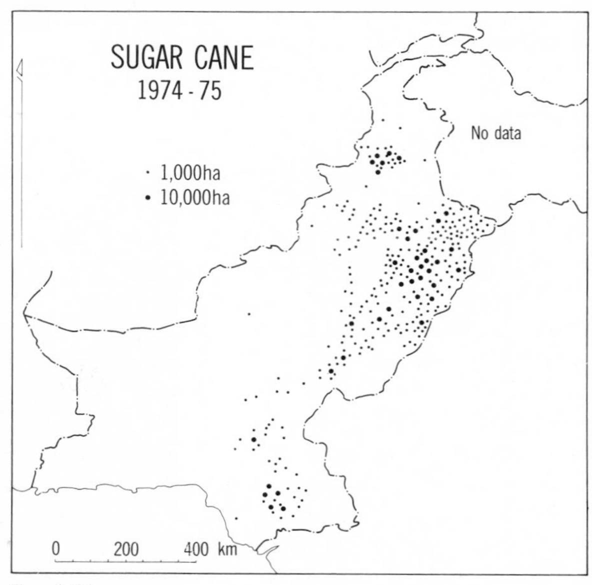

FIG. 8.20

44 While there are several modern sugar mills in Pakistan many farmers crush their own cane using simple rollers powered by bullocks as here. The juice extracted is boiled down to make gur, over a fire fuelled with crushed cane stalks.

45 A fine crop of tobacco in the Punjab. While the man on the bund smokes his hookah, his colleagues apply fertilizer to the plants.

Production is rather variable on account of fluctuating yields in response to weather. In 1976–77 output was 736,000 tonnes. The presence of a sugar mill within economic distance to transport the cane which deteriorates rapidly once cut, is a great stimulus to production. There are thirteen sugar mills in Punjab, eight in Sind, and four in NWFP. Those in the Vale of Peshawar also crush a small quantity of sugar beet. Mill capacity to crush cane ranges from 86,000 to 584,000 tonnes per year. Most handle less than 200,000 tonnes.

The fluctuation in sugar imports suggests that in a good year Pakistan is practically self-sufficient. In 1971–72, 167,000 tonnes had to be imported when mill production fell to 352,000 tonnes from 545,000 the previous year when imports were negligible.

Tobacco

Tobacco is a valuable rabi crop grown with great care under irrigation to serve both local and export markets. Local varieties are grown on a small scale in almost every district to meet the needs of hookah (hubble-bubble or water-pipe) smokers. A communal hookah is an essential item at the family hearth, or wherever men are gathered at leisure or at work. It burns slowly but continuously to give occasional solace in the daily round. Notable absentees from the list of tobacco growing districts are found in Sind, where rabi season salinity is again a countervailing factor in Tharparkar, Thatta and Jacobabad.

A remarkable concentration of tobacco cultivation is found in Mardan and Peshawar Districts which together have 52 per cent of the total

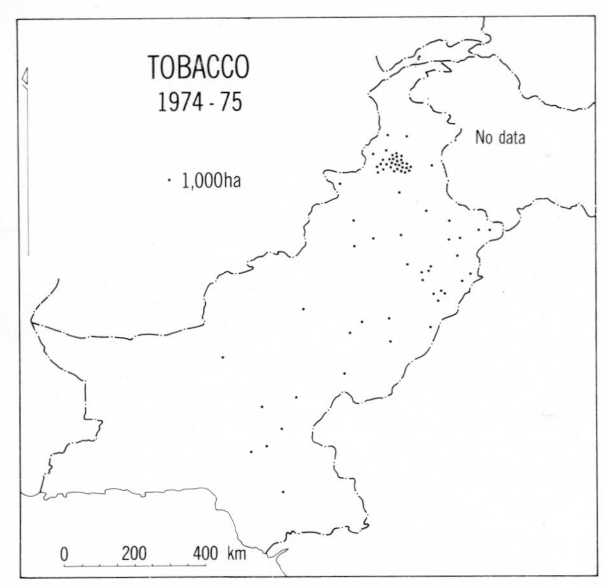

FIG. 8.21

area, and grow most of the high quality Virginia leaf. Another 30 per cent is spread between Sialkot Gujranwala and Multan through Gujranwala, Lahore, Faisalabad and Sahiwal (Fig. 8.21).

Tobacco has grown in importance by three-and-a-half times since independence reflecting an increasing interest in cash crops. Probably this has increased rural affluence, though how widely such income is spread is hard to discover.

Potatoes

Whether a tenfold increase in the area under potatoes since 1947–48 spells prosperity is open to doubt. It reflects more probably an increase in urban demand for a readily prepared starchy food. The small area concerned, 23,000 ha means more than it suggests since the crop yields a heavy weight of useable food per unit area: 10.6 tonnes per ha compared with 1.75 tonnes of HYV rice or 1.63 tonnes of HYV wheat per ha.

There are two branches in potato growing. Seed potatoes are best grown in a cool climate where pests and blights are minimal. Thus the higher country in Hazara, Peshawar, Swat and Dir Districts and Quetta and Kalat Divisions account for 35 per cent of the area, at least partly for this reason. Maincrop potatoes are irrigated in the north and central Punjab plains, in the familiar triangle subtended by Gujrat and Sialkot with its apex in Multan, where 57 per cent of the area under potatoes is found. In the rest of the Punjab and in Sind in particular, the crop is little grown.

Vegetables

Quite apart from population growth, increasing urbanization has helped develop the demand for vegetables. Production of vegetables has almost trebled since the first three years of independence (1947–48 to 1949–50), although the area used is only 72 per cent greater. Rabi vegetables have become popular, and many varieties of excellent quality are obtainable in the urban markets: cauliflower, carrots, turnips, cabbage, tomato, radish, peas, for example. Kharif vegetables are usually of the 'watery' kind: cucumber, brinjal, lady's fingers, pumpkin, etc., though various beans are also grown. Some concern is now being expressed that the traditional market garden areas close to the cities are being over-run by urban expansion, which furthermore tends to increase the demand for fodder for dairy cattle and urban work animals and so the area for such crops may be expanding at the expense of vegetables.

Chillies

Among annual crops, the rabi spice chillies, without which a Pakistani meal would appear (to the Pakistani at least) insipid, must be mentioned. The crop is widely grown, as befits a spice so universally in demand, but there are important concentrations. Tharparkar in the far southeast, otherwise little renowned for agricultural specialisation or prowess, tops the list with a third of the total crop area, to give the three southern most districts 44 per cent. The experts in polyculture, the well provided farmers of the Faisalabad, Sahiwal, Lahore and Multan block account for a further 28 per cent.

Fodder Crops

As can be seen from Table 8.4, fodder crops as such bulk large in both rabi and kharif, amounting to 12 and 19 per cent of the total cropped area respectively in 1972. These crops are discussed in Chapter 9 in context with the livestock they are grown to support.

Fruits

By far the largest share in fruit is taken up by those grown in the subtropical plains with the help of irrigation. In relation to their total area, however, the higher altitude districts with the ability to grow temperate fruits such as apples, pears, plums, peaches and apricots make a sizeable contribution. Quetta, Loralai, Zhob and Kalat in Baluchistan, Peshawar and the hills of Rawalpindi in the north come into the latter category. Quetta's grapes and apples in particular reach every city in Pakistan. Citrus fruits and guava are grown mainly in the Punjab, where Sargodha, which has the largest area of orchards of all kinds, is an important producer. Sind with its milder winters can grow bananas. Mangoes are widely grown on the plains, but dates are restricted to lower Sind and oases in Baluchistan, contributing to the 'Middle Eastern' aspect of Makran.

Fruit growing has shown remarkable growth. Taken together the area has tripled in the 15 years or so between the period 1957–58 to 1959–60 and 1974–75. The outstanding increases among major fruits are of bananas, occupying ten times their former area, and dates increasing almost seven fold. Kureshy attributes some of the increase to the exclusion of orchards from the land reform legislation of 1959.*

Crop Associations

It is well to remind oneself when considering any geographical pattern of crops and their associations viewed at a district level, let alone at a national level, that it is a generalization from a myriad of individual cases. Ultimately the decision about what crops to grow, whether one or several, rests with the farmer. He has to balance in his mind such diverse factors as the short term needs of food for his family, of cash to buy essentials or to spend on a coming marriage, his perception of the climatic environment matched with his expectations from the Irrigation Department with regard to canal or tube well water supplies, his knowledge of the alternatives open to him, of HYV and of their needs, for example, and his reading of the current straws in the wind indicative of movements in price and demand in distant markets.

The seasonal change in cropping pattern on a 5 ha farm in the rice tract of Gujranwala District in the Punjab is shown in Fig. 8.22 and the month by month requirement of water for each category of crop grown and *in toto* in Fig. 8.23. These are based mainly on a study by W. P. Falcon and C. H. Gotsch†, who are not however responsible for the interpretations placed here on those facts.

The basic fact for the Gujranwala farmer is that he has more land than water to irrigate it with. The canal has determined that in general he can expect water sufficient for him to cultivate half his land in rabi and half in kharif, and there is uncertainty even that he will receive the optimum of which that system is capable. Discounting the 0.4 ha he has permanently as orchard and so cannot be brought into his annual cropping strategy, he achieves a cropping intensity of 124 per cent. In kharif he grows mainly cash crops, cotton, sugar cane and rice. The planting of coarse and fine rice is staggered to spread the labour demand for transplanting and later harvesting. 0.4 ha is devoted to fodder crops in both seasons, for his work animals must be fed as well as his family, and there may be a ready market for any surplus in a nearby town. The large area given to vegetables further suggests a strong market orientation, as this is far in excess of subsistence requirements. The intensity of work in March and April and again in September and October particularly, when the harvest of one season overlaps sowing and planting for the next can well be imagined. The farmer has little option in the matter, however, as the flow of water in the canal is outside his control, and if he does not take it in his turn he cannot rely on it being available later. Only if he has managed to invest in a tube well can he act independently of the plans of administrators and executive engineers, and even then his financial advantage will generally lie in conforming basically to the patterns laid down by canal supply, using his tube well (which costs money to run) to supplement the supply and to tide him over critical periods of shortage.

* Kureshy K. U. *A Geography of Pakistan*, Oxford University Press, Karachi 1977, p. 111.

† Falcon W. P. and Gotsch, C. H., 'Relative Price Response, Economic, Efficiency and Technological Change: a Study of Punjab Agriculture', in Falcon, W. P. and Papanek, G. F. *Development Policy II – the Pakistan Experience*, Harvard University Press, pp. 160–185, 1971.

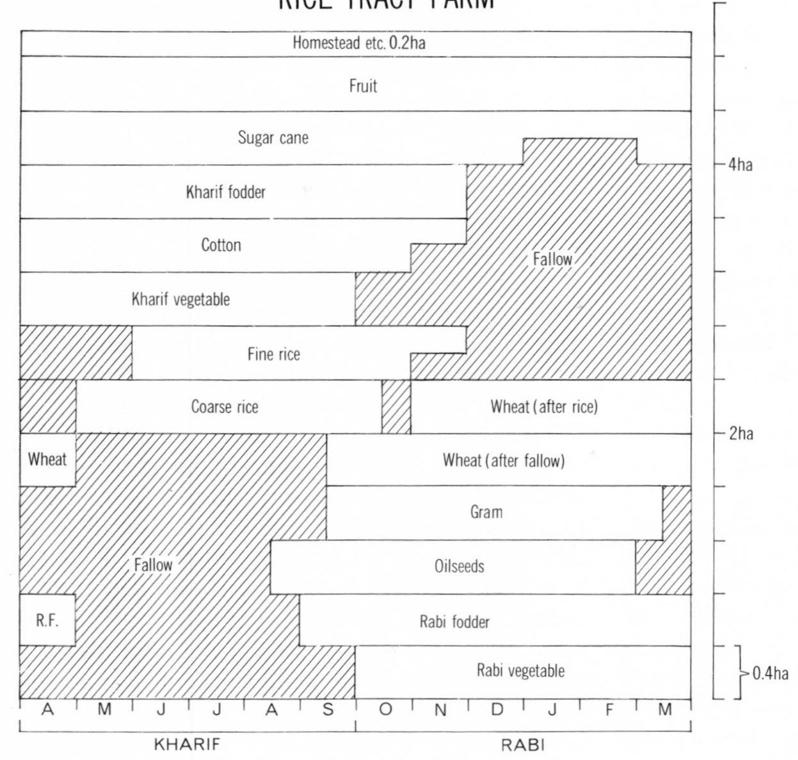

FIG. 8.22

As was seen in the gross seasonal budgeting of water supplies (Tables 8.1 and 8.2 above), kharif water is much more abundant than the supply in rabi, and necessarily so hectare for hectare, because of the higher evaporation rates during the very hot summer time. The importance of early summer watering in April-May-June is clear from the cropping calendar. While crops may not exert their maximum demand at the time of sowing and seed germination, undue shortage in April is very critical and this can be a period of annual anxiety. Once snow melt is well advanced in the Himalaya (provided snow falls have not been sparse during the winter) the rise in river levels generally guarantees the May-June supplies, after which the monsoon floods arrive. Local rainfall is not neligible in summer, and must be seen as a bonus of 100–130 mm in July and August and 75 mm in September on average, thus helping to compensate for the apparent fall off in irrigation allocations.

From the imaginary 'farm gate', in fact from the distributary channel at his boundary, the farmer has full choice as to how he allocates water between his crops. Rice puts a heavy demand through most of its growing season. Cotton on the other hand builds up its demand. Sugar cane's needs are heavy at the height of the hot weather, but fall off as the crop matures in the autumn, or early rabi, to a minimum when one crop is being harvested and the next is being planted out, in December-January. Wheat's needs are modest but may span eight months of the year, building up to a peak in February–March when the grain is swelling and maturing. Fodder crops probably take their chance and would not often take precedence over the cash crops.

The large amount of land left fallow at any one time is not entirely wasted. It can be grazed by small livestock in the care of the children, thereby adding some natural manure; the soil can 'rest' and of course can be prepared at a more leisurely pace for the seed time to come.

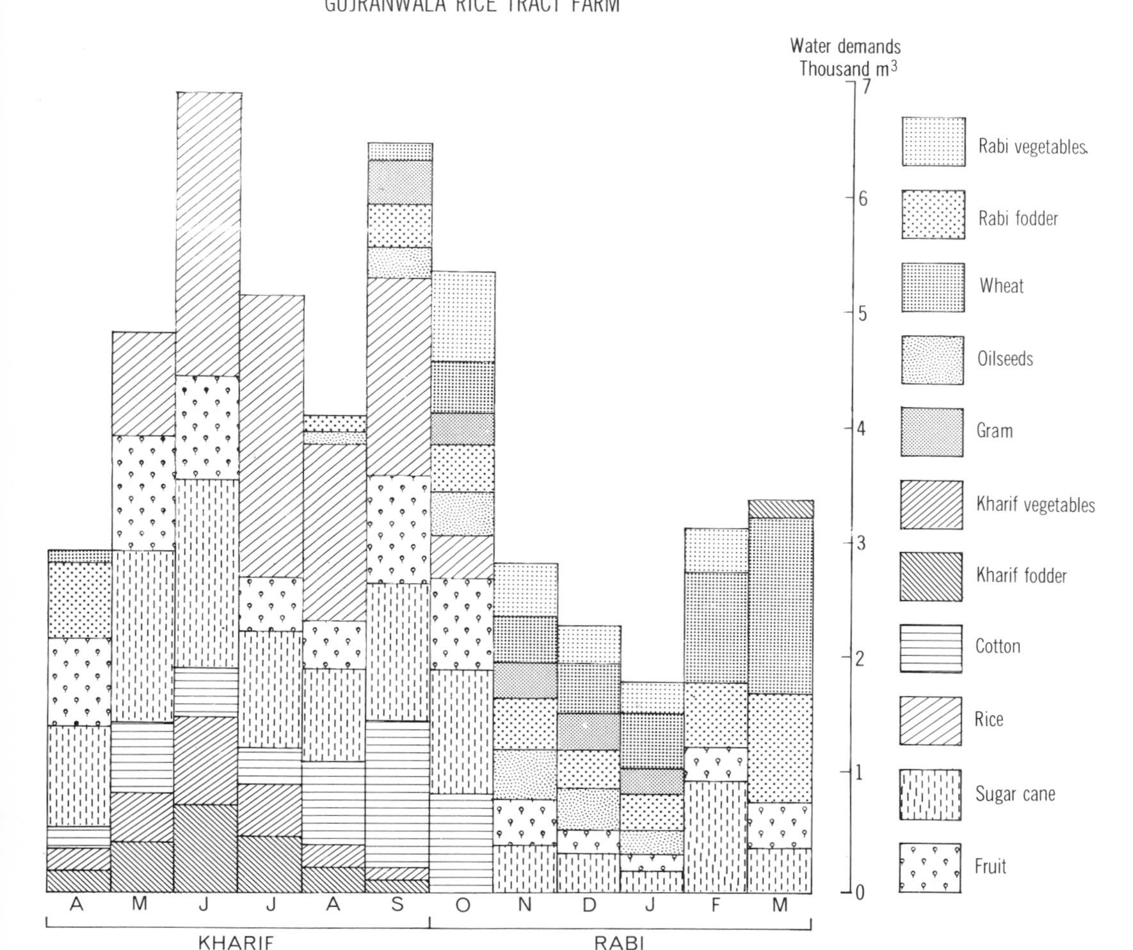

FIG. 8.23

Seasonal Cropping Patterns

From what has just been said regarding the cropping details of a single farm, it will be appreciated that Fig. 8.24 showing the cropped area by districts and major kharif and rabi crops, is highly generalised. It serves however to give an overall impression of the relative distribution of the cultivated area of Pakistan, and the basic pattern picked out by the ratio of kharif to rabi cropping based on gross areas. One important variable missing from this map is cropping intensity, indicative of the use of the same land more than once in the year. This was discussed above (see Fig. 8.5).

A simple view of the seasonal ratio of cropping is given in Fig. 8.25. Overall rabi cropping exceeds kharif in 33 of the 46 districts mapped but in many the difference is slight. Over much of southern Punjab and the mainly left-bank districts of Sind, rabi and kharif cropping are more or less in balance. Rabi cropping shows most marked dominance in a trans-Indus belt extending from Chitral through Kohat and Bannu to Dera Ismail Khan, with extension east of the river into Mianwali, Attock and Muzaffargarh, and further east through Sargodha to the Punjab piedmont in Gujrat and Sialkot. Winter rainfall is a major

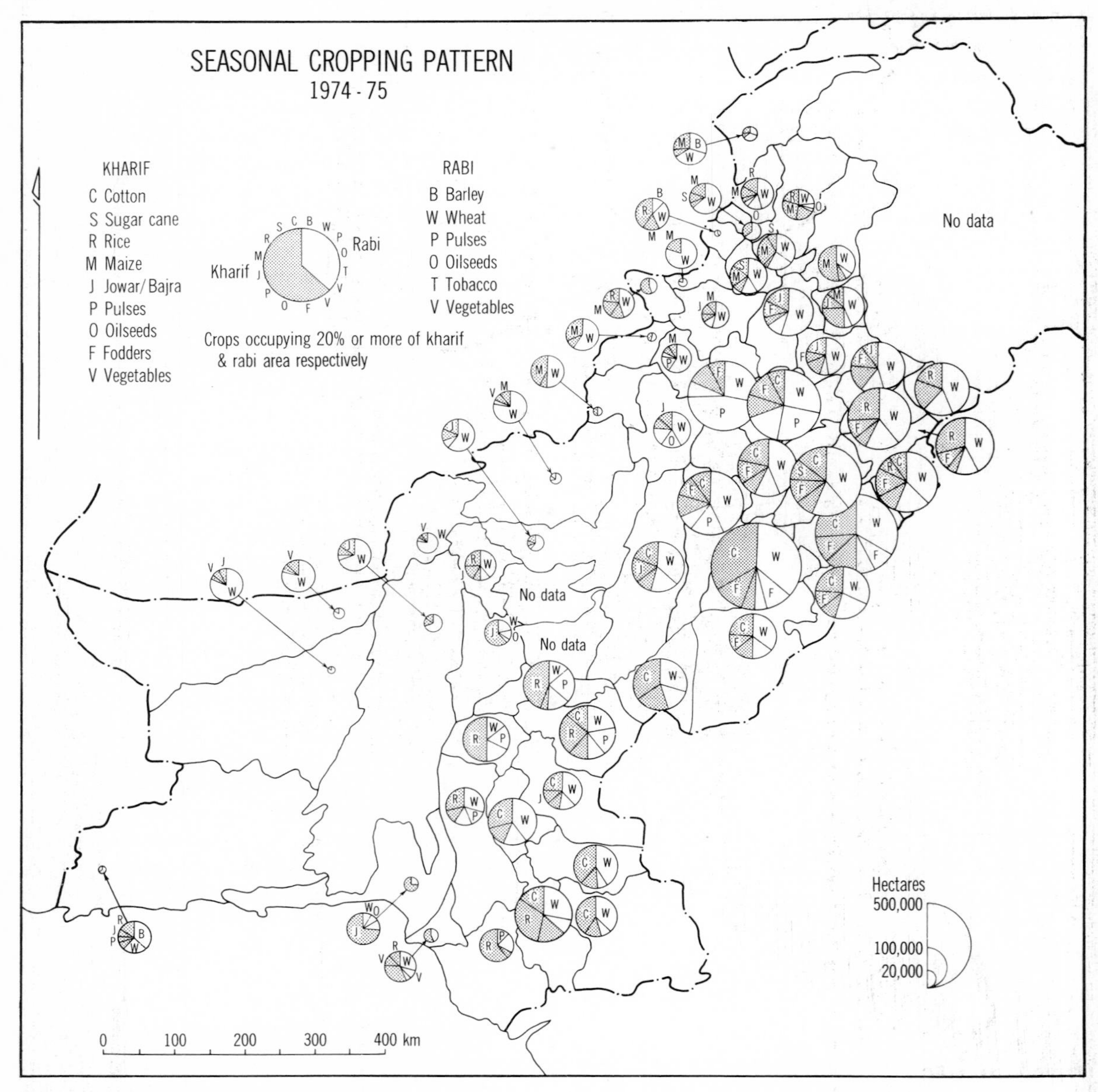

FIG. 8.24

factor in the northwest of this belt, and the greater efficiency of well irrigation for rabi cropping in the Punjab piedmont, but in Mianwali and Muzaffargarh in Thal it is at least in part the unsuitability of the sandy soils for cropping in the high evaporation season that makes rabi cropping preferable. Most of the scattered patches of cultivation in Baluchistan show very strong rabi dominance. Lasbela and Kachhi are exceptions in showing quite the opposite tendency, the first due to its receiving very occasional kharif downpours. Both tend to rely on exotic rainfall in the hill catchments to feed torrents whose waters are spread and used as soon as possible because of the very high evaporation rates. Northwards along the Sulaiman piedmont the practice is rather to carry the moisture forward for a rabi crop if at all possible. Kharif cropping is again dominant in Hyderabad and Thatta, not so much because of the slight chance of monsoon rainfall, but rather because only in kharif is canal water really plentiful and salinity can be temporarily leached out of the top soil.

When reading Fig. 8.24 it needs to be noted that in plateau Baluchistan and some of the Tribal Areas on the NW Frontier, crop totals are so small as to make some magnification necessary in mapping their kharif: rabi ratios, etc. This map summarises the analytical commentary of the proceding pages of this chapter and so is largely self-explanatory. Only crops which occupy 20 per cent or more of the kharif or rabi area respectively are identified, but this serves to demonstrate a few important points. In rabi, wheat dominates in all but Thatta where it is overshadowed by pulses. These appear in second place in some districts; in others oilseeds, barley or fodder.

The kharif share of the total cropped area is not clearly dominated by a single crop. Rice is supreme in Gujranwala, Sialkot and Sheikhupura in northern Punjab, and in the Indus right-bank Sind districts. Through much of the rest of the Punjab and left-bank Sind, 'cotton is king', while in the hilly areas and Potwar, jowar (especially in lower Potwar and Baluchistan) or maize (especially in NWFP) dominates.

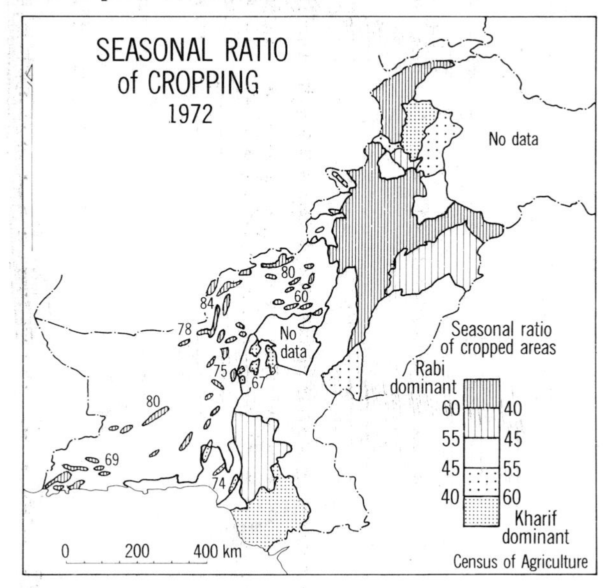

FIG. 8.25

CHARACTERISTIC TRANSECTS

The seven transects have been drawn to a standard key, simplified from the maps of landforms and land use produced under Colombo Plan by the Canadian Government (Fig. 8.26). The location of the transects is shown in Fig. 8.27. The landform units are those used by the Canadian geomorphologists and have been discussed in Chapter 4 above.

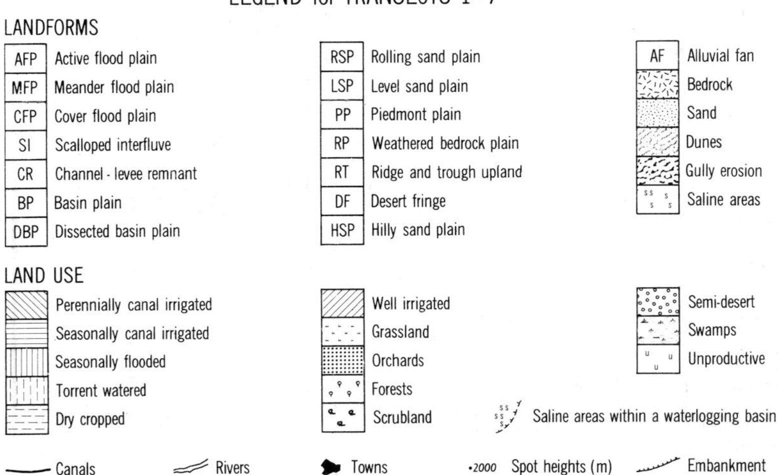

FIG. 8.26

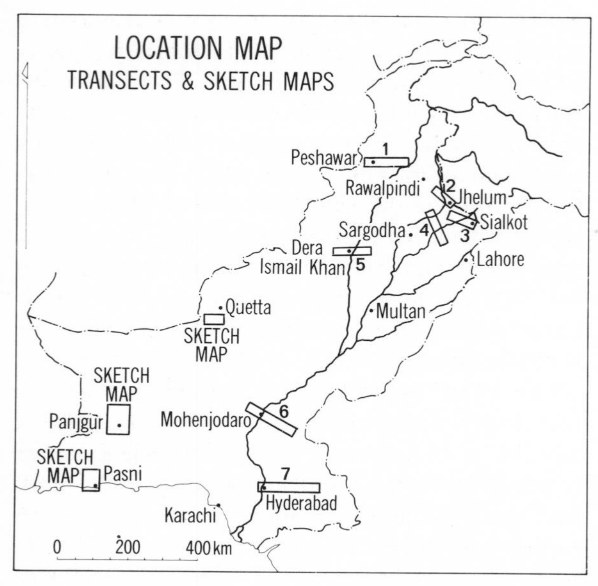

FIG. 8.27

Land-use designations distinguish types of irrigation, perennial and seasonal canal irrigated areas, well-irrigation, torrent-watered areas, dry cropped (barani) areas and orchards as well as unproductive waste, grassland, scrub forest and desert.

Transect 1 (Fig. 8.28): Vale of Peshawar

The transect traverses the southern part of the Vale of Peshawar from the Khyber Hills to the Indus near Tarbela. Broad stretches of alluvial fan built up of coarse gravels carry detritus from the mountain rim of the Vale in a multiplicity of braided torrent channels, which organize into more regular courses in the piedmont plain around Peshawar. In the northwest the meander flood plain of the Kabul bears abundant evidence of past changes of course by this heavily laden river. The braided channels of its active flood plain divide the meander flood plain into long strips. At Nowshera the flow has become concentrated into a single main channel within a narrowing

FIG. 8.28

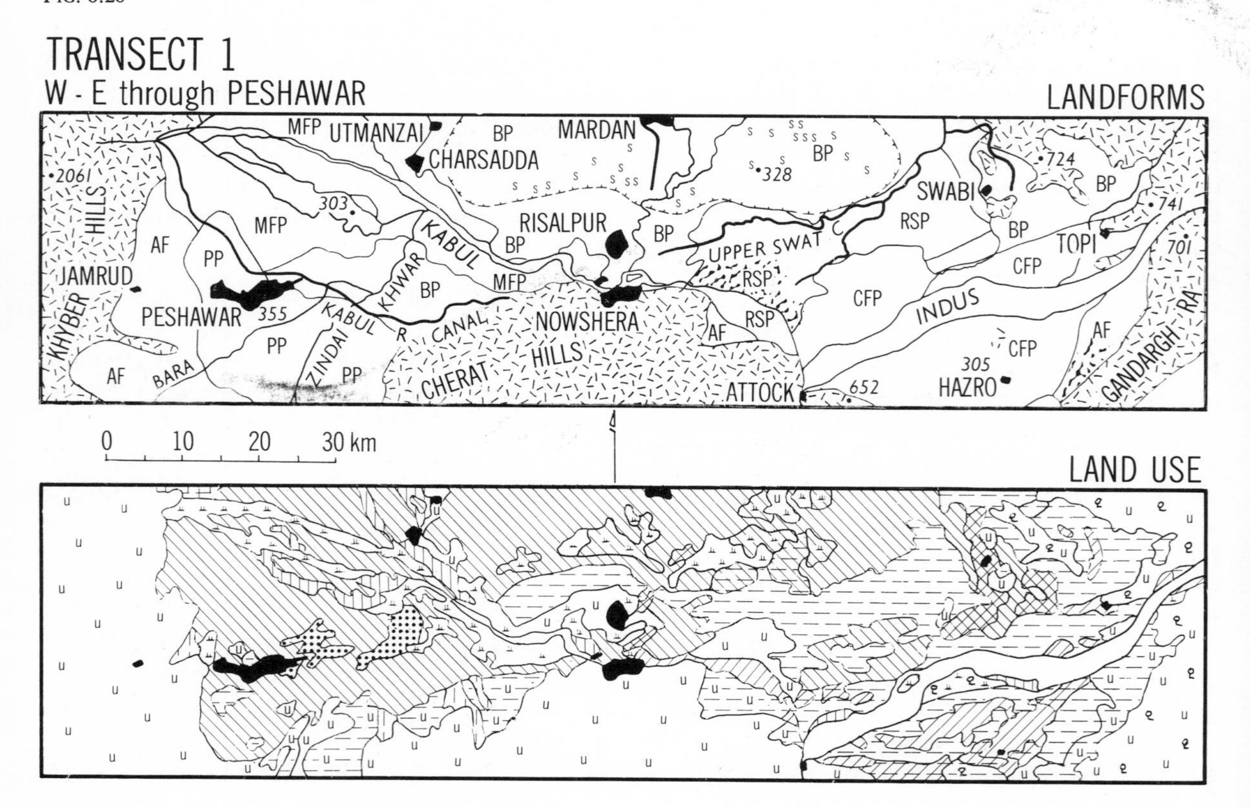

meander flood plain which impinges on the foot of the rocky hills of the Cherat Hills to the south. Northwards stretches the level featureless surface of the basin plain that is the major morphological element in the Vale. Broken ridges of the sandstone Sar-i-Maira break the smooth surface of the basin plain in the east. Gully erosion and the sculpturing by wind of sand and silt in part blown in from the Indus flood plain help give the area its rolling appearance. It gives way to the east and south to the more regular surface of the cover flood plains which flank the Indus on either bank for a distance of up to 12 km. The very fine grained sands of this plain suggest a possible lacustrine origin. The Indus here has just emerged from its long gorge through the western Himalaya, and reverts again to the constraints of solid rock banks on reaching Attock. In its open section the active flood plain of the Indus is 5 km wide, a waste of blowing dust and sand at low water, and a vast river in flood. The rocky Gandgarh Range with its skirting alluvial fans limits the plains to the east.

Land-use in the Vale of Peshawar demonstrates a great range of cropping intensity and a wide variety of crops grown. The Khyber and Cherat Hills and Gandgarh Ranges that frame the area of the transect to the west, south and east are unproductive. Over-grazed by goats and overcut by villagers in search of natural hay and firewood, the potential of the natural enviornment is never achieved. Large herds of goats and sheep forage on the gravelly alluvial fans and menace the marginal fields on the Bara piedmont plain south of Peshawar. Here cultivation is possible and very rewarding if an assured supply of water can be found. The River Kabul Canal reaches parts of the piedmont in the north, and new works are extending the supply. With irrigation, wheat and maize predominate in rabi and kharif respectively. Wheat and barley get a start from canal water in autumn and yields are good in the event of winter rains. Maize is sown in July when there is plenty of water; cotton, less thirsty, but liable to early frost damage in autumn particularly in the centre of the basin, is sown earlier in summer. Here and there torrent water is controlled to moisten fields for a rabi grain crop.

Running east from Peshawar city the small strip of basin plain known as the Tarnab plain is famous for its orchards, principally of pears. Irrigation from Warsak Dam via the River Kabul Canal supports perennial cropping, fodder for the city being important, the first flush of wheat and maize generally being cut green for sale in Peshawar. Kharif and rabi crops are pretty much in balance here, but northwards in the plains of the Kabul and Swat the Kharif crop dominates. Small private inundation canals in summer and channel bed diversions in winter make water abundantly available in many villages, while the government River Kabul and Lower Swat Canals command parts of the area. Maize, sugar cane and rice are the main kharif crop; wheat, barley and fodder the rabi crops. In the active flood plains rabi grains are grown, and a little millet in kharif where protection from flooding is feasible, but much is left to coarse grass.

The Mardan plain north of the River Kabul is the agricultural heart of the Vale of Peshwar. Perennially irrigated, cultivation intensities reach over 150 per cent. Sugar cane is widely grown for the huge mill at Mardan. Maize and wheat are the principal grains. Salinity and waterlogging have come in with perennial irrigation, affecting particularly the belt parallel to the rolling sand plain where the sandstone ridge inhibits groundwater drainage. Another menace is gully erosion which is eating headwards from the deeply incised Kalpani between Mardan and Nowshera. Only rainfed (barani) cropping is possible among the sandstone ridges of the Sar-i-Maira. Summer rainfall is generally conserved for growing a rabi wheat crop, though a little millet may be sown in kharif. The small area of basin plain round Swabi is renowned for its high grade Virginia tobacco. Well irrigation supplements canal supplies allowing cropping intensities to reach 190 per cent. Tobacco with wheat and fodder are the rabi crops. Maize is the main kharif crop, grown as much to feed the many bullocks who turn the Persian wheels at the well heads as for human consumption.

The Indus plain shows a variety of uses. Canal irrigation is at present restricted to parts of the right-bank cover flood plain, but wells are numerous locally where groundwater is near the surface. A left-bank canal from the Indus below Tarbela is projected which will command 24,300 ha in Attock District. The wells are used for intensive cultivation of the staple grains, wheat and maize, but also for crops that reward the entrepreneur for his tender care and investment in manuring: tobacco, fodders, vegetables, melons, pepper and sugar cane. Much land lying beyond the reach of irrigation carries barani crops of wheat and a little maize.

Transect 2 (Fig. 8.29): Potwar Plateau

To many travellers along the Grand Trunk Road between Rawalpindi and Lahore, the landscapes of the section across the Potwar plateau west of the river Jhelum are as bizarre as those on the moon. The steeply dipping strata of the rocks of the Salt Range, bevelled to a pediment in late Tertiary time, were unevenly 'plastered over' with a mantle of loess during the Pleistocene glacial period. Subsequent erosion, accelerated by man and his goats, has laid bare the bedrock in many areas, etching out into closely alternating 'ridge and trough' landscapes, the underlying structural alignments. The loess mantle has been fretted into a fantastic badland topography of vertically-walled gullies.

The narrow transect, 16 km across by 110 km long, cannot adequately show the marked parallelism in the surface morphology resulting from structural control. Once presumably standing as relict ridges above the late Tertiary pediment along the base of the Himalaya, several hilly bedrock areas occupy the higher ground. The old rock pediment remains a widespread feature—the weathered bedrock plain (R.P.) on the transect – but erosion has attacked it in some areas to produce the ridge and trough uplands.

Where bedrock has been covered with loess and/or alluvial deposits to a significant degree, basin plains have been formed. One north of Jhelum is 20 km across. In other parts of the mountain rim around the Indus plains, basin plains become a feature of regional extent, for example the Vale of Peshawar. In this transect where basin plains have developed within the Potwar plateau, they have been deeply dissected by gullying (areas D.B.P. on the map). Transitional between the ridges of bedrock and the basin plains are the piedmont plains, gently sloping rock pediments in effect, covered with weathered material progressing slowly down slope under gravity and periodic sheet wash. This morphological element is seen flanking the basin plain and active flood plain of the Jhelum river.

The most striking feature of agricultural land-use in this transect is its dependence on rainfall. In a narrow strip along the Soan river in the northwest, well-irrigation and seasonal flooding support more intensive cultivation. On the Jhelum's active flood plain the risks of severe flood have no doubt been reduced with the completion of Mangla dam just off the line of transect northeast of Dina. The dominance of rock structure and gullying presents itself in the parallel alignment of areas of unproductive land. The more level sections of the basin plains are widely cultivated for rabi wheat and

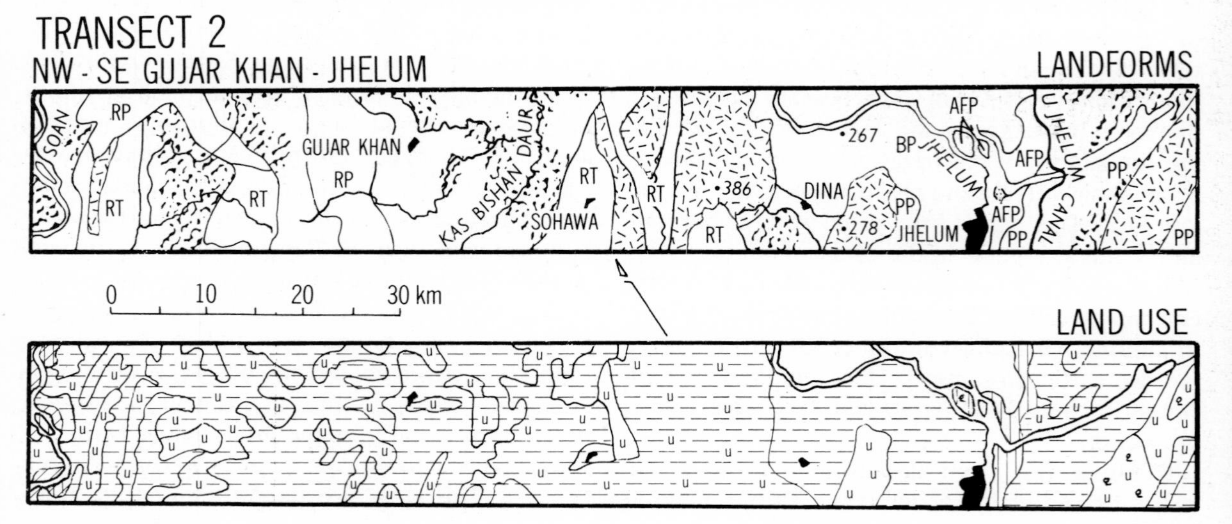

FIG. 8.29

mustard, and kharif maize and jowar. Measures to protect the fields of fertile loess against further dissection by gullying are being attempted but the continuing ravages of uncontrolled grazing on the fragile semi-arid ecosystem combined with the difficulties of persuading land holders to forego individual rights for the common good in order to re-contour the fields, make the task a formidable one.

Transect 3 (Fig. 8.30): Lala Musa to Sialkot

The sketch maps of the terrain between Lala Musa in the Chaj Doab and Sialkot in the Rechna Doab illustrate the conditions in the Punjab plains close to the Himalayan foothills which lie immediately north of Punjab's border with the disputed territory of Jammu and Kashmir. North of the Chenab the piedmont plain slopes increasingly gently to the southeast, and its cover of alluvial sediments becomes finer textured. Torrent streams have dissected the surface into gentle undulations. South of the active flood plain of the Chenab the plain around Sialkot is mostly covered flood plain, diversified only by shallow depressions in which water-logging and sometimes salinity has occurred due to canal seepage.

There is a basic similarity in the land-use either side of the Chenab. Rainfall is moderately plentiful (up to 1000 mm) and reliable by Pakistan standards, supporting barani crops in both rabi and kharif though most comes in the summer monsoon. Aquifers present near the surface are widely accessible in wells worked by Persian Wheels. Often it is the texture of the surface soil that determines the cropping pattern through its relative retentiveness of moisture.

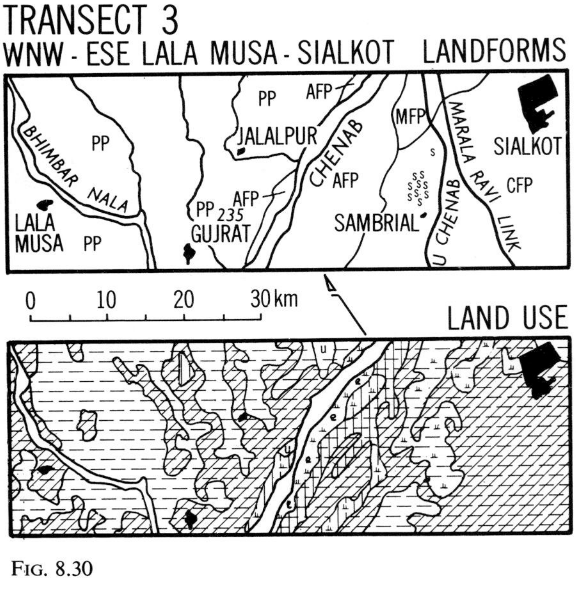

The more extensive dry cropping north of Gujrat has rabi wheat and gram (the latter on the coarser soils), pulses and oilseeds, kharif millets and pulses, in the general ratio of rabi 60 to kharif 40. The 'dofasli-dosala' rotation is common, in which rabi and kharif crops taken in succession are followed by a full year's fallow to rest the soil and conserve moisture. Where wells are available to supplement rainfall, their water is used mainly in the cool rabi season when evaporation is less and the labour of the bullocks is put to better use; wheat is then the favoured crop. A variety of kharif crops follow depending on how much water is raised: maize, sugar cane vegetables, rice and fodder being fairly thirsty crops, jowar less demanding. Settlement in the Gujrat-Lala Musa plains is particularly close, despite the absence of canals. This sub-montane zone was for centuries the major population base of the Indus plains region as a whole.

Well irrigation increases towards and southeast of the Chenab. Along the active flood plain grass and scrub occupy the coarse sand ridges close to the river. The finer sediment on either side may carry crops that benefit from seasonal inundation of the land, as well as from rainfall and occasional well irrigation. Rabi wheat is the farmer's first preference.

In the Sialkot plain proper, well irrigation is very important, though intensities of cropping seldom exceed 100 per cent. This can increase if pump and tube well irrigation proves economically more profitable than the traditional Persian Wheel system. Under the latter there is more land cultivated than can be commanded in any one season. Rabi crops exceed kharif in area. Without irrigation barani crops can be taken in both seasons. As rabi crops, wheat and gram and as rainfed kharif crops jowar, pulses and in low-lands, rice, can be grown. With irrigation, water-demanding crops are added to these staples: vegetables, fruits and fodder in rabi; maize, sugar-cane and cotton in kharif. The city of Sialkot stimulates intensive fruit and vegetable cultivation in its immediate

Fig. 8.30

vicinity, the limit of its influence still being set however by that of the pony-drawn rehra.

Transect 4 (Fig. 8.31): Salt Range–Rech Doab

The transect southeast from the Salt Range through Malakwal traverses the Chaj Doab between the Jhelum and Chenab and the broader Rech Doab. Most of the morphological elements of the plains are illustrated here, as are their different potentialities as realised by man and as rendered useless, at least temporarily, through lack of foresight in the design of irrigation schemes. (see also Fig. 4.5.)

The top of the Salt Range scarp stands at about 700 m above sea level almost 500 m above the plain below. A strip of particularly coarse gravel fans, gullied by torrents, lies along the foot of the scarp merging into a piedmont plain of gentler slope but relatively coarse surface material. Between Jhelum and Chenab the classic sequence of flood plain units is found. A narrow active flood plain along the present course of the Jhelum is followed southwards by a strip of meander flood plain still bearing the traces of former meander scrolls and depressions, the latter picked out in patches of swamp. The oldest terrace, dubbed by geomorphologists the 'scalloped interfluve' because of the characteristic line of its margins, presents a degraded bluff to the lower meander flood plain. To the south this bluff is strengthened by the channel levée remnant at its foot, a pronounced depression with sandy rises along old levees and marshy in former backswamps. A strip of meander flood plain and the active flood plain of the Chenab complete the symmetry of the Chaj Doab. Southeast again the Rechna Doab presents a very wide meander flood plain with occasional channel remnant features, and the transect ends in the older scalloped interfluve surface west of Sheikhupura. Had the transect been taken further west, much more of the scalloped interfluve would have appeared but in the Rech Doab this feature presents such an abrupt boundary to the meander flood plain lying immediately to its north as to suggest that tectonic movement may be responsible.

As to land-use, the south-facing, sun-scorched scarp of the Salt Range is practically barren rock dotted with hardy xerophytic scrub with an occasional patch of dry cultivation of rabi wheat and mustard. The alluvial fans at its base are similarly neglected by man because of the extreme ruggedness of the stony surface. The piedmont plain is better favoured as the aquifers in its alluvial mantle are readily accessible in shallow wells, supporting rabi wheat and tobacco, kharif millets, cotton, sugar cane and vegetables.

The sandy banks in the active flood plain of the Jhelum are flooded every summer but carry a rabi wheat crop irrigated by diverting the braided channels at low water. Immediately south of the river its meander flood plain is commanded by the lower Jhelum canal nominally perennial but supplemented in the rabi season by well irrigation of the more valuable crops following their establishment with a late kharif or early rabi watering

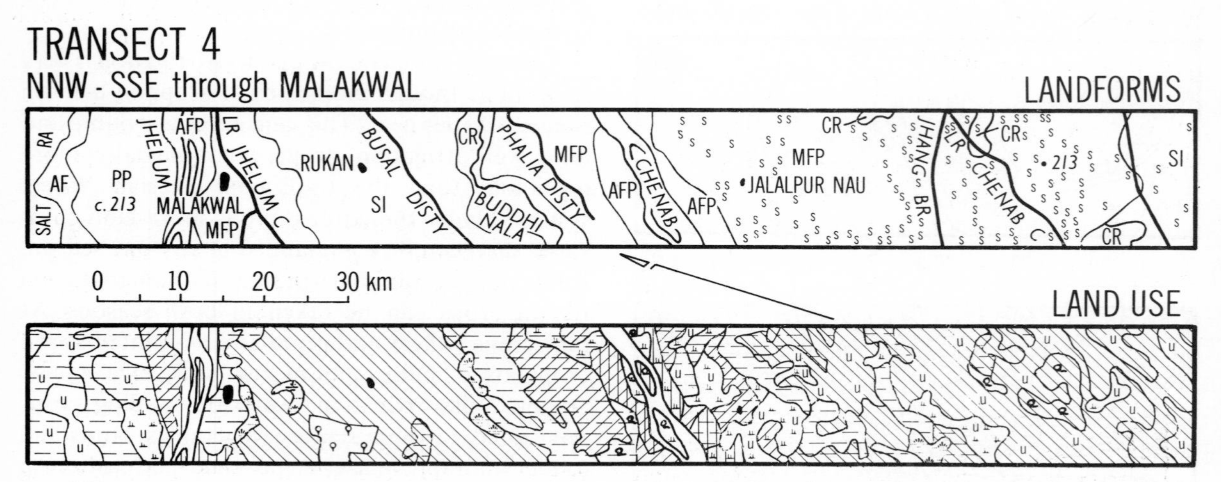

FIG. 8.31

from the canal. A wide range of crops is cultivated: wheat, oilseeds, gram and fodder berseem (the latter needing regular watering) in rabi, and in kharif cotton, jowar and maize. Low sandy ridges near Malakwal are above the level reached by irrigation and so carry barani crops of wheat and gram in the cool season when the little rainfall gives maximum benefit.

The scalloped interfluve, the Kirana Bar, is similar in its canal commanded crop association. Rabi crops slightly exceed kharif, and well irrigation is relatively unimportant. There is a little barani cropping on sand ridge remnants. Of particular interest is the extensive irrigated reserved forest.

Along the right bank of the Chenab, the meander flood plain and channel levée remnant provide a more complex environment. The highest levels, generally sandy textured, grow barani wheat and gram or are left in scrub. Depressions are severely waterlogged, supporting at best rice as the crop most tolerant of the saline conditions, some wheat and much 'sarkanda' scrub. Between these extremes the lands at intermediate levels receive water from inundation canals in kharif and early rabi, to grow millets, cotton and rice while water is plentiful, and wheat in rabi. Wells are used for more intensive cultivation of sugar cane, vegetables, fruits, and fodder.

The active flood plain of the Chenab is more extensive than that of the Jhelum and its floods are too strong to allow permanent structures like wells near its changing bed. Rabi wheat, gram and other pulses are grown on the annually changing shoals, but much is left under scrub and clumps of tough grass.

Almost a third of the transect consists of the Hafizabad plain, a meander flood plain commanded by the Lower Chenab Canal system. The region has suffered extremely serious deterioration in productivity owing to salting and waterlogging. The activities of SCARPs (Salinity Control and Reclamation Projects) discussed in Chapter 6 above are making headway to combat the problem. Perennial irrigation supports the usual range of kharif and rabi crops, the latter tending to dominate. Where rabi water is short, rice is the crop irrigated in kharif, with wheat and gram as barani crops started perhaps with a late kharif watering.

Transect 5 (Fig. 8.32): Sulaiman Range–Dera Ismail Khan–Thal

This 136 km transect from the Sulaiman Ranges eastward through Dera Ismail Khan on the west bank of the Indus and on into the central Thal clearly indicates the very restricted development of riverain alluvial features in this reach of the Indus. From the base of the alluvial fans that apron the mountain wall a barely sloping pediment or piedmont plain extends for 58 km to the edge of the active flood plain (see Fig. 7.4). Intermittent torrents crossing the plain have incised their courses slightly. The Indus flood plain is about 16 km wide, with braided channels and sand banks occupying up to 6 km of this. Crossed by a bridge of boats in winter time, a paddle steamer ferry operates in summer to link D.I. Khan to Darya Khan on the east bank. To the east the rolling and hilly sand plains, to a small extent levelled by man (in the Level Sand Plains), are characteristic of vast tracts of Thal.

The region is semi-arid with limited availability of irrigation water. The torrents of the piedmont plain yield occasional summer floods which to the extent that they can be diverted into bunded fields provide moisture for a subsequent rabi crop or wheat or pulses. The rod kohi system has already been described in Chapter 6. Rainfed cropping is similarly hazardous: depending on its incidence the rains may support a low yielding kharif crop of jowar or bajra, or a rabi crop of wheat or grain.

Close to the Indus, perennial cropping on the piedmont plain is made possible by the Paharpur inundation canal and intensive well irrigation of sugar cane, vegetables and wheat. Perennial irrigation is on the way from Chasma barrage. The active flood plain is flooded in summer but can be irrigated by temporary channel diversion in the low water months, the area watered varying from year to year. Uncultivated sections provide rough grassland and scrub grazing.

The flood plain ends abruptly in the high bluffs that mark the edge of the Sind Sagar Doab. Here seasonal and well irrigation is used near the river, while along the line of the Indus Branch Main Line Lower of the Thal Canal perennial cultivation is possible. Rabi wheat and grain dominate under irrigation with millets in summer, when sand storms militate against cotton growing. Away from

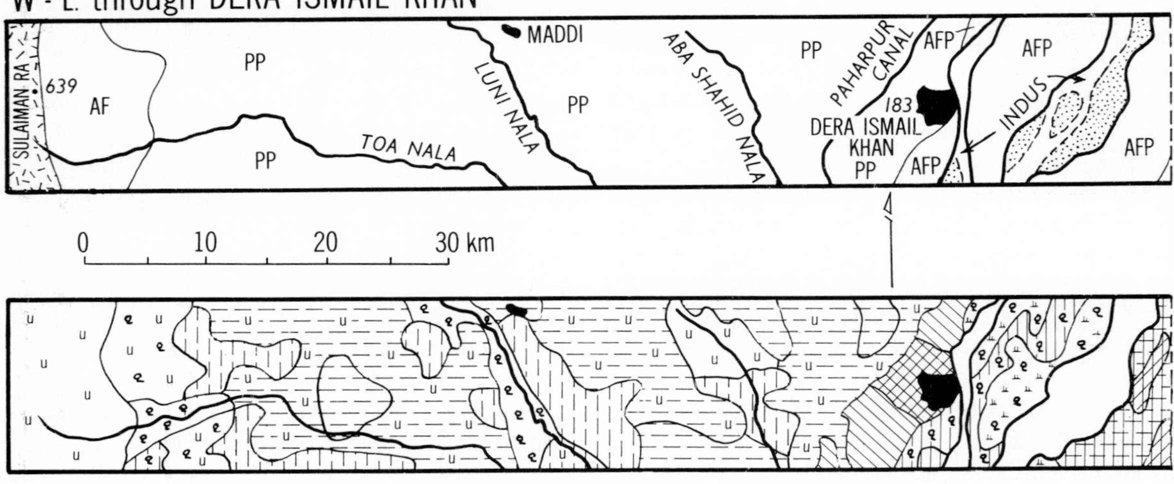

FIG. 8.32

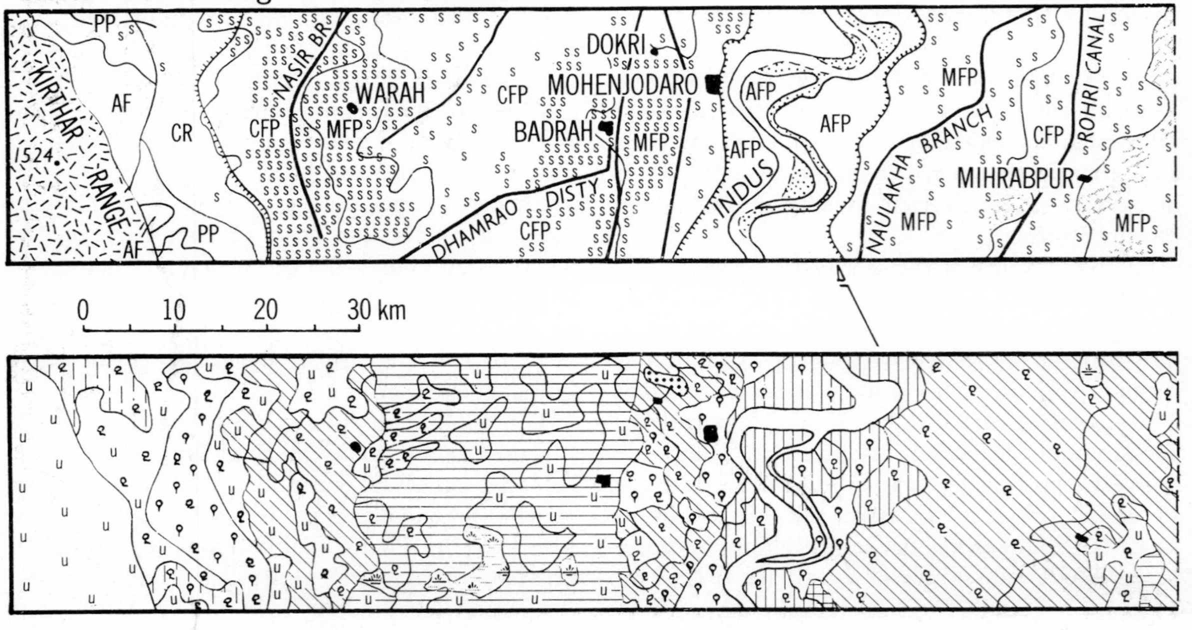

FIG. 8.33

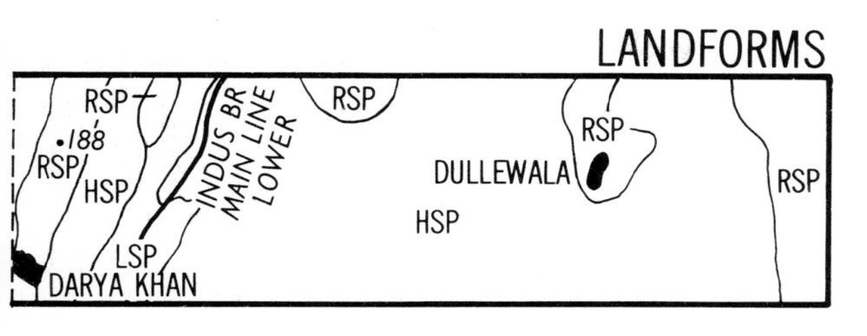

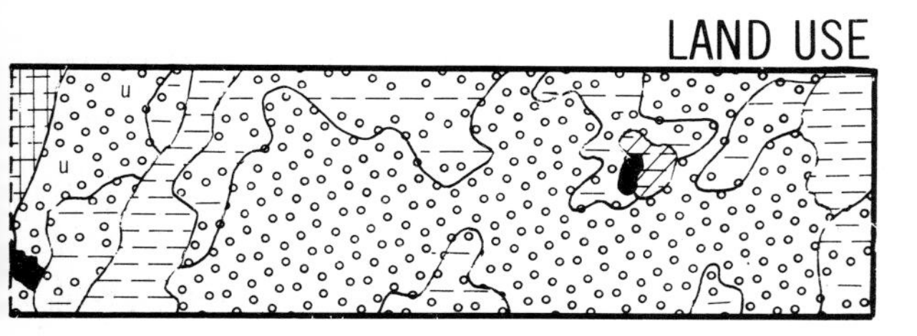

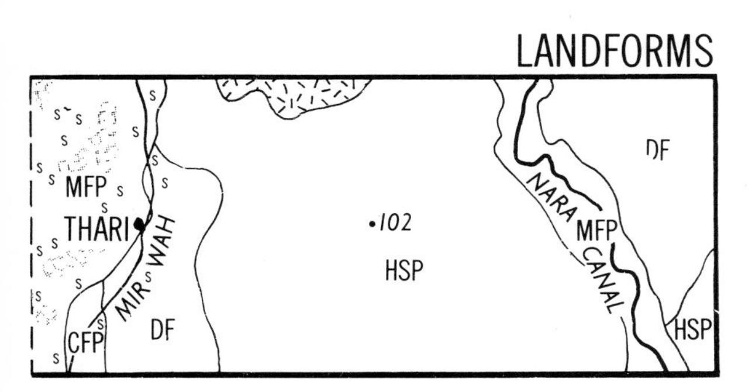

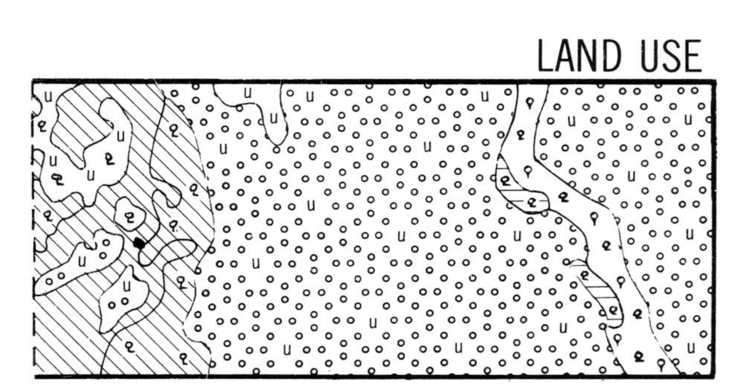

the areas commanded by canal distributaries, traditional wells worked by camels, or modern tube wells powered by diesel engines reach sweet ground water in the narrow depressions or 'patti' between the dunes. The water raised is intensively used to grow rabi crops of wheat and barley, oilseeds and pulses. Gram may also be sown without irrigation if there has been rain. In kharif, bajra is grown, and near to the wells sugar cane and cotton on a small scale.

Much of the hilly sand plain of central Thal is semi-arid grazing for sheep and goats whose herdsmen lop trees to supplement the poor forage (Plate 37).

Transect 6 (Fig. 8.33):
Kirthar Range – Mohenjo–Daro

By contrast with Transect 7, alluvial morphological elements dominate in the Indus plains in Sind downstream of Sukkur (see Fig. 4.5). At the foot of the Kirthar Range alluvial fans merge into a very narrow discontinuous zone of piedmont plain. At some time the Indus, or at least one of its braided channels, has swept close to the mountain foot leaving evidence in the form of channel levée remnants. 150 km to the east the channel of the Nara, now used by a canal, is another formerly active channel of the Indus which has deposited and moulded the basic features of the whole intervening area. Cover flood plains, meander flood plains and active flood plains represent successively younger stages in river development. On the cover flood plains soil wash has practically obliterated the traces of former channels which, however, are still clearly seen in the meander flood plains. In the active flood plain, channels are still in periodic use and the surface undergoes constant change in successive floods. The older flood plain elements are repeated in broad belts either side of the present active flood plain. East of the Indus wind plays a major role in determining the pattern of the surface and sand dunes become increasingly common. In the hilly sand plains of the Nara tract alluvium has been reworked into dunes and patti depressions.

Two major areas of unproductive land mark either end of the transect: the rocky Kirthar Hills in the west, the hilly sand plains and desert fringe to the east, into which a small area of the rocky

Rohri ridge penetrates from the north. It is this ridge that separates the Nara from the Indus above Sukkur. Although unproductive from the point of view of cropping, these tracts provide sparse grazing for goats, sheep and camels. Grazing is particularly abundant in the channel remnant depression at the lower edge of the alluvial fans and piedmont plain where torrent water accumulates and is prevented by an embankment from invading the well-ordered irrigated fields on the cover flood plain. Torrent watered cultivation on the alluvial fans is restricted to kharif millets.

The cover flood plain strip west of Warah is severely afflicted by salinity and waterlogging, and much land has been lost to agriculture. Irrigation here is now mostly perennial, supporting more rabi crops (wheat pulses, oilseeds) than kharif (rice). On seasonally (summer) irrigated land the kharif crop is more important.

East of this belt, salinity is less of a problem except in depressions which provide swamp grazing. Irrigation is principally seasonal, supporting rice, which may be followed by oilseeds as a 'bosi' crop, or pulses as a 'dubari' crop. A 'bosi' crop gets one autumn watering to establish it after which it depends on soil moisture; a 'dubari' crop is sown immediately after the harvest of an irrigated kharif crop, and relies entirely on remanent moisture; winter rain is a bonus not to be relied upon. Where perennial irrigation is available, the rabi crop, mainly wheat, exceeds the kharif.

Immediately adjacent to the Indus active flood plain the narrow strip of meander flood plain is perennially irrigated from the Dadu canal and supports rabi and kharif in equal proportions. Rice and millets with some sugar cane and cotton in kharif are matched by irrigated wheat and 'dubari' crops of pulses and oil seeds. Fields lying beyond the regular command of the canals tend to revert to scrub of tamarisk and acacia.

The Indus flood plain is too deeply awash in summer time to carry a crop. In winter, channel

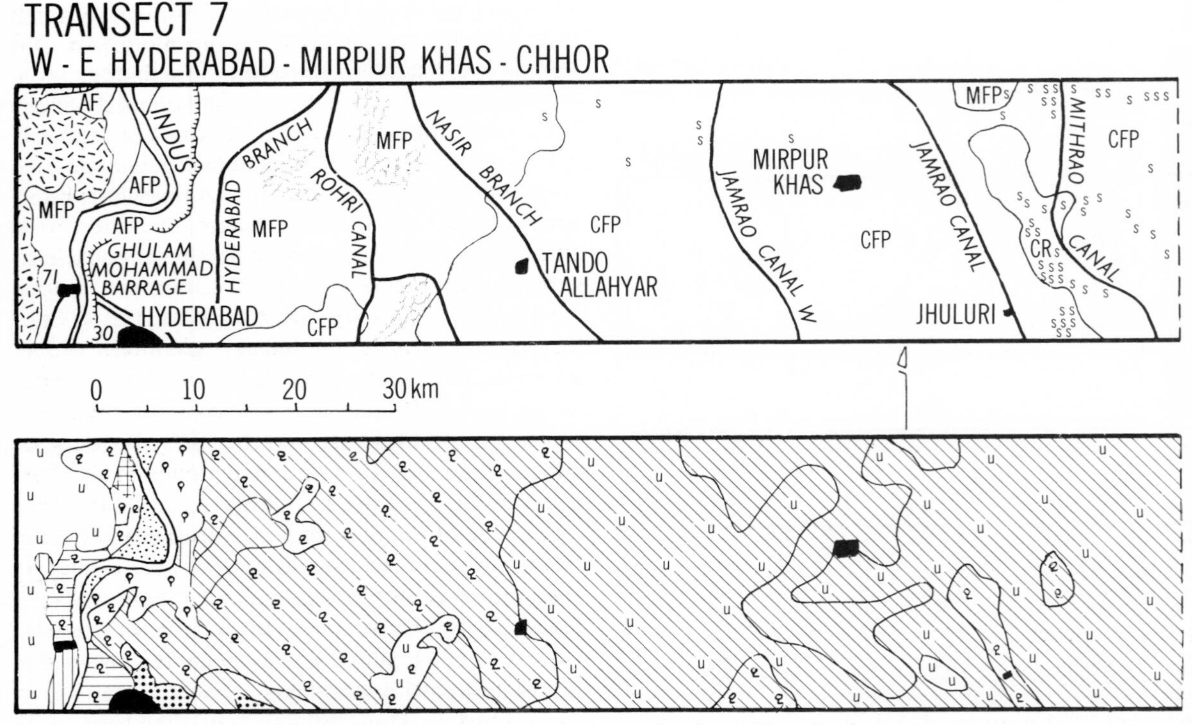

FIG. 8.34

diversion or 'jhalari' (raising water to fields by Persian Wheels) supports the standard association of wheat, oilseeds, gram and other pulses. Considerable areas of reserved forest grow shisham in addition to tamarisk and acacia and provide firewood for the towns. East of the Indus the cropping patterns of the perennially irrigated cover and meander flood plains west of the river is repeated. Salinity and waterlogging are insignificant. In the broad Nara Tract further east, sand dunes sometimes rising 30 m above the intervening patti lands, bear witness to the proximity of the Thar Desert. Jhalari irrigation from the Nara Canal (occupying the course of Nara Dhora anabranch of the Indus) waters small patches of kharif, cotton, rice and fodder and rabi wheat.

Transect 7 (Fig. 8.34): Hyderabad–Chhor

At the latitude of Hyderabad, the active flood plain of the Indus impinges almost directly onto the bedrock of the mountain wall with only a narrow strip of meander flood plain, and an occasional alluvial fan, separating them. For kilometre upon kilometre the train eastwards from Hyderabad to Chhor, close to the Indian border, runs across featureless flood plains, the monotony relieved only by the occasional belt of sand dunes. Even these have often been levelled to extend cultivation. Towards the east, a channel level remnant followed in part by the Mithrao canal marks the beginning of fairly extensive salinification due to seepage from the canals and poor natural drainage in such flat terrain. Much land has been lost to farming in the commands of the canals that parallel the desert fringe.

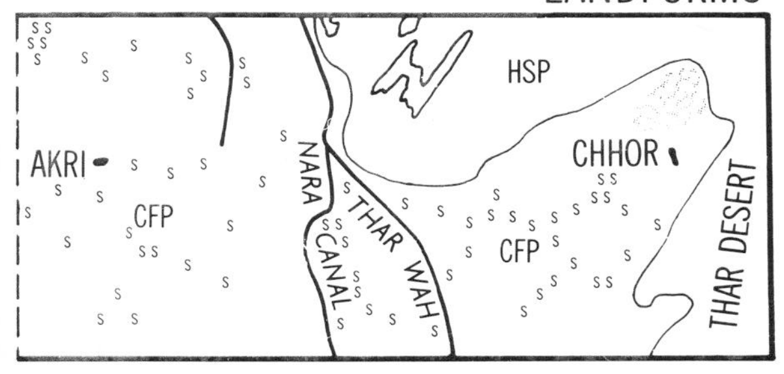

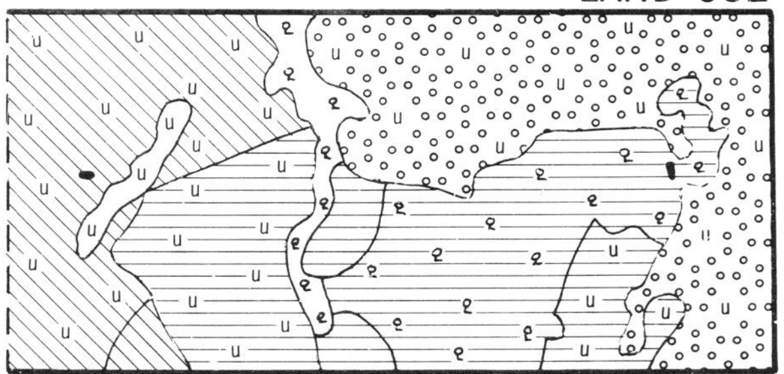

Characteristically the unproductive wastelands of the mountain rim to the west give place to sporadic torrent watered cropping at its foot. Seasonally flooded and irrigated cultivation on the west bank is mixed with scrub. The active flood plain is occupied mainly by reserved forest. Protected by the continuous flood embankment on the east bank, the meander flood plain is perennially irrigated by the Rohri canal and its branches supplemented by well irrigation of intensively farmed fruits and vegetables close to Hyderabad. Kharif crops have a slight edge on rabi crops, millets being the main kharif crop, with some cotton, wheat the major rabi season crop with vegetables in association. Cotton becomes dominant in the cover flood plain commanded by the perennial Jamrao canals, with millets in a secondary position in kharif, and wheat almost the sole rabi crop. A similar pattern appears in the salinity affected areas further east. Despite nominally perennial irrigation, rabi, wheat and oilseeds are often grown as 'bosi' or 'dubari' crops, and even kharif millets may be sown as a 'barani' (rainfed) crop. The region is far enough south to benefit from summer monsoon rainfall, unreliable though it is. Traditionally depressions among the dunes are bunded to trap any rain that eventuates, and whether a rabi or kharif crop is sown depends on the incidence of the rainfall.

The desert fringe provides forage for camels, sheep and goats, but it is a harsh environment, the day temperatures extremely high and oppressive for all but two or three months of the year.

CHAPTER NINE

LIVESTOCK FOR WORK, MILK AND MEAT

That 15 per cent of Pakistan's total cropped area is devoted to raising fodder crops is indicative of the importance livestock have in the economy. Apart from the several grain crops that may also be cut green for fodder, there are specific fodder crops among which 'berseem' or Egyptian clover is probably most common as an irrigated crop yielding up to five cuttings between December and May. Guar, now also being grown for its protein-rich seed, is widespread as a kharif fodder, either rainfed or irrigated. Pakistan is heavily dependent on animal muscle power to till the fields, to turn the Persian Wheels, and to carry the farmer and his produce to market, and no small fraction of intra-urban transport is still carried on the backs of donkeys or pulled by horse rehras and tongas, bullock or camel carts. Milk, 'lassi' (yoghurt), and for the more affluent, butter, are important elements of daily diet, while meat, particularly mutton of sheep or goat is a desirable if not often obtainable food, and well nigh essential for celebrating the great Muslim festival of Eid. In recent decades, with improved strains and more scientific management, poultry has come to play a more significant role. At the time of independence, eggs were never better than 'sub-standard' in size by western grades, but now rival the world's best in size, and excel in flavor, being from hens fed largely on maize.

To take the smallest first, poultry numbered almost 14 million in 1972. About half were in the Punjab plains, Faisalabad with 875,000 heading the list. Karachi with 786,000 had nearly 6 per cent of the total to feed its urban population. Next came Peshawar, 754,000, which with the other northern districts from Bannu to Jhelum—all important maize growers, perhaps not incidentally—accounted for 25 per cent. No district recorded was without poultry. By 1976–77 there were in excess of 32 million birds, 8.3 million broilers, 3.5 million layers and 20.6 million Desi (local) poultry.

The numbers of quadrupeds are given in Table 9.1.

TABLE 9.1
Quadruped livestock, 1976–77

	('000)		
Goats	20,881		
Sheep	18,131		
Cattle	14,372	in milk	2,947
		for work	7,383
Buffaloes	10,418	in milk	4,125
		for work	556
Donkeys	1,939		
Horses	382		
Camels	694		

Source: Livestock Division, Ministry of Agriculture.

Statistics on the distribution of livestock are not available after 1972, and the much lower numbers of the Census of Agriculture in that year must serve to indicate the general patterns for the major types.

Goats and sheep are often herded together. Fig. 9.1 maps their spread which is universal. No clearly meaningful pattern can be discerned. While the districts with extensive rough grazing in hills or rolling sand plains have quite high numbers, for example 254,000 in the three Thal districts, the intensively cultivated canal colonies have similarly high figures, Multan, Faisalabad and Sahiwal totalling 294,000. These small stock do a useful job grazing stubble and weeds and adding manure, a task from which the shepherds of the transhumant flocks from Baluchistan and the western hills make a partial living.

46 Sheep and goats for sale in the market at Lahore at the season of Bakr-Id when every Muslim family strives to have its own animal to kill for a feast. Prices become exorbitant by normal standards. The ubiquitous charpoy is a seat by day, a bed by night.

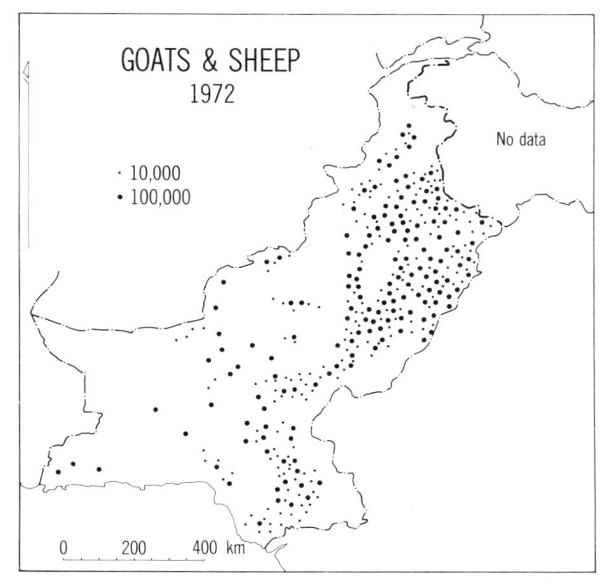

FIG. 9.1

The quality of wool is coarse, but well suited to carpet making. For fine woollen fabrics wool has to be imported. There are numerous regional breeds of sheep. Fat-tailed sheep are favoured in the dry mountain country of NWFP and Baluchistan and one, the Hashat Nagri, is also popular around the large cities for sacrificial purposes. Thin-tailed breeds are found in the more humid mountains and on the plains, several being used as milkers by their generally very poor owners. Goats are bred for mutton or milk, some breeds also providing fine under-coat fibre.*

*Khan, A. A. and Ullah, M. H., *Agronomy and Horticulture*, Caravan Book House, Lahore 1962, pp. 239–252.

Cattle and buffaloes furnish the agricultural work force and the dairy herds. Buffaloes, although strong for work, are slower than bullocks and more expensive to feed. They are bred mainly for milking, producing milk rich in butter fat. Cattle, of the *Bos indicus* species are relatively poor milkers, unless crossed with *Bos taurus* of European ancestry. Sahiwal and Red Sindhi breeds are the best of the local milk producers. Sahiwal bullocks are regarded as too lethargic to make good plough animals but can be fattened for beef. The Red Sindhi is tolerant of hard conditions of forage and high temperatures. They are also hardy workers at the plough or bullock cart, and are bred for that purpose primarily.

In the north the stocky spotted Dhanni breed is not a good milker but has a good reputation for work in barani lands and mountains, while the Bhagnari breed originating in Kachhi and Sibi can tolerate extremes of heat and coarse fodder, is a strong load puller and has adapted satisfactorily to the submontane areas of the north. It is a good all-purpose animal as the male is a good plough beast. Even hardier is the Tharparker breed from the margins of the Thar Desert, which has the reputation of being a speedy ploughrer. Naturally the distribution of work animals ties closely to the level of agricultural activity (see Fig. 9.2). Relatively high figures for Sahiwal, Faisalabad and Multan (1,420,000) may reflect the raising of the Sahiwal breed for sale, and Hyderabad's 250,000 the fame of the Sindhi breed, but this is conjecture.

Buffaloes show a more distinct regional pattern. Entirely absent from upland Baluchistan and in very small numbers in Sind and the drier and

47 A bullock team, Dhanni breed, levelling with a board on which the driver stands. A crop of maize has been drilled in the furrows. A walled village may be seen in the background. A scene in late winter on the Maize and Millets Research Farm, Pirsabak, in the Vale of Peshawar.

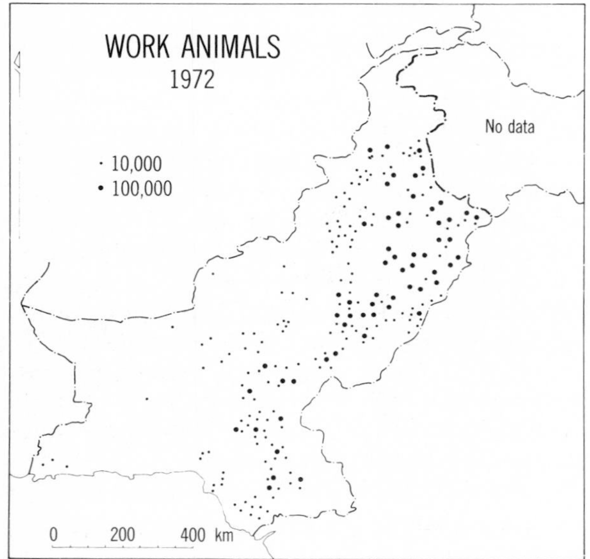

Fig. 9.2

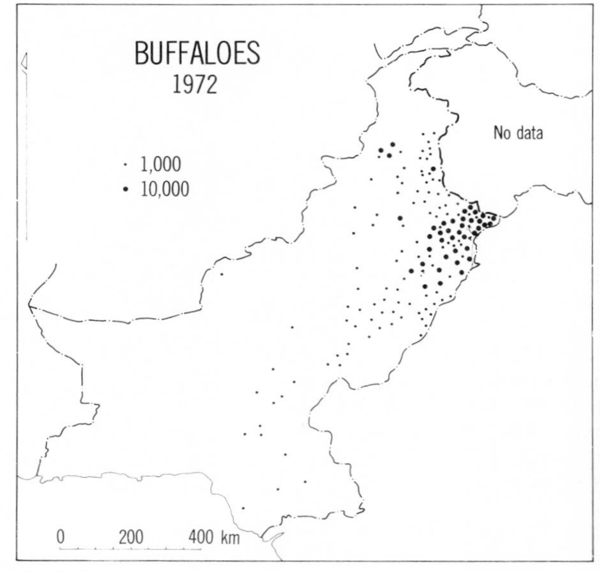

Fig. 9.3

hotter south and west Punjab, they are remarkably concentrated in Faisalabad, Sahiwal and the districts to the northeast, Gujranwala, Sheikhupura, Lahore and Sialkot (Fig. 9.3). These six districts had 74 per cent of the total buffaloes for work. Their absence in the intensely hot and arid regions is explained by the buffalo's sensitivity to prolonged dryness. They need a daily wallow for health. It is the buffalo's acceptance of stall feeding and weaning that makes it so popular as a town dairy animal. The most notable breeds are the Nili from the Sutlej valley originally, and the Ravi from the 'bar' lands of Sahiwal, Faisalabad, Sheikhupura and Jhang. The Kindi buffalo is the local breed tolerant of the extreme conditions of Sind.

Milch cattle and buffaloes together show a more even distribution, but these are also very sparse in Baluchistan, and not numerous in Sind (Fig. 9.4.). The same six Punjab districts pre-eminent for working buffaloes had only 26 per cent of the total, but the triangular block enclosed by Multan, Sargodha, Gujrat, Sialkot, Lahore and Sahiwal accounted for 53 per cent.

Milk supplies for the urban population present a problem in a community still mainly dependent on traditional systems. Apart from some notable exceptions such as the Australian designed dairy farm for Islamabad, and some government and military dairy farms, urban supplies still come mainly from buffalo herds installed in the cities themselves and necessitating the transport of fod-

48 Buffalo hauling a rubber-tyred cart along the Grand Trunk Road near Lahore, carrying sugar cane to a mill for crushing. The man on foot guards the load against pilfering children who like to chew cane.

49 A mixed milking herd of buffaloes and cows on the banks of the main drainage canal that cuts through Lahore. Fodder has to be brought in from the countryside, but efforts to force such herds out of the city failed when the supply of fresh milk deteriorated as a consequence. The upper middle class residents have to put up with their bovine neighbours.

Fig. 9.4

der to maintain them, cut grass being less perishable than milk in the conditions of Pakistan. It need hardly be stressed that such dairy herds kept on vacant lots lacking appropriate drainage or water supplies constitute a land use incompatible with modern hygienic city life! A valuable by-product, cattle dung for use as domestic fuel, is collected and dried in circular cakes plastered on any convenient wall. A vivid mixture of economic and cultural levels is seen in the most elite residential suburbs, where a herd of milking buffaloes may be seen in occupation of a building site between luxury mansions.

It is to support working cattle, milch animals, and the much smaller numbers of donkeys, camels and horses, that so much fodder has to be grown, year in, year out. Figs. 9.5 and 9.6 show the percentage of the total cropped area occupied by fodder crops in 1972. In addition to these green fodders, the chopped straw from the wheat, rice, jowar and bajra crops and the trimmings of the sugar cane are available. Kharif fodder crops totalled 1.51 million ha, rabi fodder 1.22 million ha. In five districts of the Punjab, kharif fodder occupied more than 30 per cent of the kharif sown area, and almost the whole province exceeded 20 per cent. No district in Sind, NWFP or Baluchistan reached even 20 per cent and a substantial area in right-bank Sind and beyond had percentages below three.

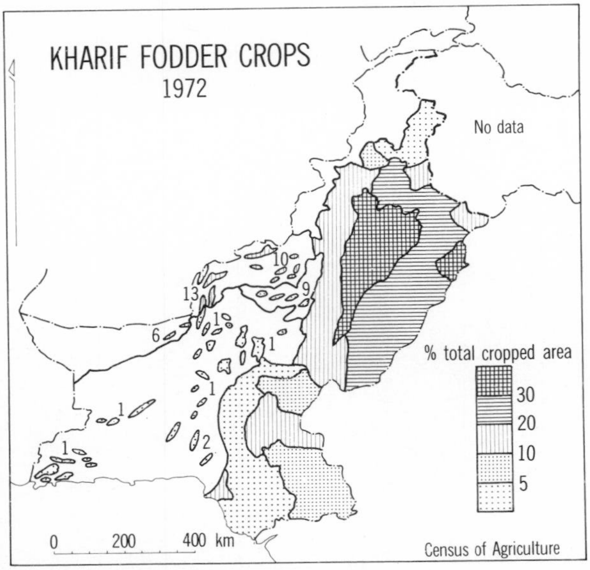

FIG. 9.5

Sind makes some amends to its cattle in the rabi season, with three districts using 15 per cent or more of the cropped area for fodder. In Punjab the fodder crop area contracts in the rabi season, the important districts all with over 15 and two over 20 per cent being the triangle of districts mentioned above.

Camels, like cattle and buffaloes are dual purpose animals, used for milking by their herding families but reared principally as transport animals. Camels drawing four-wheeled, tyred carts are not uncommon, but the more usual sight throughout the length and breadth of Pakistan–and most common in the western Punjab–is a string of camels under loads of firewood, of chopped fodder in bulging nets, or prodigiously bulky stacks of straw (Fig. 9.7).

Donkeys constitute the lowliest of the animals' working class, but do a great amount of work carrying saddle sacks of earth or bricks on building sites, or of anything else that needs carrying for a pittance. Harnessed to small 200 litre tanker carts they may be seen delivering kerosene in the towns. They outnumber horses by almost three to one. The latter are mainly in the shafts of passenger carrying tongas in the city street or drawing rehras

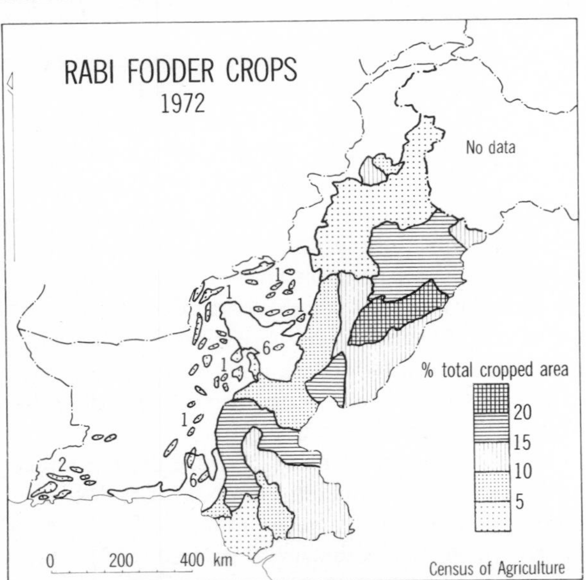

FIG. 9.6

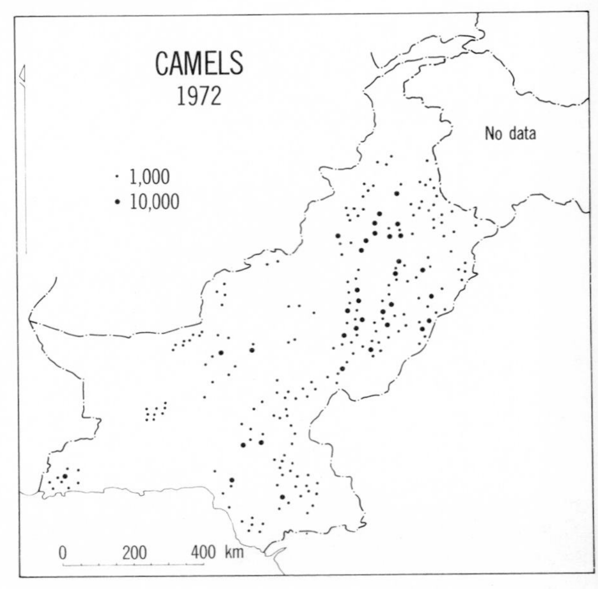

FIG. 9.7

carrying farmers' produce and family to market. Occasionally the landowner still rides a mount, but more often nowadays he aspires to a jeep, a tractor or a comfortable car.

Meat, hides and skins. These terminal products of livestock rearing make an important contribution to the economy. Livestock slaughtered by kind and by province in 1974–75 are shown in Table 9.2.

As mentioned goats and sheep are kept mainly for slaughter. Beef on the other hand is rather a by-product of rearing for work and milking.

Hides and skins, converted by tanning to leather, currently make up 5.6 per cent of Pakistan's exports by value. Bones weighing over 26,000 tonnes in 1974–74 (136,000 tonnes in 1972–73) are also a source of foreign exchange.

TABLE 9.2
Livestock slaughtered, 1974–1975 ('000s)

	Punjab	*Sind*	*NWFP*	*Baluchistan*	*Total*
Cattle	152	217	58	13	440
Buffaloes	229	182	78	0.4	489
Sheep	705	233	111	34	1082
Goats	1259	1130	167	89	2645
Camels, etc.	1.1	—	—	—	1.1

Source: *Agricultural Statistics of Pakistan*, 1975.

50 Rubber-tyred camel-carts in convoy near Dulle Wala in the Thal region. The old fashioned iron clad cart wheels are generally forbidden on bitumen surfaced roads because of the damage they cause. The tyres are lorry cast-offs.

51 The patient overloaded donkey is a common sight on Pakistan's roads. Singly or in caravans these little beasts do a vast amount of the country's carrying.

CHAPTER TEN

TECHNOLOGICAL AND INSTITUTIONAL CHANGE IN AGRICULTURE

INTRODUCTION

This chapter might alternatively be entitled 'who works whose land and with what?' Two distinct aspects of modernization are involved. They are basically, on the one hand the application to previously traditional agriculture of a scientific approach to the cultivation of crops and drawing upon the world's accumulated knowledge and experience to find the appropriate technology to achieve that application, and on the other the change in the institutional structure under which individuals work the land, requiring fundamental changes in the traditional social order. How far the two aspects are interdependent is open to argument; that they are both inevitable there is little doubt. Technological change goes on apace; land reform by fits and starts and legislation tends to be 'more honoured in the breach than in the observance'.

Pakistan has often been held up as a glowing example of successful achievement in the 'Green Revolution'. The fortunate combination of an efficient system of irrigated agriculture, a tradition of wheat cultivation accessible to penetration by the new high yielding varieties becoming available in the 1960s, and a generally economically progressive land-owning class, were prerequisites and guarantees for success. Some of the key elements in the process of agricultural modernization in Pakistan are reviewed to discover the extent to which the scientific method is being applied: fertilizer use, adoption of HYV, chemical control of plant pests and diseases, and mechanization. The advantages to the entrepreneur in adopting in whole or in part the methods implicit in the Green Revolution are generally sufficiently evident for him to require no coaxing to alter his traditional ways. If the market price is right and the inputs are available, the profit motive in man will do the rest! Government's concern has rather been to meet the demand for the inputs and to steer them in the direction of the nationally most desirable crops.

The question of reform in the land-holding system is part of a potential revolution of a quite different character. Governments have introduced several land reforms since independence in 1947. But many landowners have been able to avoid the impact of legislation by manipulating titles to land among relatives or nominees in advance of the promulgation and implementation of the reforms.

Land reform is a part of a more general process of social change. The world over, the currents of opinion demanding a more egalitarian society and social justice are running strongly. Land reform has always been resisted by the established elite who hold their power through the sometimes semi-feudal traditional system of landlordism. In this Pakistan is no exception. Since independence, land reform has been used as an instrument in the struggle to gain political power through the ballot box. When democratic processes are allowed to run, the mass vote carries weight. In Pakistan the masses are rural, underprivileged, somewhat cowed by centuries of feudalism, less difficult in this age of mass media to reach than they were, and very susceptible to political promises and slogans, such as 'land to the tiller'. The politicians for their part, have found it difficult to put such promises into effect. Much real power and influence still lies with the conservative property-owning classes. Recent political events in Pakistan suggest that the day may still be a long way off when every worker in the fields will be tilling his own soil, but at least few will now be unaware that the possibility exists.

THE GREEN REVOLUTION

There has been debate among academic economists about whether the Green Revolution has succeeded, but much of the discussion, because it is in economic terms, fails to appreciate its meaning. The term 'Green Revolution' implies the application in practical agriculture of the knowledge and discoveries of modern international agronomic science. Specifically it means the use of high yielding varieties of plants and their cultivation under optimal conditions of water supply, nutrition through fertilizers and protection from pests and diseases. The revolution can be said to have happened when the farmers know what can be done and have a will to adopt the necessary means to achieve it. That government agencies, economists included, fail to make adequate provision for the essential inputs or to manipulate the structure of their costs and the prices to be given for the product, cannot be judged 'failure' of the Green Revolution, failure though it may be in some aspect of the functioning of administration in an underdeveloped country.

In the case of Pakistan the evidence for success is quite impressive. Table 8.5 records the farmers' conversion to HYV. Mexican high yielding varieties were first widely used in rabi 1967–68 when 16 per cent of the total wheat area was sown. By 1974–75, 63 per cent was in HYV and the target for 1976–77 was 74 per cent, in effect all non-barani wheat. Although the very high yield of 2.34 tonnes per hectare achieved by HYV in 1969–70 has not been maintained, the general level since 1971–72 has been held at 1.61 to 1.65 tonnes per ha. The table and Fig. 10.1 demonstrate the impact of HYV wheat on overall yields which rose suddenly from a low plateau of around 0.8 tonnes between 1958 and 1967, to more than 1 tonne in 1967–68 and progressively upwards to 1.45 tonnes in 1976–77 as more and more farmers took up HYV in preference to traditional varieties. It is doubtful if any substantial body of farmers decline to cultivate HYV wheat for other than sound technical and/or

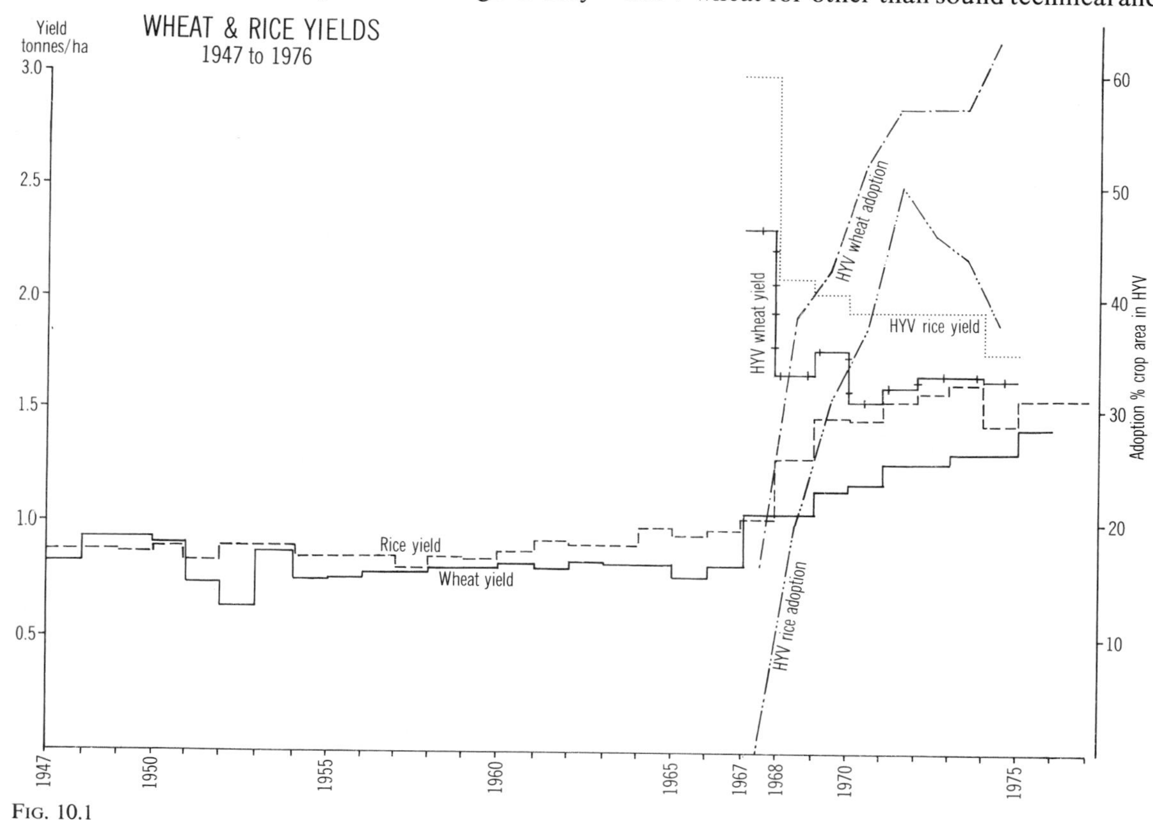

Fig. 10.1

economic reasons. About one-quarter of the total wheat area is barani and so unsuited to planting with Mexi-Pak seed which must be adequately irrigated if it is to benefit from the heavy fertilizer applications prescribed. Improved yields in the barani lands await the development of non-irrigated HYV to replace the well-trusted Desi varieties.

Fig. 10.2 shows the distribution of wheat yields by district, the pattern clearly reflecting the association of irrigation with high yields. The barani wheat areas are easily distinguishable in the low yielding areas in the north and northwest.

How close to the theoretically optimum applications of fertilizer the farmer comes depends on a number of economic rather than technical factors, the relative prices of fertilizer and wheat being paramount. The cost of water and fertilizer have been subsidized and support prices fixed for wheat and rice, so that between 1964–65 to 1966–67, and 1967–68 to 1968–69, that is before and after the introduction of HYV wheat and rice, the profitability of growing these crops increased 37 per cent in Sargodha, 69 per cent in Peshawar District and 46 per cent in Hyderabad. The comparative increase for sugar cane in the same districts was 13,4.5 and 0.9 per cent and for the wheat-cotton combination, 13, 6 and 2 per cent.*

The adoption levels for HYV rice fall well short of those for wheat (Table 8.5) but for very good reasons. All rice in Pakistan is irrigated and it can be said that, unlike in India or Bangladesh, the artificial environment for cultivating HYV as a kharif crop could hardly be bettered. The importance of fine 'Basmati' rice in the export trade cuts out almost a third of the rice area from consideration for planting with HYV. IRRI-Pak rice† occupied 39 per cent of the total area in 1974–75 (it reached as much as 50 per cent in 1971–72) so it could be argued that a potential of about 30 per cent of the total rice area remains to be exploited. Market preference has been more difficult to satisfy in the case of HYV rice than with wheat and this may have deterred more wholesale adoption, though the export price strongly favours IRRI-Pak against local 'Joshi' rice (see Chapter 8 above). The explanation probably lies in the relative degrees of tolerance by local and HYV rice to conditions of salinity.

The pattern of improvement in rice yields seen in Fig. 10.1 follows wheat with a delay of a year or so. Rice has always tended to yield better than wheat, largely because of the greater uptake of water, and this lead was increased for a while with the introduction of HYV of both crops. The decline in the yields of HYV wheat and rice from the levels achieved at their introduction is a feature experienced elsewhere. An innovation tends first to be tried out in the best available conditions, on the best land, cultivating and fertilizing 'by the book'. With more widespread adoption less than optimal areas are brought into use, less than the ideal financial outlay is available, and less than scrupulously scientific husbandry is practiced. Inevitably yields level out but on a higher plateau than before.

FERTILIZER USE

In the 1950s fertilizer use was negligible in Pakistan, and to judge from Fig. 10.3 which graphs the consumption from 1959–60 onwards, the big boost came with the introduction of HYV in the late 1960s. Production of fertilizer has become a large industry, using natural gas as a major feed stock to make urea, but much has still to be imported to meet the shortfall. The world petroleum crisis from 1974 onwards has caused fertilizer prices to soar, and has led to a flattening of demand. Further subsidization from 1975–76 has raised consumption which may reach 555,000 tonnes in 1977–78. The manufacture of fertilizer is examined in Chapter 11 below. Despite the impressive increase in fertilizer use, Pakistan still ranks low in the world table. In 1972–73, 22 kg per hectare was the apparent consumption, comparing with 85 kg/ha in USA and 126 kg/ha in Italy. The comparison is not quite valid since the cropping patterns and climates of these countries are very different from Pakistan's. Nevertheless, even assuming an increase has since occurred in

*Afzal, M., 'Implications of the Green Revolution for Land Use patterns and Relative Crop Profitability under Domestic and International Prices', *Pakistan Development Review*, VII, 1973, pp. 135–147.

†IRRI stands for International Rice Research Institute, the Philippines-based source of new rice varieties for the Green Revolution.

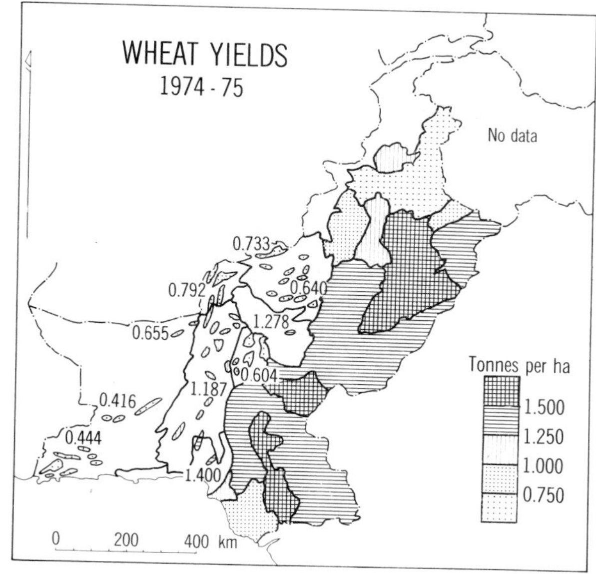

FIG. 10.2

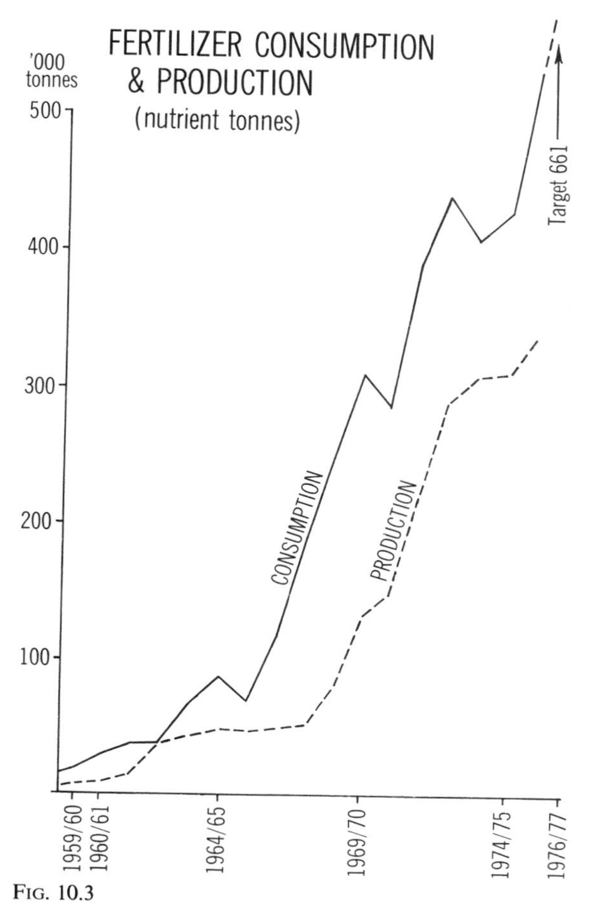

FIG. 10.3

Pakistan, there is probably great scope for further expansion, provided as has been said, the price obtained for the extra product brings adequate reward to the farmer.

The use of fertilizers is not restricted to the HYV wheat and rice crops by any means. Fig. 10.4 shows the percentage of the cropped area fertilized in 1972 according to the Census of Agriculture, which also reported on the crop-wise use of fertilizer. The map suggests a strong correlation between canal irrigated agriculture and fertilizer use, the barani areas making a poor showing. In Table 10.1 are listed for each of the major crops the number of districts in which fertilizer is applied to (i) over 80 per cent, (ii) 60 to 79 per cent, and (iii) under 20 per cent of the area of the crop concerned.

Such data tells us nothing about how much is applied, but it is indicative of the neglect of certain groups of crops like the pulses, oilseeds and fodders, and the concern lavished on cash crops like cotton and sugar cane. The food grains occupy a middle position. While nitrogenous fertilizers have been by far the most used to date, it is now realised that a better balance of nitrogen and phosphate is needed. In 1975–76 the target ratio was 100 phosphate to 455 nitrogen, whereas in 1970–71 it was 100 to 824.

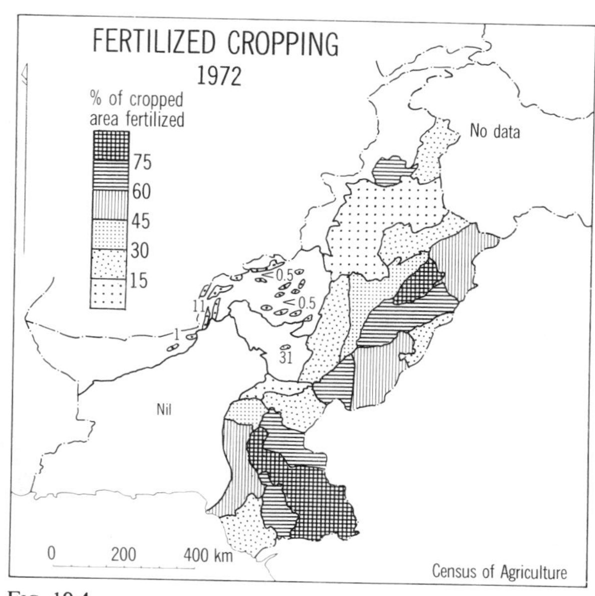

FIG. 10.4

TABLE 10.1
Use of fertilizers

Crops	Total districts	Number in which fertilizer is used on 80% or more	60–79%	less than 20%
Rabi crops				
Wheat	38	4	10	8
Pulses	34	–	–	31
Oilseeds	35	–	1	23
Fodder	38	–	4	12
Vegetables	37	5	8	1
Tobacco	29	6	3	6
Kharif crops				
Cotton	39	10	7	5
Sugar Cane	38	14	11	2
Rice	38	5	13	7
Maize	38	5	5	10
Oilseeds	35	1	1	16
Fodder	38	–	3	14
Vegetables	38	5	8	1

PLANT PROTECTION

This is much less widespread than is fertilizer use and is applied almost exclusively to kharif crops and orchards, with the exception of one rabi crop, tobacco. Only five districts record plant protection to more than 10 per cent of the cotton crop, to sugar cane, 5 to rice, 3 to maize, and 8 to vegetables. Orchards are treated by spraying in every district, and usually at least a quarter of the area is protected. The four main tobacco districts in Peshawar Division and D.G. Khan spray their crops, 80 per cent of the specialised Mardan area being treated. These data suggest much remains to be done to protect crops against diseases and pests, a fact that was brought home sharply to cotton growers when the crop was heavily attacked in 1973 by boll weevils and army worms.

AGRICULTURAL MECHANIZATION

Tractors

Tractors are now no longer a source of wonderment in the countryside. As Fig. 10.5 shows, over 10 per cent of farm households in 15 districts used tractors in 1972. Together with diesel and electric powered pump units for raising well water, tractors represent a potential displacement of animal power from the farm. At present the total impact of mechanical energy is not great. In terms of how agricultural manpower time is allocated it seems that less than 10 per cent is linked to powered equipment, 25 per cent to animal-powered equipment and the rest to using hand tools for digging, harvesting, threshing, etc.

There is much inconclusive debate about the effect of mechanization, particularly tractors, on the utilization of labour. The evidence suggests that when tractors are used in combination with plentiful water to raise the intensity of cropping, labour demand is also increased. When however

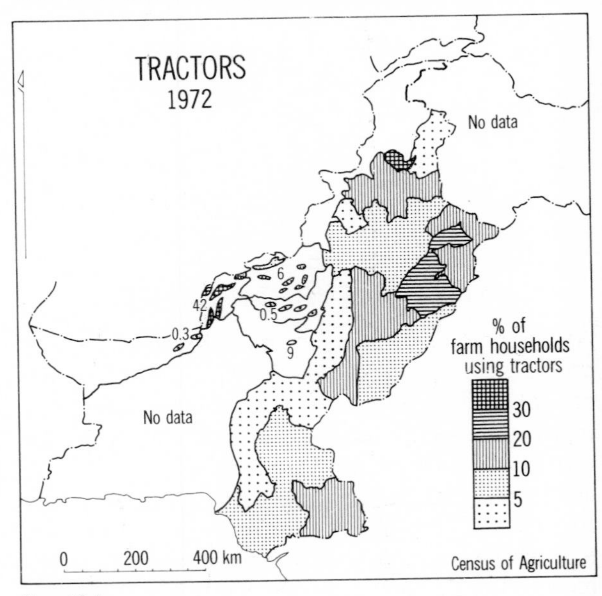

FIG. 10.5

they are used in drier areas they tend to displace tenant occupiers, as owner farmers find themselves able to handle more of their land than previously and will do so extensively rather than intensively. It has also been suggested that social reforms that were being introduced in the years immediately prior to the current (1978) period of Martial Law Administration tended in the direction of increasing the bargaining power of labour and tenants with the landlords, and has had the effect of accelerating mechanization as a move to reducing the labour needed on the farm. In favour of mechanization it can be argued that speedier cultivation promotes intensification, that tractor drawn implements do a better job than those drawn by bullocks, and that a reduction in livestock numbers by diminishing the need to grow fodder, frees land for alternative crops and more income. Among the counter arguments is the powerful one that tractors represent a cost in foreign exchange both at the time of purchase and when refuelling with imported petroleum products.

Meanwhile Government is allowing tractors to be imported at an increasing rate and has kept duty on them lower than on other vehicles. Over the years since 1970–71 the imports have been as in Table. 10.2. Allowing a life of eight years, some 54,000 would have been in service by 1977.

TABLE 10.2
Tractors imported

1970–71	4,021
1971–72	3,571
1972–73	2,679
1973–74	5,216
1974–75	7,190
1975–76	10,900
1976–77	15,098

AGRICULTURAL CREDIT

An essential element in modernization is the provision of credit, by which a farmer can borrow capital to finance inputs to improve output against repayment from that output. Without access to credit the farmer can modernize only slowly, investing from savings made in previous years. In 1975–76 only 20 per cent of the fertilizer used was bought on credit. The Agricultural Development Bank disbursed Rs 395,500,000 in advances in 1974–75, and planned to lend Rs 520,000,000 in 1975–76. Items of mechanization made up 55 per cent of the total as Table 10.3 shows. Most of the credit went to farmers with more than the average sized holding, as Table 10.4 indicates.

52 Disc cultivation using tractors in the Vale of Peshawar, NWFP. Mechanisation of tilling operations and of farm haulage is increasingly common.

53 Tractor-drawn combine harvester at work near Sahiwal, Punjab, cutting wheat.

TABLE 10.3
Agricultural development bank loans, 1974–75

Purpose	*Rs million*
Seeds, seedlings	13.1
Fertilizers	80.1
Tractors, etc.	130.4
Tubewells and repairs	86.3
Surface wells	3.3
Plough animals	18.7
Poultry, dairying etc.	6.5
Fisheries	4.0
Cold stores	2.7
Other	50.5
Total	395.5

TABLE 10.4
Distribution of credit by farm size, 1974-75

Category	*Rs million*
Landless	58.8
1 ha	4.6
1–5 ha	40.6
5–20 ha	183.0
20–40 ha	84.5
40 ha	28.9
Total	395.5

Another source of credit may be a co-operative society of which there were 23,293 in 1973–74 with 1.2 million members, and Rs 135 million out on loan to individuals, though not necessarily for agricultural purposes. Commercial banks also lend money (usually against security) for agricultural purposes. Loans for production totalled Rs 291 million in the first eight months of 1975–76, and for purchasing tractors and tube wells, Rs 57 million.

LAND REFORM

Probably the most fundamental expression of the *real* structure of a mainly agricultural society is its land holding system. The distribution of wealth within the community is a product of the land holding system. The power of the landlord derives from his ability to channel to himself the products of the land, even though, as is so often the case, he resides in the city, and may not be in a position to take decisions about what happens on his land, except at second hand through his steward, let alone take any personal part in working the soil. A small or medium landowner may well live and work close to his land, using his own, his family's and some hired casual labour. Often, however the land is let out to tenants or to share croppers who in return for a very large share of the product of their toil, maybe a half, are allowed to work it. Rarely does the landlord contribute proportionately to the costs of the necessary inputs, though this is changing to some extent. The tenants, the share croppers and the landless labourers are in a weak position to bargain with the landlord. Even though distant legislators may have sought to improve their lot at the landlord's expense, the latter is backed by the traditions that have permeated rural society for centuries. Tenants, share croppers and labourers must behave deferentially towards the landlord and their disinclination to kick against the *status quo*, even when legislation seems to justify such reaction, is understandable. The village is a long way from the law courts in more senses than mere distance, and for the poor peasant the paths to social justice are tortuous and uncertain of anything but their cost. Corruption at every gate upon the way gives little confidence of ultimate success against the vast vested interests of 'the establishment'.

Pakistan inherited at independence a variety of land-holding systems. In Sind the system was 'raiyatwari'. Government held permanent title to the land which was worked by tenants (and by tenants' tenants, to the n'th degree). Punjab was a stronghold of private ownership in effect, the landlords were zamindars with legal rights to transfer their land even if in theory they held it from the state. In the Northwest Frontier Province landlordism operated widely, though in the more traditional tribal areas very democratic systems obtained through which periodic redistribution of the tribal lands ensured a measure of 'fair shares for all'. Baluchistan, except in the irrigated tracts of the Kachhi and Sibi plains, was thoroughly feudal under the 'sardari' system which gave the ruling sardars powers to command the forced labour of their people, to imprison them in private jails, to lead them in private wars and generally to behave like medieval European barons.

54 A government tube well, electrically powered, on a SCARP area in the Rechna Doab west of Lahore. Such wells are used to control the level of groundwater and so to counter waterlogging and salinity.

In the first flush of independence the authors of the first Five Year Plan (which incidentally had also to apply to East Pakistan where the abolition of the zamindari system was an essential part of continuing policy) made bold recommendations to give security to tenants against unjustified eviction, to abolish illegal exactions by the landlord, and to reduce rents. There was also talk of establishing ceilings for land holdings and of redistributing surplus land.

The situation confronting the planners was one of extreme inequity. In the Punjab 21.5 per cent of the land was owned by a tiny fraction, 0.6 per cent, of those owning land. In Sind three per cent of the owners, each having more than 405 ha owned 49 per cent of the area cropped, and 7 per cent owned 70 per cent of the total. There were 294 landlords owning more than 2025 ha and some with over 24,300 ha. About half the cropped area in NWFP was held by large owners. Against this background of landlordism, the larger part of the area was cultivated not by the owners, who were often absentee landlords, but by tenants-at-will, who worked 56 per cent of the Punjab, 80 per cent of Sind and 50 per cent of NWFP.

With the backing of the Five Year Plan, the then Prime Minister in 1954 could speak of the ideal of putting the land in the hands of the tiller, to give tenants security and to break up the large estates. Understandably, Sind was notably cool in its reception of such thoughts. It was not until 1959 that land reform legislation was introduced under the Martial Law Administration of the time. It set ceilings to land holdings, but at very generous levels when compared with the size of holding regarded in the same legislation as 'economic' at the subsistence level. A land-owner could retain 36,000 'produce index units', i.e. not more than 203 ha of irrigated land or of unirrigated land or some combination of the two. Orchards were excluded as also were stud farms and lands used as reserves for shooting game, all of which suggest that the well-to-do landlord was comfortably cushioned against the legislation. Also he could gift land to his heirs thereby postponing any impact during his own lifetime. The Land Reform Commission calculated that almost one million hectares were resumed by government for redistribution by sale, to tenants and others. The recipients of such land could take up only 5 ha in most areas, 6.5 ha in Hyderabad and Khairpur Divisions, though this minimum could be multiplied up to four times if there were up to four adult males in the family. Some 200,000 small-holder tenants benefitted.

Clearly the 1959 reforms were a hollow gesture, barely touching the fundamental issues of rural inequality. The next round of reforms came in 1972 following the Bangladesh debacle, when Pakistan's ruling party under Mr Bhutto were trying to come to grips with the nation's problems and its need for a new co-operative spirit.

Certain facts about farm size and tenancy distribution are available from the Census of Agriculture of 1972, and although they may have altered somewhat as the reforms of the same year began to take effect, they are still indicative of the quality of the land holding situation, if not its precise quantitative aspects. Since a farm may be operated by its owner or by a tenant with sole occupation or by some combination of owner and tenant (as often happens when an owner lets out a part of his land on a share-cropping arrangement) the statistics may sometimes appear confusing. Thus 40 per cent of farmland was operated by the owner alone, 30 per cent by tenants alone and 31 per cent by owner-cum-tenant. If the latter area is divided between the owners' and the tenants' share, the owner-operated lands are 54 per cent of the total and the tenants' lands 46 per cent, the latter being separable into 39 per cent under share-cropping agreements, and only 7 per cent leased on a fixed monetary rent. The share taken by the owner is generally one half of the produce. Of almost 2.5 million contracts reported, 79 per cent were on a 50/50 basis, nearly 4 per cent were for two-thirds or more going to the owner, and nearly 20 per cent for two-fifths or less (14 per cent being for one-third). Figure 10.6 shows the percentage by districts of the farmland operated by the owners. The high percentage for plateau Baluchistan is accounted for by the Sardari system in which the owner chieftain directs the agricultural operations of 'his' serfs and claims possession of the product of their labours. The barani districts of Potwar and Hazara, with Kohat, show more than two-thirds of the land worked by the owners. It is in the canal irrigated plains that tenant operation is strong, reaching its highest levels in two-thirds of the districts of Sind, where the very high tenancy rate in 1947 has already been noted.

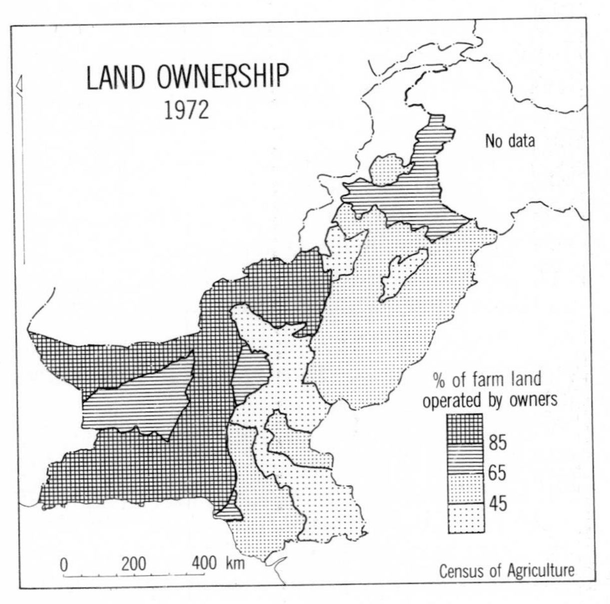

FIG. 10.6

On paper the land reform of 1972 was quite hard on the landowner, but since he had plenty of warning of what was coming, evasive action could be taken to blunt its severity. The core of the reform was to reduce ownership ceilings to 61 ha of irrigated and 122 ha of unirrigated land with concessions of up to 61 ha acres of irrigated land 122 ha unirrigated for family dependants, and a bonus equivalent to 10 ha of irrigated land for those who had purchased tractors or tube wells before December 1971 and who, presumably, had invested thereby in a capacity to farm efficiently more than the ceiling limits. Tenants were given enhanced protection and the landlord became responsible for land revenue taxes. Small holders (having below 5 ha irrigated or 10 ha unirrigated) were exempted from land revenue, while that of the larger owners had increased to double in the case of those with over 20 ha irrigated or 40 unirrigated land. In Baluchistan sardari rights were totally abolished, the sardars losing thereby their traditional powers to arrest, judge, and jail, and to exact tribute and free labour.

Up to December 1977 land resumed by government under the reform amounted to 1.4 million ha (much of it in the Punjab) of which 567,000 ha had been redistributed to 136,600 farmers at the expense of 2555 landlords. In 1977, before his fall from power, Mr Bhutto abolished land revenue entirely, substituting income tax, and cut the land ceilings still further to 40 ha of irrigated and 81 ha of un-irrigated land. The concessions regarding orchards and gifts to heirs had been removed in 1975.

It remains to be seen how energetically these recent land reform measures will be implemented. While they may not have a very great immediate impact on the present generation of landlords, it seems likely that large rural properties will become a thing of the past. Absentee landlords may well disappear as tenants flex their muscles and decline to pay up to the old levels of exaction. Farm units will become smaller, and typically there will be owner-operated medium sized holdings employing a little seasonal labour, and owner and tenant-operated smaller units making ends meet by selling their surplus labour to the larger farmers. Opportunities for casual employment by landless workers are likely to diminish with increasing mechanization.

Some idea of the variation in farm size within Pakistan is given in Fig. 10.7 of median farm size – that is the size category above and below which half the farms lie. The farms are smallest in the northern hills and Peshawar Vale, and in the well-irrigated, densely and long settled Punjab piedmont of Gujrat and Sialkot. Upper Sind, where tenancy has cut farm size to near the minimum is in the same category, with the median less than 5 ha. The irrigated plains otherwise show remarkable consistency with the median 5 to 10 ha with only the districts with appreciable non-canal irrigated land rising above that level – Sargodha, Mianwali (Thal), Dera Ghazi Khan and Sibi. In the western hills and Baluchistan (except for Kharan) farms probably have to be over 10 ha or even 20 ha or possibly even 60 ha to be able to support just a subsistence level of living.

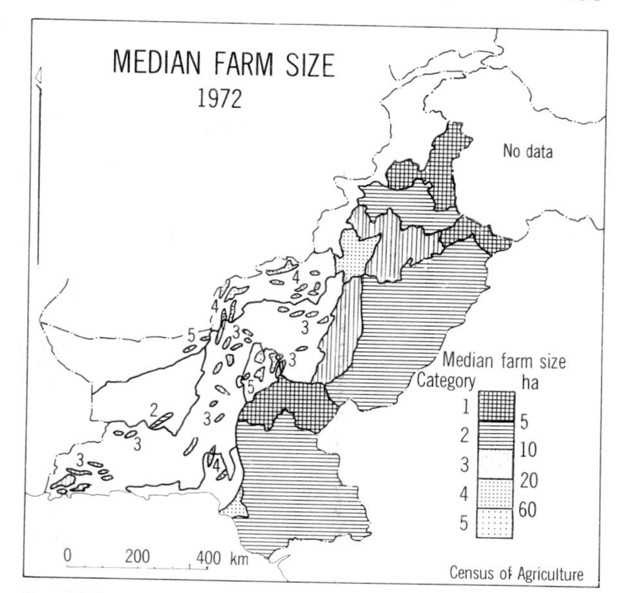

FIG. 10.7

CONSOLIDATION OF HOLDINGS

A common characteristic of peasant agriculture reinforced in Pakistan by the Muslim laws of inheritance whereby land is divided equally among the sons, is the extreme fragmentation of the land. Thus an operational farm may be in a number of scattered parcels within a village or even in several villages. Fig. 10.8 shows the percentage of farmland fragmented in four or more parts. The incidence of such fragmentation is more severe in the western hills, the barani tracts and the Punjab piedmont, all areas of long continuous settlement. In the plains not only has time been shorter for fragmentation to develop, but also a more modern outlook has helped expedite the process of consolidation, a process begun before independence. By 1974–75, 6.8 million ha had been consolidated so that a farmer's lands became contiguous. It is a necessarily meticulous operation, requiring patience and tact on the part of the administrators responsible. 0.4 million ha more were billed for consolidation in 1976–77 and the task will take many more years to complete.

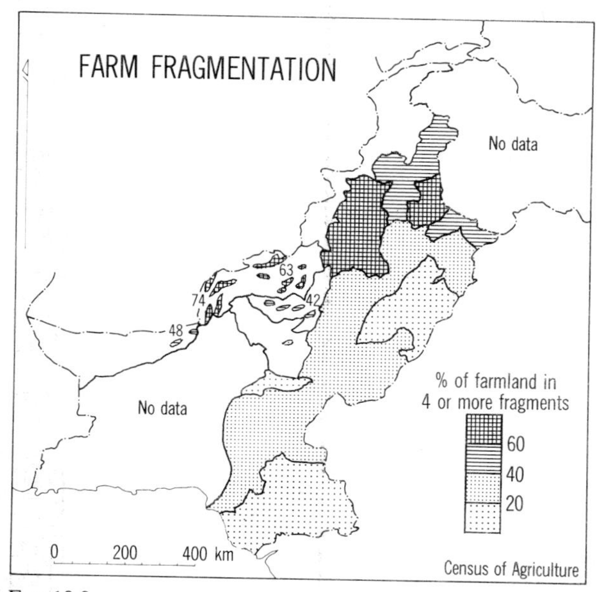

FIG. 10.8

INTEGRATED RURAL DEVELOPMENT

To recognize that all is not well with the complex organism that constitutes the rural economy together with all its social attributes is one thing. To prescribe and to administer a cure is quite another. From time to time efforts have been made to project into rural life plans for its general uplift or development usually thought out by theoreticians living outside the system. One early attempt to formulate a policy was made by F. L. Brayne in the years before and during the World War Two.* As an Indian Civil Service officer, Brayne had lived in various districts of the undivided Punjab and knew at first hand the conditions and the character of its people. His original advocacy of rural reconstruction well merits recognition, for subsequent schemes lean heavily on his ideas. Rural reconstruction is the regeneration and revivifying of the countryside by the people themselves working in co-operation with each other and with government. The objective is better villages and healthier, happier homes. The movement embraces every activity of village life and every villager from the cradle to the grave. Brayne explored every activity and comments on the needs and opportunities for improvement in every one.

After independence in the middle 1950s a major effort on similar lines was undertaken in the Village Agricultural and Industrial Development (AID) scheme, largely under US initiative and staffing of the training centres for Village AID personnel. Admirable in intent, the scheme seems to have foundered on two rocks: one was the quality of the village workers who tended to be college graduates with little real experience of village life and who could too easily find themselves in the situation of 'teaching their grandmothers to suck eggs'. The other was that the comprehensive developmental aspect of the scheme trod on too many administrative toes by urging self-reliance, and perhaps excessive boldness in the underprivileged in the semi-feudal society. At any rate the scope of Village AID was curtailed. Lack of co-operation among government ministries often frustrated the scheme. Somewhat similar ideas are now embodied in the Integrated Rural Development Programme which, to quote a World Bank report 'is the vehicle by which the government hopes to create a modernized countryside with greatly increased productivity per worker and thriving towns based on industrial and service activity'. The approach is at root more economic than social in its objectives, with the tacit assumption that all the things Brayne wrote of would follow. Basic to the programme is the establishment of 'marakaz', centres to serve several villages in order to

1. integrate all services to agriculture (extension, credit supplies and marketing);
2. to integrate government department programmes at the local level (Agriculture, Public Works, WAPDA, etc.);
3. to integrate villagers more fully into relationships with the rest of the economy through associations or co-operatives;
4. to institute a system of local government based on elected rural councils.

Pilot projects are working near Lahore and Peshawar. As was the case with the village AID programme, the risk is ever present that conservative forces within government or within the rural community will deliberately frustrate the efforts of the project leaders, for their programme is potentially revolutionary and radical to a degree. As the World Bank report puts it:

> To be successful the program must: provide access to credit, supplies and markets for all (especially small) farmers: free small farmers from exploitation by large farmers and from government officials such as the canal 'patwari'; build village and project area experience among farmers of ability to meet their needs through co-operation; bring about land improvement and consolidation of small holdings and perhaps co-operatively planned patterns of cultivation; strengthen the rights and security of tenants; and provide the institutional structure to make mechanization available to small farmers.

While one must hope against hope that such policies will prevail, the programme aims so directly at the institutional roots of rural inequalities that one fears that vested interests will succeed in crushing it as they did its forerunners. To tackle the social institutions that delay development is more difficult than to adopt economic policies. A recent article in Pakistan Times, 20 December 1977, comments that, 'economic policies are . . . seen as softer options than social policies. They help in by-passing the frustrations of challenging vested interests, violating deepseated inhibitions, offending cherished traditions and beliefs, and working against the heavy weight of social inertia.'

* 'Rural Reconstruction in the Punjab', in Chatterji, G. C. and Brayne, F. L., *The Punjab Past and Present*, Indian Science Congress Association, Lahore 1939, pp. 172–181.

CHAPTER ELEVEN

MANUFACTURING INDUSTRY: RESOURCES, DEVELOPMENT AND POTENTIAL

INTRODUCTION

Land, water and a warm climate are Pakistan's main natural resources, directly supporting a large proportion of the population. Before being used for irrigation some of the water is put to good purpose in generating hydroelectricity. The rocks of the country contain large reserves of natural gas and some liquid petroleum; both are exploited and the search for further productive structures actively continues. In other minerals Pakistan cannot be said to be richly endowed. There is some coal, in difficult structures and rather remote from industrial centres. Limestone and rock salt are reasonably abundant, and valuable deposits of chromite have long been mined for export. Exploration has identified workable ores of copper and substantial quantities of rock phosphate. Iron ore resources so far discovered are inaccessible and small (if of good quality), or abundant but low grade and remote from coal of adequate quality for an iron industry.

The products of agriculture are the main basis for industrial development. Most agricultural raw materials have to be processed into a form suitable either for consumption, export or further manufacture into useful articles. Thus there is a wide range of processing industries: flour and rice mills, cotton ginneries, oil mills, sugar mills, leather tanneries. Secondary manufacturing is dominated by the cotton textile industry, through which much foreign exchange is earned by exporting the value added by making yarn and cloth, rather than the raw cotton. Similarly leather goods are exported rather than raw hides and skins, and represent a better income for Pakistan's workforce.

Another group of industries supports the economy by producing substitutes for manufactures previously imported, and there is plenty of scope for further development in this area, using if necessary imported raw or semi-processed materials to be worked up or assembled by Pakistani labour. Of Pakistan's imports 57.5 per cent are machinery and other manufactured goods and commodities, including fertilizers. Starting from a very low position at the time of independence, when Pakistan inherited a meagre share of the industry of British India, manufacturing now contributes 14.4 per cent of Pakistan's Gross National Product, 10.9 per cent from large scale and 3.5 per cent from small-scale industry. This compares with 34.5 per cent from agriculture. Industry's share of GNP is much greater proportionately than its share of the economically active population for which it provides 13.25 per cent of employment compared with 54.3 per cent in agriculture. Its contribution to foreign trade is increasingly important, accounting for 45.3 per cent of the value of exports (leaving aside processed raw materials).

55 Hewing a thick coal seam in a mine near Quetta, Baluchistan.

56 Timber being weighed for charcoal burning in the kilns in the background, at Dera Ismail Khan, NWFP. The bark drying in the foreground will be used in leather tanning. Stones and a brick substitute for weights. Charcoal is widely used for domestic heating and cooking.

ENERGY RESOURCES AND DEVELOPMENT

Inanimate energy is an essential ingredient for industrialization and for development generally. As living standards improve the demand for energy increases, and rural electrification is an important part of any programme to reduce urban-rural disparities in the quality of life. It may be noted in passing that of 42,730 villages in Pakistan, only 7879 had been provided with electricity up to 1975–76, and a further 1000 were to be linked to the grid in 1976–77, bringing to 21 per cent the proportion of the 'haves' in this connection.*

In 1976–77 the major sources of energy were estimated to be as in Table 11.1 and the uses to which energy was put as in Table 11.2.

Natural Gas

Gas has been an enormous boon to Pakistan (Fig. 11.1). In 1955 it provided less than 1.2 per cent of the energy used, now it tops the list and will go higher. Proved reserves now total 4×10^{12} cubic metres, three-quarters of this in the Indus basin. Currently production runs at about 15.3 million m^3 per day (5577 million m^3 in 1976–77), and this could easily be doubled if the demand warranted.

*The provincial proportions by 1976–77 should become Punjab 19, Sind 38, NWFP 27 and Baluchistan 3 per cent. The great distances between settlements in sparsely populated Baluchistan will long delay electrification.

TABLE 11.1

Sources of energy, 1976–77 (percentage of total)

Natural gas	38.0
Oil	37.0
Hydroelectricity	18.0
Coal	5.0
Nuclear power	1.7
Low pressure gas	0.3

TABLE 11.2

Energy use, 1975–76 (Total 386.32 $\times$ 10^{12} kilojoules)

	per cent.
Industry	35.2 (42.3)
Transport	17.3
Electric power[a]	16.9
Fertilizers	7.9
Government	7.7
Domestic	7.4 (9.6)
Agriculture	5.3 (9.7)
Commerce	1.4
Bulk	0.9 (4.1)

[a]This list is of first users. Electricity is subsequently distributed among sectors thus:
Industry 42%
Agriculture 26%
Public lighting and bulk supplies 19%
Domestic and commerce 13%
Redistribution among these heads gives the proportions shown in brackets above.

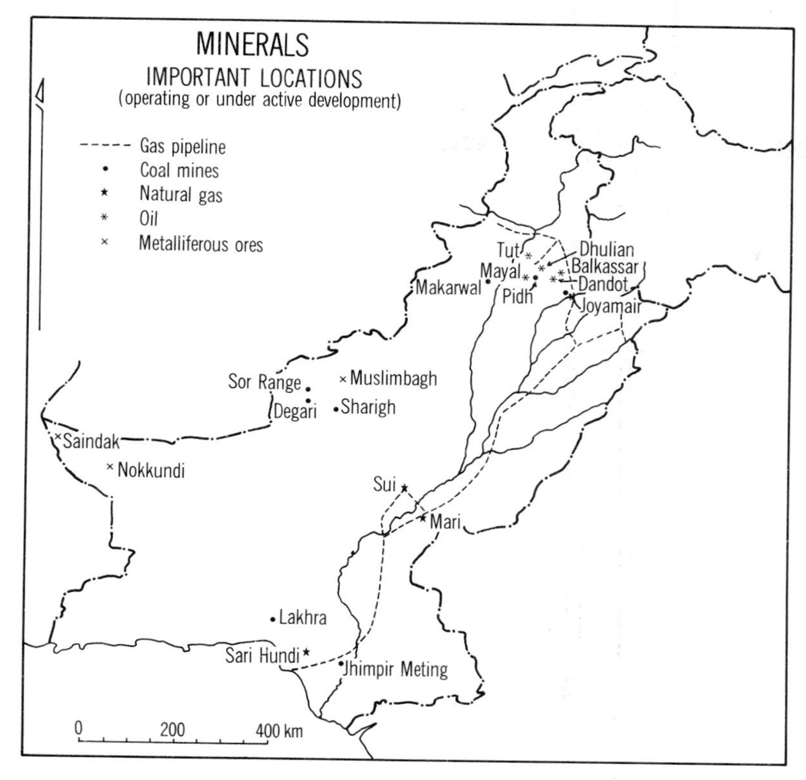

FIG. 11.1

The major gas field is at Sui in Sibi District where gas was first discovered in significant quantity. The output from Sui alone in 1976–77 was 4488 million m^3, equivalent to 5.43 million tonnes of furnace oil. The gas is piped the length of the country to Lahore and Peshawar District in the north, and to Karachi in the south. Important single users are the new gas-fired electricity station at Gudu, and the Multan fertilizer plant that uses the gas as feedstock. Close to Karachi the Sari-Hundi gas field is yielding 0.34 million m^3 daily for that city. The Mari field northeast of Sukkur produces 0.92 million m^3 daily to provide power and feedstock for the Daharki fertilizer plant at Sukkur and will also supply a plant being built at Mirpur Mathelo and another projected at Sadiqabad.

In addition to these purely gas fields, natural gas is also produced at the oil wells in the Potwar plateau. Dhulian and Mayal fields supply 1.67 million m^3 daily, and when the Tut field is linked to the pipeline the output will rise to 2.8 million m^3.

Several other gas fields are known but not yet exploited: Zir, Uch, Jacobabad, Khandkot and Mazarani in the Sibi re-entrant, and Khairpur on the left bank of the Indus south of Sukkur. Others include Kothar, Rodho and Pirkoh. Dhodak is a new find of natural gas and condensate in Potwar.

Gas of course needs to be piped to its market. Karachi is fed by the 40.6 cm Indus left-bank line and a new 45.7 cm line on the right bank. The pipe lines running north from Sui have had to be increased to meet continually expanding demand, particularly from large users like the Faisalabad gas turbine electricity station, and the Pak-Arab fertilizer plant at Multan. Overall natural gas utilization by sector in 1975–76 was as follows: power 28 per cent, general industry 28 per cent, fertilizer manufacture 22 per cent, cement 16 per cent, domestic use 4 per cent and commercial use 3 per cent.

Petroleum

There are hopes that intensified exploration may make Pakistan self-sufficient in liquid petroleum.

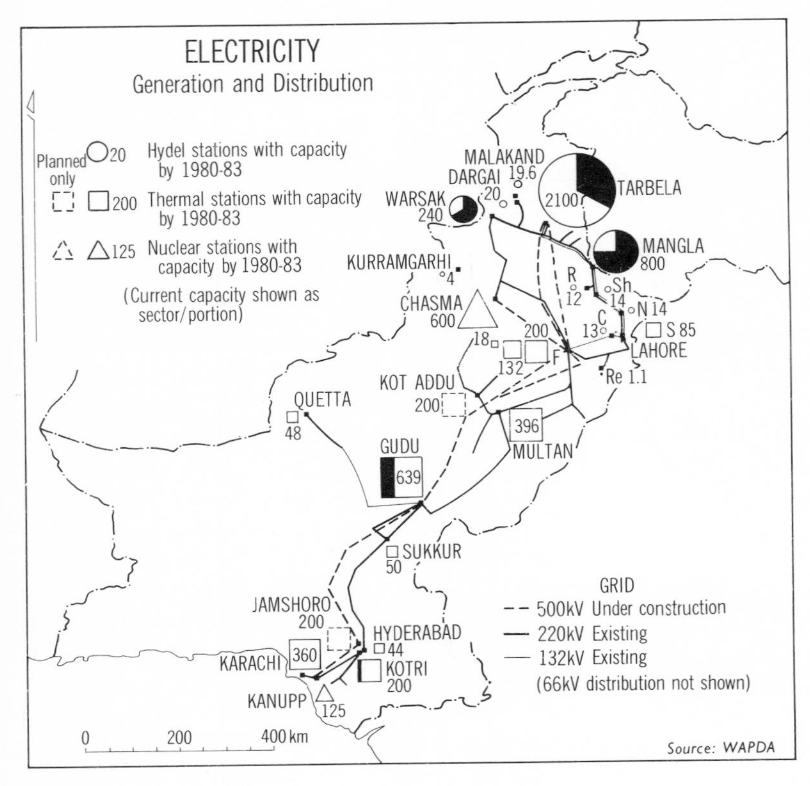

FIG. 11.2

Electricity generation and distribution: locations not named are: R., Rasul; Sh – Shadiwal; N – Nandipur; S – Shadara; C – Chichoki Mallian; Re – Renala; F – Faisalabad.

At present up to 15 per cent of needs are met from the several fields in Potwar: Dhulian, Joya Mair, Balkassar, Mayal and Tut. Output is around 13,000 barrels per day to which Dhodak will add up to 80,000 b/d by 1981. Imports totalled about 3.85 million tonnes of crude oil in 1975–76 which was refined in two refineries in Karachi. The National Refinery has a capacity of 2.1 million tonnes. A smaller unit near Rawalpindi handles one million tonnes and a new refinery to process 2 million tonnes piped from Karachi is being set up at Kot Adu, Multan.

Hydroelectricity

As a country in which most of the population live in semi-arid to arid plains, Pakistan may be regarded as fortunate in possessing some well-watered mountainous terrain in the north where hydroelectricity can be developed. Between 25,000 and 30,000 megawatts is the estimated potential, of which maybe 10,000 are utilizable in the sense of being concentrated at places not impossible to contemplate developing under foreseeable technological and economic conditions. Much of this country is however extremely inaccessible, and although it is its intense state of dissection that creates the potential, it also means considerable instability of slopes mantled in screes and this poses acute engineering problems.

To date 1567 MW have been harnessed but the total will rise to 2500 MW from the three largest producers by 1980 and when Tarbela is fully developed by 1983, hopefully, to about 3550 MW. The major stations (Fig. 11.2) have all been built since independence.

Warsak on the River Kabul, built by Canada under the Colombo Plan in 1960, is a run-of-the-river station, its output subject to the vagaries of seasonal flow. Its installed capacity of 160 MW is being increased to 240 MW.

Mangla, associated with the irrigation storage on the Jhelum was completed in 1967 and yields 600 MW at present, rising to 800 MW by 1980.

Tarbela dam on the Indus in well on the way to displacing Mangla as the keystone in the hydroelectric system. Its ultimate capacity is 2000 MW from 12 units of 175 MW each. Four such units (700 MW) are in operation (1978) and all are planned to be producing by 1983.

There are a number of smaller HEP stations mostly utilising concentrated falls in the upper reaches of the canal system where slopes are appreciable and consequently suffering similar variation in output as the farmers suffer in their receipt of water. Two such schemes had been developed before independence, at Renala (1.1 MW) 112 km from Lahore, and Malakand (19.6 MW), using water diverted through tunnels from the River Swat principally for irrigation purposes in Mardan District. Since independence the following stations have been built (in order of construction):

- Rasul (12 MW) on the Upper Jhelum canal in Gujrat;
- Dargai (20 MW) using the tail race from Malakand;
- Kurram Garhi (4 MW) on the Kurram near Bannu;
- Chichoki Mallian (13.2 MW–4 MW available in winter) on the Upper Chenab canal 30 km from Lahore;
- Shadiwal (13.5 MW–3 MW in winter) on the Upper Jhelum canal near Gujrat;
- Nandipur (13.8 MW) on the Upper Chenab canal near Gujranwala.

In addition several small units of up to 1 MW capacity have been set up in remote valleys where grid supplies are unlikely to penetrate, and where demand is small. Such units have been built in Gilgit, Hunza, Skardu and Chitral and others are planned in Swat, Dir, Hazara and Azad Kashmir. One large scheme is being considered on the Jhelum at Kohala near Muzaffarabad in Azad Kashmir where a tunnel through a wide incised meander loop in the river could yield 3760 MW at peak flow, and 306 MW at low water.

Coal

Output of coal was about 1.13 million tonnes in 1976–77. The coal is of mediocre quality. While there is scope for some expansion, reserves are not large and most energy needs can better be met from natural gas, oil or electricity. The estimates of reserves of Lower Tertiary lignitic to sub-bituminous coal range between 449 and 478 million tonnes. Coals of coking quality for metallurgical use without blending are lacking, and what is mined finds use in foundries and brick kilns, the latter consuming 90 per cent of the output, the rest going to ceramics, railway engine use, lime burning, and so on.

The coal mining industry is inevitably characterised by a large number of small operations, on account of the structural conditions. Multiple seams are rare, and the coals generally do not exceed 1.5 m in thickness. Adit mining from the outcrop in a scarp face is common practice.

Development is the responsibility of the Pakistan Mineral Development Corporation, which has mines at Makarwal (4.5 per cent of its total output) in the trans-Indus Salt Range west of Kalabagh, and in the Quetta region in Baluchistan, at Sor Range, (20 per cent), Degari (26 per cent), and Sharigh (8 per cent). Smaller mines operate in the main section of the Salt Range at Dandot in Jhelum District and Pidh, and this whole field east of the Indus is being investigated to establish its capacity for expanded output. In Sind the Lakhra lignite field 50 km north of Hyderabad is being surveyed with the possibility in view of supplying a thermal electric power station proposed for Jamshoro to generate 250 MW. This lignite might also become the basis for upgrading by desulphurisation, hydrogenation and distillation for various chemical products.

There are minor deposits of coal 120 km north of Karachi; at Jhimpur-Meting, in several parts of Baluchistan, NWFP and Azad Kashmir but they are unlikely to become of more than local small scale use. Currently interest focuses on the Sharigh field in mountainous country near Harnai on the branch railway line running north from Sibi. Here Pakistan's best quality coals are found which after washing can be blended with imported coal to make metallurgical coke. Plans are in hand to increase output to 100,000 tonnes of washed coal, most of which will be used by the Karachi Steel Mills now under construction at Pipri.

This rather extended discussion of small coal resources should not blind us to the fact that Pakistan has other more abundant sources of energy and will not have to regard coal as other than of marginal advantage, even to a resource-poor country.

ELECTRICITY GENERATION AND DISTRIBUTION (Fig. 11.2)

Pakistan is moving towards a unified power grid into which will be fed supplies from WAPDA's (Water and Power Development Authority) hydro- and thermal stations, those of the Karachi Electric-Supply Co., and of the Karachi Nuclear Power Project (KANUPP). The stations are:

- Faisalabad, Abdullahpur (18.2 MW): an old diesel station now standby for National grid;
- Faisalabad, Nishatabad (132 MW): gas-fired steam turbines;
- Faisalabad Gas Turbine (200 MW): quick start units;
- Shahdara Gas Turbine Lahore (85 MW);
- Multan Natural Gas (395.7 MW);
- Hyderabad Thermal (43.7 MW): steam and gas turbines;
- Sukkur Thermal (50 MW);
- Kotri Gas Turbine (30 MW; 80 MW by 1980; 130 in 1981; rising to 200 MW);
- Gudu Thermal (229 MW; 439 MW by 1980; later 639 MW);
- Quetta (48 MW): coal and gas turbine;
- Karachi Thermal Stations (360 MW): Karachi A and B, Landhi and McLeod Road stations;
- Karachi Nuclear Power Project (125 MW).

Major stations in prospect are Kot Addu Tehsil (200 MW) to use surplus furnace oil produced by 1986 from the Pak-Arab Oil Refinery planned for construction near Multan, and the Jamshoro (200 MW) station to use Lakhra lignite. A second nuclear power station with 600 MW capacity is to be located near Chasma barrage and become operational by 1983. Not all areas will be within economic reach of the ultimate grid, and many remote townships in Baluchistan – about 55 in all – are being equipped with diesel units and local high and low tension distribution lines.

By 1983 when Tarbela's full contribution should be available, it is expected that a 500 kV transmission system will link Tarbela to Karachi through Faisalabad, Multan, Gudu and Jamshoro. At present the country is divided into northern and southern grids, the 'power shed' lying just north of Gudu. The super-grid will allow the most advantageous switching of power from hydro- to thermal stations depending on the seasonal fluctuations in the former. Till then a 220 kV system links Tarbela, Mangla to Lahore and Faisalabad, and Gudu to Sibi, with 132 kV lines joining Faisalabad, Multan, Gudu and Hyderabad–Kotri. Karachi meanwhile remains a separate system.

57 Machine-cutting rock salt in the Khewra salt mine in the Salt Range.

MINERAL RESOURCES

Apart from the energy sources discussed above, Pakistan's major minerals of economic value as exports are chromite, which has been mined near Muslimbagh in Zhob Districts for many years, and copper, now being developed at Saindak in Chagai District, both in Baluchistan. The Saindak copper deposit near the Iranian border 640 km west of Quetta has been assessed as containing 342 million tonnes of ore containing over 1 million tonnes of metallic copper as well as gold and molybdenum. A plant to process ore yielding 15,000 tonnes of blister copper annually is thought feasible. At Nokkundi 150 km nearer to Quetta, iron ore reserves are being surveyed and may be suitable for a mini-steel plant with a capacity of 100 to 200,000 tonnes annually to be set up with Chinese assistance. A major constraint for all mineral development in Baluchistan is the difficulty of finding adequate water for thirsty industrial plants and their workforce. Rock salt is mined in the Salt Range mainly at Khewra west of the Indus, at Kalabagh, Warcha and other places along the scarp face.

Output of a miscellany of minerals in 1976–76 was as follows (in '000 tonnes, provisional figures):

Rock salt	341	Marble	36
Silica sand	59	China clay	7
Chromite	8	Fireclay	40
Gypsum	292	Fullers Earth	12
Limestone	3341	Magnesite	1

MANUFACTURING INDUSTRY

Like most underdeveloped countries Pakistan shows in its industrial structure clear evidence of the workings of a dual economy. There is a strong substructure of rural cottage and urban craft industries carrying on mainly traditional manufacture of home spun and woven fabrics in cotton, wool and silk, of carpets, footwear, wood and cane products, metal, pottery, ropes and cord. There has been no full survey of cottage industry since 1960 when it employed 151,000, probably a minimum figure for many rural dwellers occupy their 'spare' time in some kind of constructive handicrafts for home use, barter or sale. The number of workers employed in more formal factory industry is in excess of 555,000, the figure that can be approximated from the official lists of registered factories. Registration is not absolutely obligatory so an unknown number of industrial units and their workers may be left out of the count. The Economic Survey 1977–78 estimated there were two million workers in small scale industry and 300,000 in large scale industry, but these categories do not correspond to 'formal' and 'cottage' industry.

Pakistan entered independence with an extremely poor inheritance of manufacturing industries from British India. Successive governments have endeavoured in different, and ultimately sometimes mutually contradictory ways, to stimulate manufacturing as a way of creating work for large numbers of unemployed and underemployed, and in order to improve the nation's income and its ability to pay its way in the world. Early policies helped private enterprise, critics would say it amounted to 'feather-bedding', by the Pakistan Industrial Development Corporation sharing the burden of capital investment which the private entrepreneur could purchase when the concern became viable. The pendulum swung violently the

58 BECO bicycle factory, Lahore. The push-bike is the average Pakistani's first and only vehicle for self-propulsion, so the market is a huge one.

other way when soon after assuming power Mr Bhutto introduced in January 1972 measures to nationalize basic sectors of industry, even though in some instances there was little to nationalize. The ten corporations which are controlled under the Board of Industrial Management (B.I.M.) are as follows:

- Federal Chemical and Ceramics Corp.
- Federal Light Engineering Corp.
- National Fertilizer Corp.
- Pakistan Automobile Corp.
- Pakistan Industrial Development Corp.
- State Cement Corp.
- State Heavy Engineering and Machine Tools Corp.
- State Petroleum Refining and Petro-Chemical Corp.
- Pakistan Steel Mills Corporation
- National Design and Industrial Services Corp.

In July 1976 a further round of nationalization took place to try to ease the hold that middlemen in the agricultural processing industries were considered to have over the small farmers. Cotton ginning (555 units), rice husking (2072) and the larger scale flour mills (125) were taken over. In May the year following, the rice millers received a partial reprieve with the denationalization of 1523 mills. A policy of phased denationalization of much of industry was announced in October 1978.

Private industry receives some assistance from government through the Pakistan Industrial Credit and Investment Corporation and the Industrial Development Bank of Pakistan which provide credit facilities, share participation arrangements and the underwriting of public share issues. Government has assured industry that further nationalization would not take place, and in 1976 was trying to woo foreign investment by legislation specifically to protect it. In the atmosphere of 'wait-and-see' under Martial Law Administration following Mr Bhutto's downfall, the industrial investment scene is understandably sluggish. Table 11.3 shows the output of major industrial products in 1971–72 and in 1976–77. It will be noted that apart from industries making cotton yarn and cloth, cement, sugar and fertilizers, most industries are aimed at the low income market for bicycles, tyres, sewing machines, etc. The motor vehicle industry, so dominant in developed economies, together with the machine tool and other essential ancillary trades, are absent from the list, though some development is taking place. Table 11.4 showing the Quantum Index numbers for manufacturing since 1971–72 reveals the present depressed state of the key export industry, cotton textiles, and the very modest growth considering the time span involved, in several other areas. The most cheering aspect is the growth in fertilizer production.

In 1976–77 the installed capacity of the cotton textile industry was 3.49 million spindles of which 2.57 million (74 per cent) were working, and 29,000 looms, 21,000 (72 per cent) at work. Of the 441 million m² of cloth produced 15 per cent was of fine fabric, 56 per cent of medium, and 28 per cent coarse, with almost one per cent of blended cloth.

TABLE 11.3

Manufacturing industries output, 1971–72, 1975–76 and 1976–77

Item	*Unit*	*1971–72*	*1975–76*	*1976–77*
Cotton yarn	'000 tonnes	336	350	283
Cotton cloth	million m²	628	520	408
Jute textiles	'000 tonnes	30	42	34
Mild steel products	'000 tonnes	184[a]	231	270
Cement	'000 tonnes	2,876	3,093	3,090
Particle board (straw, paper, chip)	'000 tonnes	38	21	22
Paper	'000 tonnes	27	23	23
Fans	'000s	223	148	145
Electric bulbs	'000,000s	11	17	15
Gramophone records	'000s	999	1,082	n.a.
Bicycles	'000s	212	211	212
Sewing machines	'000s	66	64	58
Cycle tyres	'000s	2,542	3,180	3,461
Motor tyres	'000s	100	166	148
Sea salt	'000 tonnes	240	151	138
Sugar	'000 tonnes	375	630	736
Cigarettes	'000,000,000s	28	27	28
Fertilizers	'000 tonnes	573	833[b]	824
Soda ash	'000 tonnes	77	79	55
Caustic soda	'000 tonnes	34	38	24
Sulphuric acid	'000 tonnes	35	46	45
Paint	'000 litres	5,983	7,128	7,193

[a] 1972–73.

[b] Urea 605, Superphosphate 59, Ammonium Sulphate 98, Ammonium Nitrate 71 (all '000 tonnes).

Source: *Monthly Statistical Bulletin*, May–June 1977, and *Pakistan Economic Survey*, 1977–78.

DISTRIBUTION OF INDUSTRY

Fig. 11.3 shows the distribution of 554, 895 factory workers by districts as estimated from the register of 1976, with the proportion in cotton textiles, the largest group accounting for almost 39 per cent of the total. This may be read in conjunction with Table 11.5 showing the breakdown of the total into 15 major industries for the seven districts having more than 17,000 workers. These seven contain the major industrial cities, and together account for 68.5 per cent of the total industrial work force.

The basic pattern, as Fig. 11.3 shows, is bi-polar, except that the southern end of the axis is heavily dominated by Karachi with Hyderabad alone in significant support. At the northern end Lahore is the largest, but not far behind in size are Faisalabad and Multan. Twelve others in the north and northwest have over 5000 workers, while in Sind only four in addition to Karachi and Hyderabad, reach this level.

The proportion of workers in textiles varies widely, from nil in six districts to over 60 per cent in nine. The seven leading districts show interesting differences. Faisalabad and Multan are heavily dominated by textiles, Karachi and Hyderabad to a lesser degree. In the latter the concentration of the tobacco industry brings its workers above those in textiles. In Karachi the industrial mix outside textiles is quite diverse, food and a variety of metal and engineering industries each having in excess of 10,000 workers. In most industries, in fact, Karachi's workforce exceeds that in any of the other six with the exception of footwear, iron and steel, engineering and electrical goods. Lahore has the best balance reflecting its maturity as the former capital of the British Punjab. Iron working, engineering, some textiles, printing, chemicals, food and footwear all exceed 3 per cent of the total.

Faisalabad is an important milling centre and makes agricultural machinery, while Multan has a large concentration in cotton ginning. Sialkot

TABLE 11.4
Quantum index numbers of manufacturing (1969–70 = 100)

	Manufacturing overall	*Cotton cloth*	*Cotton yarn*	*Cement*	*Cigarettes*	*Fertilizers*
1971–72	106	104	123	98	97	160
1972–73	115	97	138	108	124	201
1973–74	122	98	139	118	123	218
1974–75	121	92	129	125	120	227
1975–76	120	86	128	120	123	235
1976–77[a]	117	68	104	116	127	234
1977–78[b]	123					

Source: Monthly Statistical Bulletin.
[a] Calculated from Pakistan Economic Survey 1977–78.
[b] Preliminary estimate, June 1978, for manufacturing overall only.

59 Interior of the Machine Tools Factory at Taxila, near Rawalpindi. It provides the finely engineered parts and builds machines for a variety of manufacturing industries.

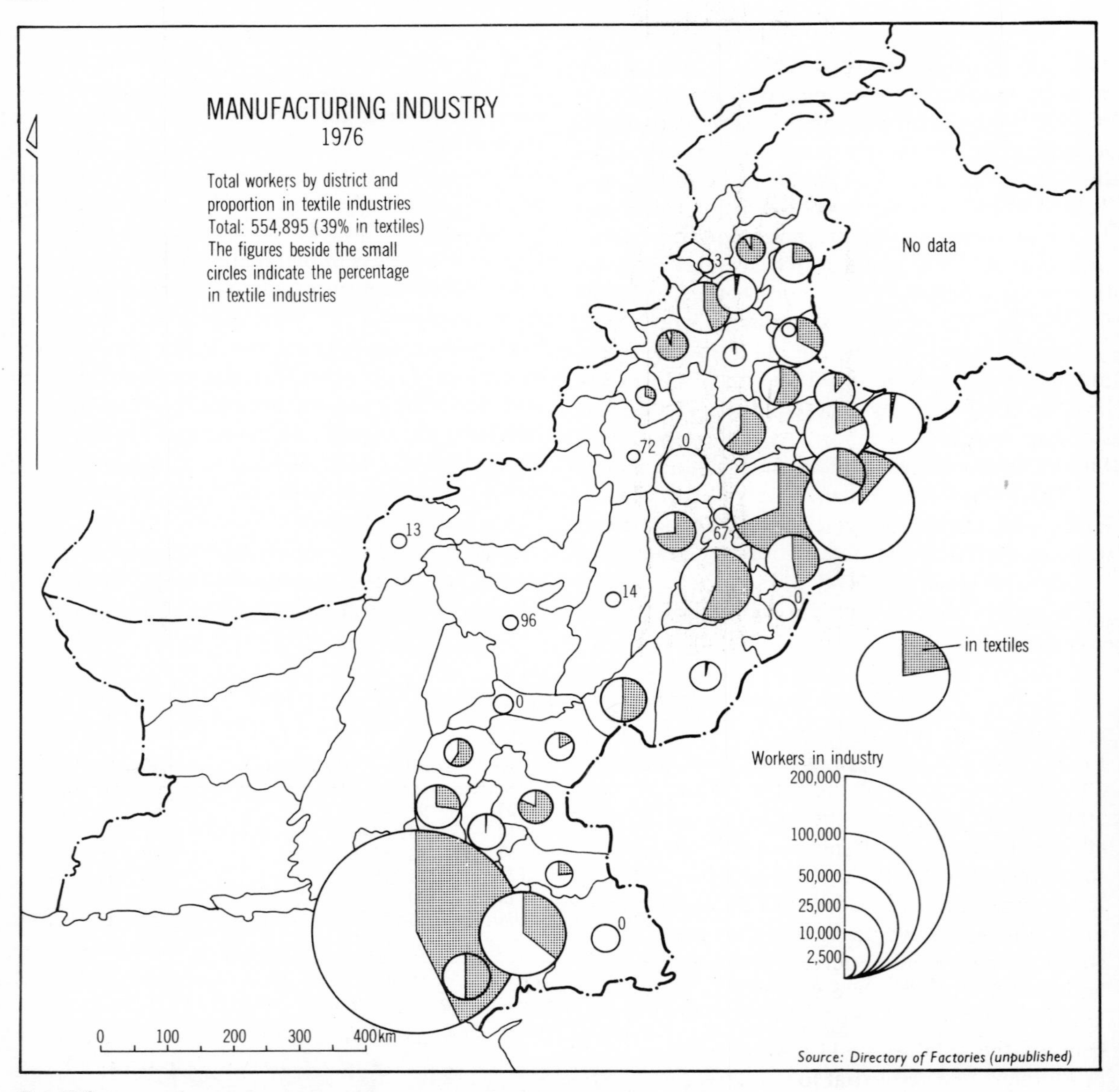

FIG. 11.3

is unlike the rest in its small textile sector but has world renown for its stainless steel surgical instruments and its sports equipment. Gujranwala spreads its interests between the metal engineering and electrical goods industries, and textiles particularly bleaching, dyeing and finishing. The somewhat atypical industrial mix of Sialkot and Gujranwala may well reflect their relatively rapid development after independence following the influx of refugees bringing with them a variety of skills and experience. The same is true of Karachi to a degree, though it already had much of such industry, particularly textile, as existed at the time of partition.

The 1970–71 *Census of Manufacturing* found 427,411 workers in 3549 factories. There seems no reason to believe that the size distribution of industrial units will have changed much since then. The factories with less than 100 workers numbered then 2952, 83 per cent of the total, but

TABLE 11.5
Major industrial groups in seven major industrial cities, 1976 ('000s and %)

strial group	*Karachi*		*Lahore*		*Faisalabad*		*Hyderabad*		*Multan*		*Sialkot*		*Gujranwala*	
d and drink	18.0	9.3	2.9	5.3	6.1	15.7	3.4	10.2	2.9	11.8	0.1	0.5	2.1	12.6
acco	3.6	1.9	0.2	0.4	–	–	13.0	39.0	–	–	–	–	–	–
iles and carpets	83.6	43.0	6.2	11.2	26.9	69.5	12.2	36.4	13.6	55.7	0.3	2.0	3.3	19.2
her	2.7	1.4	0.5	0.9	–	–	0.4	1.0	0.5	1.9	t	0.3	0.1	0.7
twear	1.4	0.7	1.7	3.0	–	–	0.3	0.9	–	–	0.3	1.6	–	–
od working	1.3	0.7	0.7	1.3	0.3	0.6	t	t	t	t	–	–	0.1	0.5
ting, etc.	6.0	3.1	2.8	5.1	0.2	0.6	0.2	0.4	t	0.1	–	–	–	–
ical, chemical	8.0	4.1	3.1	5.7	0.9	2.3	0.1	0.2	1.5	6.3	–	–	0.1	0.5
ber	3.4	1.8	1.8	3.3	0.1	0.3	–	–	–	–	0.3	1.7	0.3	1.7
& steel, foundries, lling mills	10.3	5.3	8.8	16.0	0.3	0.7	t	0.1	t	0.2	t	0.2	0.5	2.9
al working	10.0	5.2	1.8	3.2	t	t	0.1	0.2	0.4	1.6	1.8	10.1	6.4	37.4
ineering	3.5	1.8	5.7	10.3	2.9	7.4	0.2	0.5	0.4	1.6	0.6	3.5	1.5	8.7
trical goods	4.5	2.3	6.0	11.0	t	0.1	0.1	0.3	–	–	t	t	1.8	10.4
nsport equipment	10.1	5.2	4.8	8.8	t	t	t	0.1	0.1	0.3	0.1	0.5	0.2	1.2
ision equipment	1.1	0.6	0.7	1.3	t	0.1	–	–	–	–	3.1	18.0	–	–
er specialities									4.5	18.4[a]	1.0	5.5[b]		
otal (inc. others)	194.5	100.0	55.0	100.0	38.8	100.0	33.4	100.0	24.4	100.0	17.4	100.0	17.0	100.0

< 0.1% or < 0.1 total.
Cotton ginning.
Sports equipment.

accounted for only 19.25 per cent of the workforce. Factories of over 100 workers numbered 597, or 17 per cent of the total, and those over 500, 179 or 5 per cent; these size ranges covered 80.75 per cent and 60.16 per cent of the total workforce respectively. If the cottage industrial work force were added (almost entirely in very small units) the scales would be tilted, as the under 50 worker units would increase enormously, perhaps ten fold, to become 90 per cent of the total and the workforce in such units would be 40 per cent of the total. Fig. 11.4 shows some of the more important industrial locations.

COTTAGE AND SMALL-SCALE INDUSTRY

The 1960 Census of cottage industry gave the following breakdown of the total estimated workforce of 150,953 into major groups of which 42 per cent were rural and 58 per cent urban:

Handloom, cotton textiles	87,864 (58%)
Handloom, woollen textiles	688 (0.5)
Handloom silk textiles	1,899 (1.3)
Carpets	3,650 (2.4)
Leather and footwear	6,626 (4.4)
Wood, cane, bamboo, etc.	14,756 (9.8)
Metal	5,020 (3.3)
Pottery	5,316 (3.5)
Ropes, cord, etc.	10,864 (7.2)

The range of products is closely related to local raw materials, local demand and tradition, and in several cases to the local specialization in the organized factory sector.

Thus the making of reed screens or 'chatais', so essential a part of domestic furniture where privacy is important yet outdoor living and sleeping in the hot weather is preferred, is a big craft industry particularly in Dera Ismail Khan, Muzaffargarh, Kalat and Makran. Village needs for rough earthen pots for cooking and storage are met by the local potter, shoes are made from local materials in local designs for local use, and simple metal-ware such as lamps, cooking vessels, milk churns, etc. are made by the local craftsman from materials brought in from the city. The cottage metal workers are most numerous however in the shadow, as it were, of the factory industry to which they relate.

60 The modern Fauji Textile Mills at Jhelum, Punjab, designed with north lights to minimise sunshine entering the factory. A shed for workers' cycles is seen to the left of the minaret of a small mosque in the foreground.

FIG. 11.4

The industrial specialities of the centres named on the map (other than those already discussed and tabulated) and of others not named on the map (marked * with district in brackets) are:

Saidu Sharif	silk textiles
Mardan	sugar
Charsadda	sugar
Kohat	sugar
Jauharabad	sugar
Khanpur	sugar
Larkana	sugar
Nawabshah	sugar
* Mirpur Khas (Thar Parkar)	sugar
* Tando Mohammed Khan (Hyderabad)	sugar
* Chistain (Bahawalpur)	sugar
* Leiah (Muzaffargarh)	sugar
* Darya Khan (Mianwali)	
Bannu	woollens and sugar
* Shahdadpur (near Sanghar)	sugar
* Takht-i-Bhai (Peshawar)	sugar
Jaranwala	jute, sulphuric acid and sugar
* Tando Allahyar (Hyderabad)	sugar
Sargodha	sugar
* Rahwali (Guiranwala)	sugar
* Gujrat	china ware
Lala Musa	sugar
* Sibi (Sibi)	tanning
Sukkur	sugar
Taxila	cement
Wah	cement
Daudkhel	cement chemicals
Khewra, Dandot	cement soda, ash and salt
Jhelum	cement
Rohri	cement
Karachi	cement paper, china and tanning
Kotri	wool, jute
* Thatta (Thatta)	wool, paper
* Muzaffargarh (Muzaffargarh)	wool, cement
Harnai:	woollens
Jhang	woollens
Rawalpindi	woollens, silk
* Campbellpur	woollens (Attock)
Quaidabad	woollens
Nowshera	paper, drugs
Sheikhupura	drugs
Okara	cotton
Burewala	drugs
Rahimgar Khan	drugs
Khairpur	drugs
Liaquatabad	drugs

61 In the shadow of a minaret of the Badshahi Mosque, Lahore, cotton piece lengths are being dried after bleaching and dyeing. Two pairs of men can be seen airing lengths of cloth before final folding. In the background is a sports field with cricket pitch and volleyball court marked out.

62 Making reed screens, or chatais, for shading verandahs is a widespread craft industry, seen here on the outskirts of Lahore.

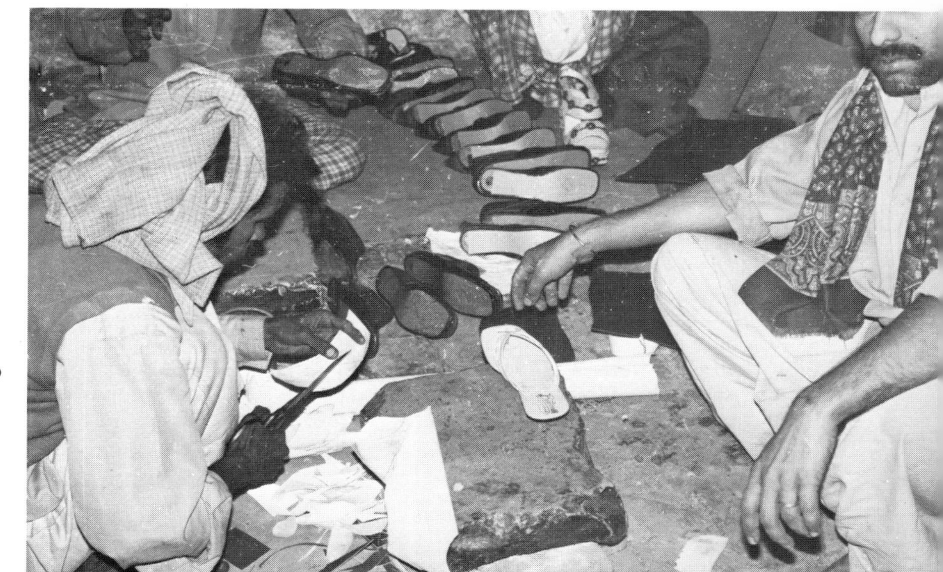

63 Making ladies' shoes in a workshop in the Old City, Lahore. Although the craft is still organised in small units the designers follow the latest fashions.

64 This Punjabi weaver powers his hand-loom by pressing down with his foot to throw the shuttle between the threads he separates with his hands. He is making a patterned cotton durri, a type of light cotton carpet.

65 The charpoy is the basic piece of furniture, in the Pakistani home. In this openair workshop in the Old City, Lahore, a charpoy is being strung. Finished charpoys and stools can be seen behind.

66 In the NWFP it is common for a Pathan to go about armed with a rifle or shotgun. The rifles on the wall are exact replicas of British Lee-Enfield .303 army weapons of World War I and II vintage. The village of Durra, between Peshawar and Kohat, specialises in making guns.

67 Brick-making in the Punjab. The donkey-cart is being loaded from a recent firing. Local clay and firewood are used. The overseer's dignity and comfort are assured by his chair, the donkey's by his sack of chaff.

Thus Sialkot, Gujranwala and Karachi had each over 780 craft workers in metal. Carpets are more for sale outside the immediate community; Sheikhupura, Gujranwala, Lahore and Karachi are outstanding in this craft. Like the metal workers, the handloom industry clusters near its factory counterpart. Faisalabad district had 25,000, Multan 12,000 and Gujranwala 10,000 cotton weavers.

The problem of protecting the cottage industry section against the uneven competition from the factory sector is ever present and it appears that many workers are drifting into unemployment or to seek jobs in urban factory industries. To the extent that cottage industry produces exportable articles of unique design it can be assisted to find markets and will not suffer unduly from large-scale competition. The future for small-scale industry generally is probably away from the traditional crafts except where these cannot readily be replicated by factories (e.g. chatai making) and into modernized production appropriate to the small entrepreneur. One cannot however easily translate village cobblers and tinsmiths into their sophisticated factory equivalents. Rather do small scale entrepreneurs come up from the ranks of organized industry when men of initiative see opportunities to do better for themselves on their own. There are Small Scale Industries Corporations and Boards under Provincial Government auspices trying to encourage initiatives of this kind by providing industrial estates in which basic services are established, and by giving guidance, credit and training where necessary. At the small town level, small entrepreneurs may be able to use the skills of the cottage workers in more modernized units and so help their adjustment to a higher economy.

The value of production in the small-scale industries sector in 1974–75 was estimated as follows (in Rs. millions):

Carpets	456
Sports goods	205
Ready-made garments	245
Surgical instruments	129
Cutlery	23
Handicrafts	105

Compared with the large-scale sector, small industries probably accounted for about 30 per cent of the value added in the textile trades, 15–20 per cent in engineering and 15–20 per cent in the food industry. About one-fifth of cotton cloth output came from the non-mill sector.

INDUSTRIAL DEVELOPMENT

To round off this survey of the state of development of manufacturing by looking at the other end of the industrial spectrum, mention of some of the major projects in hand under Federal government auspices indicates important growth points.

Several large fertilizer projects will greatly enhance Pakistan's ability to support its own agricultural needs in this essential input for the Green Revolution. Some OPEC names are evident in these enterprises which require large capital investment. Pak-Saudi Fertilizers at Mirpur Mathelo in Sind to produce 587,000 tonnes of urea will start in 1978. Pak-Arab Fertilizers at Multan will add to the several plants there. At Haripur the Hazara fertilizer complex will make urea and phosphates. These plants are in addition to several which have been in production for several years at Daudkhel (Mianwali), Faisalabad and nearby Jaranwala, Multan, Sukkur and Jhelum. Natural gas is the feedstock at most of them.

Cement has been a successful development for Pakistan which can use abundant sources of limestone and gypsum. New plants, at Kohat and the Pak-Iran Cement works in Baluchistan are to be added to the list which already includes Daudkhel, Wah and Hattar near Rawalpindi, Jhelum, Rohri, Hyderabad and Karachi.

Expansion in the textile industry is imminent with the two Pak-Iran plants at Lasbela and Bolan in Baluchistan, each with 50,000 spindles and 1100 looms. Another at Shahdad Kot in Larkana District has 25,000 spindles and 550 looms, and with Chinese aid a yarn mill is to be built at Tarbela.

Oil-refinery capacity has now reached 5 million tonnes with the completion of additional plant at the National Refinery at Korangi, Karachi, bringing the Karachi total to 4.5 million tonnes. The refinery near Rawalpindi handles 0.56 million tonnes.

68 Cow-dung cakes are used here to fire the shrinking-on process by which an iron hoop is fitted to a wooden wheel. A solid rubber tyre will be fixed to the hoop. In the background a joiner is making a wheel.

69 A stonemason puts the finishing touches to a marble screen. Memorial plaques and grave stones are inscribed here also (Old City, Lahore).

Two interesting developments for the consumer market are the mass production of wholesome roti (chapattis) in 16 factories in the major cities, and the setting up of three ready-made garment works in Karachi, Lahore and Peshawar. These are signs of modernization indeed, for these industries will invade the preserve of the small bread-maker and the man who sits with his sewing machine in a shop one metre by two in size.

In heavy industry, Taxila already has a machine tools factory and will soon inaugurate a Chinese-built heavy foundry and forge with capacity to make special steel castings (6500 tonnes), steel ingots (38,000 tonnes), iron castings (5000 tonnes), alloy press forgings (4600 tonnes) non ferrous castings (50 tonnes) and forged balls of carbon steel (2500 tonnes).

Finally, the Pakistan Steel Mills Corporation, with technical and financial assistance from USSR, is having built at Pipri, about 40 km east of Karachi on Gharo Creek, the Karachi Steel Mill, the first integrated iron and steel plant in the country. The site layout for the plant and its associated Port Qasim is shown in Fig. 11.5. The tidal creek is being deepened. By 1983–84 it is expected that the steel mill will be producing 1.032 million tonnes of steel billets, hot and cold rolled sheet, galvanized sheet and formed sections. Pig iron and coke will also be made in excess of the mill's own needs. Apart from some 75,000 tonnes of Sharigh coal for blending, dolomite from Jhimpir-Meting in Sind and electricity from the grid, coking coal and iron ore will have to be imported. Nevertheless it is hoped to save Rs. 2255 million of imported steel and to give direct employment to 15,000 workers. In addition, the spread effects might give work for 200,000 more.

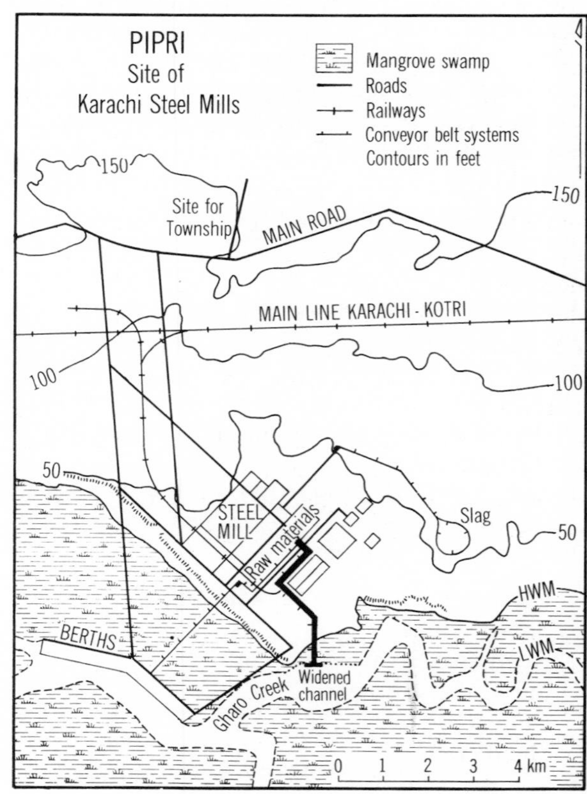

Fig. 11.5

CHAPTER TWELVE

THE SPREAD AND CONCENTRATION OF THE POPULATION

SUMMARY

A study of the changing pattern of population density since 1881 provides an overview and a perspective of development in Pakistan. The impact of canal colonization is seen in the changes in the relative ability of districts to support an agricultural population, and in recent decades the process of urbanization has further altered the pattern. Urban growth and the distribution of towns are other indicators of progressive development associated with the considerable expansion of the secondary and tertiary sectors of the economy. As the basically self-sufficient structure of traditional rural life with small towns and a few cities providing for the trading, administrative and cultural needs of the population gave place to a more fully commercialized cash economy the need for urban centres expanded, and with it the necessity of linking the increasingly interdependent parts of the economic, social and political system, joining town to countryside, city to city and region to region. The patterns of communication provided by road, rail, pipe-line and air demonstrate some of the fundamental problems of spatial integration in a country where the areal distribution of its population is constrained by the physical limitations of water and cultivable land.

POPULATION DENSITY

Five maps (Figs. 12.1–12.5) show population density by districts for the present area of Pakistan in 1881, 1901, 1921, 1951 (the first census following independence) and 1972 (the latest census). The same density values have been used throughout the series to make comparision easier, the only variation being the addition of an upper grade for 1972. In the discussion the names of present districts are used even though not all have been in existence all through the period. Some changes in the boundaries of districts have occurred usually on account of a large district being subdivided, a part keeping the original name while a new name appeared for the other. Because of this some areas may appear to drop in density from one census to the next but this would almost certainly be due to statistical accident of grouping and not to any real reduction of density at the village level.

In 1881 the barani lands of the northwest and the submontane Punjab, in a belt extending out into the plains as far as Lahore and then northwest through the Potwar plateau to Peshawar was the most closely settled. Sialkot just failed to top 200/km^2 and Gujrat had 130/km^2, both, as today, districts where shallow well irrigation supplemented a reasonably high rainfall. For the rest of the belt densities ranged between 50 and 100/km^2. Probably it was the presence of the 'small' city of Lahore with almost 150,000 that brought the district as a whole to the top of this density range. In the pattern of the rest of the Punjab plains some persistent features were already apparent. The Sutlej forms a boundary between a more heavily populated right bank (25–49 per km^2) and the left bank districts of former Bahawalpur State on the edge of the Thar Desert with under 25/km^2. To the west Mianwali containing the Thal tract with Dera Ismail Khan across the Indus from it had less than 25/km^2. In the central Punjab however there was a remarkable sameness in the range 25–49/km^2. Inundation canals were an important feature in the lower Punjab districts compensating them for the lesser rainfall in the south compared with the north. There was thus an equalizing effect. In Sind

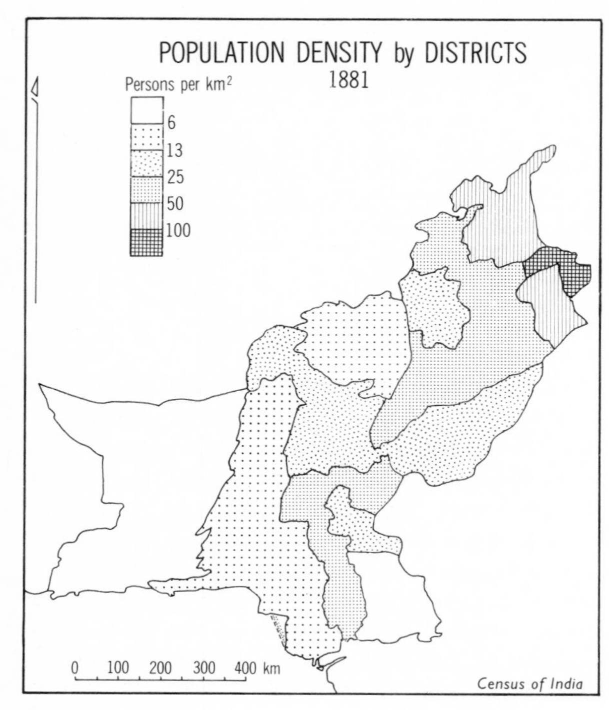

Fig. 12.1

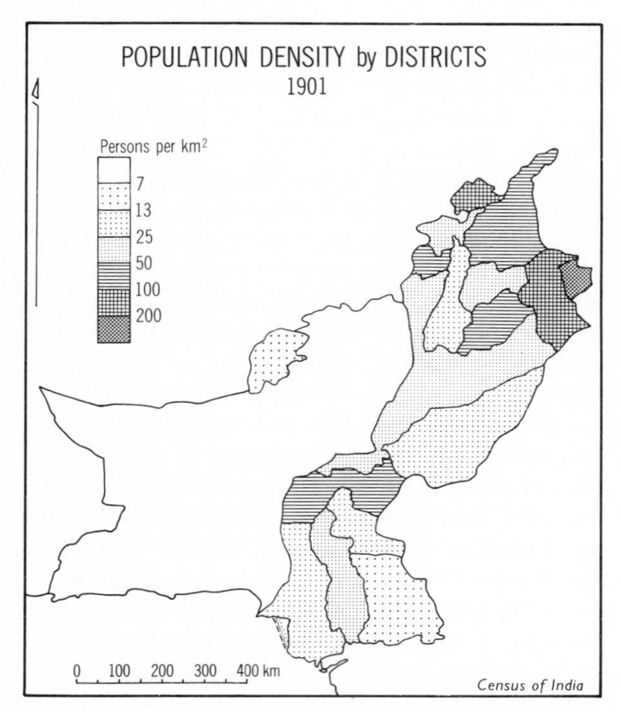

Fig. 12.2

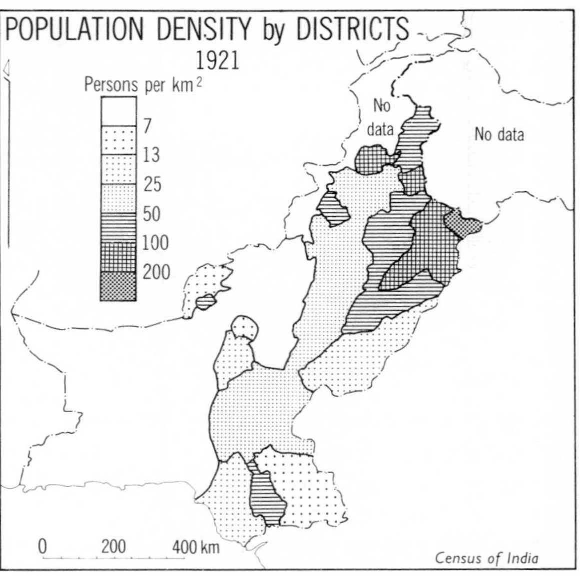

Fig. 12.3

only the immediate riverain belt from Sukkur through Nawabshah to Hyderabad compared with the central Punjab. Inundation irrigation canals and associated water raising mechanisms were essential here. Sanghar and Tharparkar were still desert country, and on the right bank of the Indus, Dadu and Thatta were only a shade better populated. At this date Karachi was of significance only to Lower Sind and had a population of 74,000. Westwards of the Indus plains only Quetta, already of strategic importance, had a density in the middle range. The remainder of upland Baluchistan fall into the lowest two categories of density. 1881 was the first occasion a census was attempted in Baluchistan, and it was admittedly more of an estimate than a head count. It will be noted nonetheless that throughout the series these Baluchistan districts retain their status in the lowest grades. Even a doubling of a low density like 2 or 3/km² fails to move it into the next higher category.

By 1901 colonization of the Punjab by perennial canals had begun. In the Chenab Colony 56 per cent of the population were immigrants, the extreme situation being in Jhang district where 83 per cent had come from other districts, mostly from the overcrowded piedmont areas of Sialkot and the belt running east into Amritsar, Jullunder, etc. On Fig. 12.2 Jhang stands out from the rest of the central Punjab. Lahore as capital of the emerg-

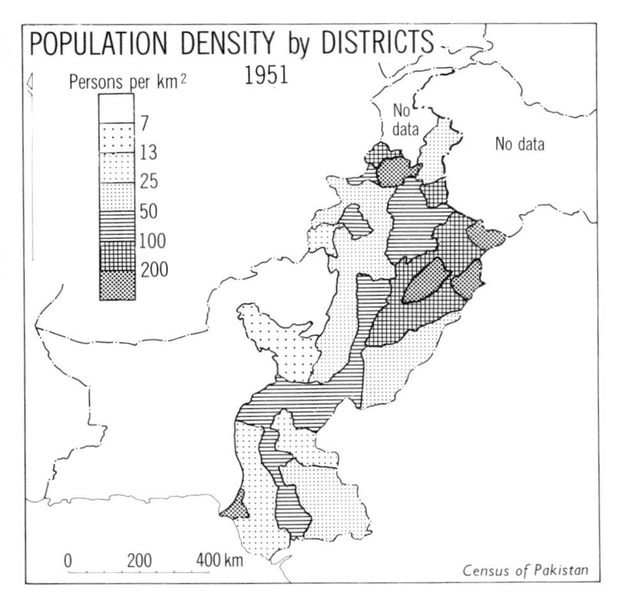

FIG. 12.4

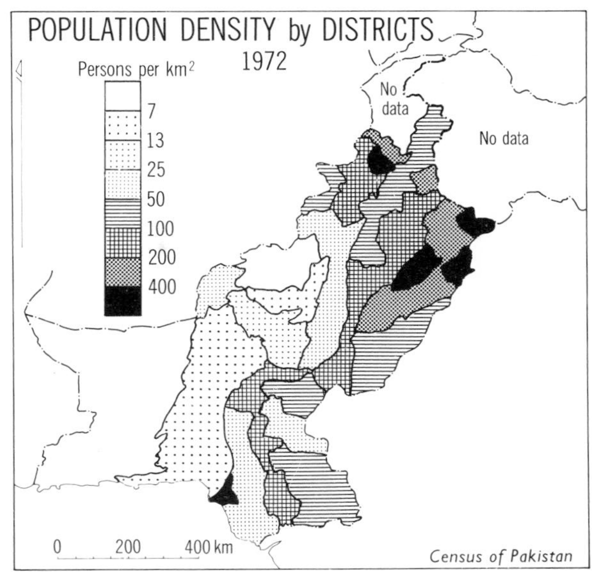

FIG. 12.5

ent Punjab on the threshold of boom had grown 36 per cent to 203,000 since 1881. The Punjab piedmont zone from Lahore to Gujrat rose one step up the density ladder to exceed 100/km². Sialkot, despite its losses to the Chenab Colony breaking into the 200/km² category, was still ahead of all others. Peshawar at the western end of the barani belt also rose a step, but the differentiation of Potwar and Hazara from the districts able to exploit irrigation had begun. Bahawalpur and Mianwali remain the the sub 25/km² class.

Like Jhang, the Upper Sind districts Sukkur, Larkana and Nawabshah advanced a step, leaving Hyderabad below 50/km². Sanghar and Tharparkar achieved a rise to over 7/km² while Dadu, Thatta and Karachi by rising above 13/km² became more distinctively different from the Baluchistan districts beyond.

The two decades from 1901 to 1921 saw much development of perennial irrigation in the central Punjab which now distinguished itself from most of Sind and from its less fortunate flanking districts. The piedmont remained in the same category as in 1901 as did Peshawar and most of the barani belt with the exception of Rawalpindi, which no doubt under the stimulus of the Great War (1914–18) had increased its population to reach a density of 114/km². Jhang remained at its 1901 level, but was joined in the 50–99/km² range by Jhelum, Sargodha, Multan and Sahiwal. Faisalabad went one step further to 122/km². These advances are all the more impressive when it is remembered that the world wide influenza epidemic had struck hard at the intercensal natural increase between 1911 and 1921. The Indus valley districts now formed a continuous belt with 25–49/km² from Attock and Kohat through Dera Ismail Khan and Mianwali to Dera Ghazi Khan and Muzaffargarh, joining the districts in upper Sind. These areas at best still had only seasonal irrigation. Hyderabad by a small change in twenty years rose by a grade. (There were some changes in district areas in the period which distorts the picture somewhat. Fluctuations in density were in fact slight but occurred around the 50/km² boundary). In Baluchistan Quetta continued to prosper.

In the 30 years from 1921 to the first post-independence census of 1951 probably the single most significant change that occurred was the completion of the Sukkur barrage in 1930 giving perennial irrigation and more certainty of kharif supplies to a large area of Sind. In the Punjab elaboration of the existing canal system continued. Population increase had been at a rising rate so it is to be expected that most districts would have advanced. It is interesting to note the laggards, for these are the areas where the ceiling to development seemed to have been reached for the time being at any rate. Hazara suffered a set back, while Kohat, Mianwali, and the two Deras remained almost static. Upper Sind and the Indus banks went ahead, as did Sanghar and Tharparkar but at a lower level. Dadu and Thatta revealed their limits to growth staying at the same level, but Karachi which suddenly became the capital and the nation's sole port naturally raced ahead. Urbanization had by 1951 become a more substantial factor in raising district levels of density above the general run. Thus, apart from Karachi which is almost entirely an urban district, Lahore, Faisalabad and Peshawar are with Sialkot in the top category with over 200/km^2. The influx of refugees to the Punjab cities was an important factor. The changes over the next 21 years to 1972 were not generally out of the normal run due to natural increase. The move of the national capital to Rawalpindi and then to Islamabad brought that district into the over 400/km^2 class which now had a solid block of districts in the highly developed east-central Punjab from Sialkot-Gujrat in the piedmont to Multan. Improvements in canal irrigation following completion of new barrages on the Indus helped Muzaffargarh and Rahimyar Khan in the Punjab and enabled Sind to more than keep pace with its natural increase.

Fig. 12.6 shows the 1972 density pattern for tehsils and clearly indicates the major urban centres. In Baluchistan the effect of the scale change merely accentuates the near emptiness of vast areas.

A sharper focus on change since independence is provided in Fig. 12.7 where an attempt is made to show simultaneously the percentage change in a district's population and the size of the population involved. The strong stimulus to growth in Sind is marked. As was mentioned in Chapter 2, some over-enumeration of population for political reasons probably took place in Sind but corrected data are not yet available and the map uses the census of 1972 as it stands. Urban growth was strong in the four Punjab districts with growth between 90 and 119 per cent. Multan, Lahore, Gujranwala and Faisalabad. The comparative failures merit attention. Sialkot, Sahiwal and

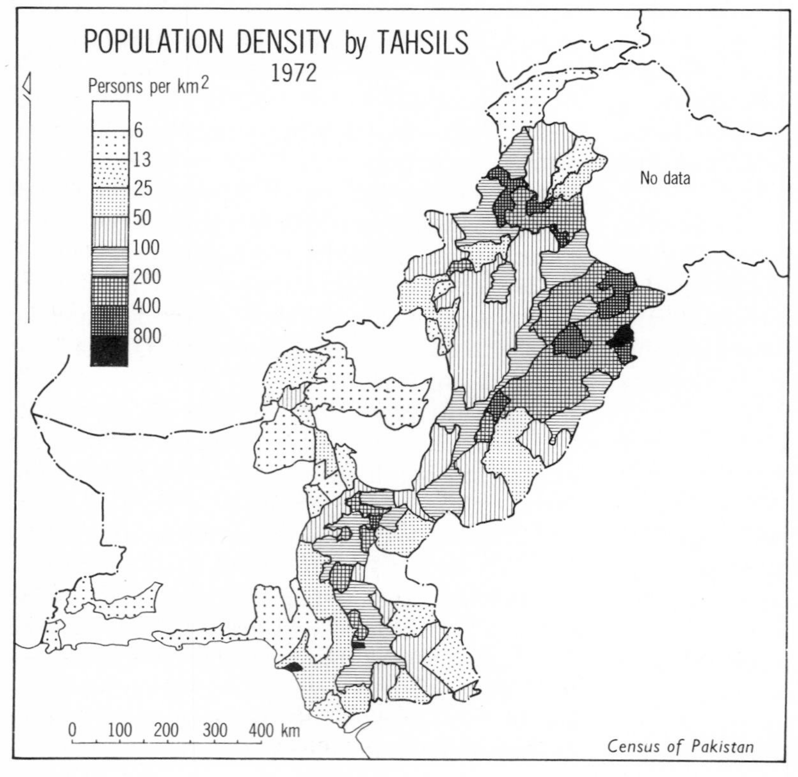

Fig. 12.6

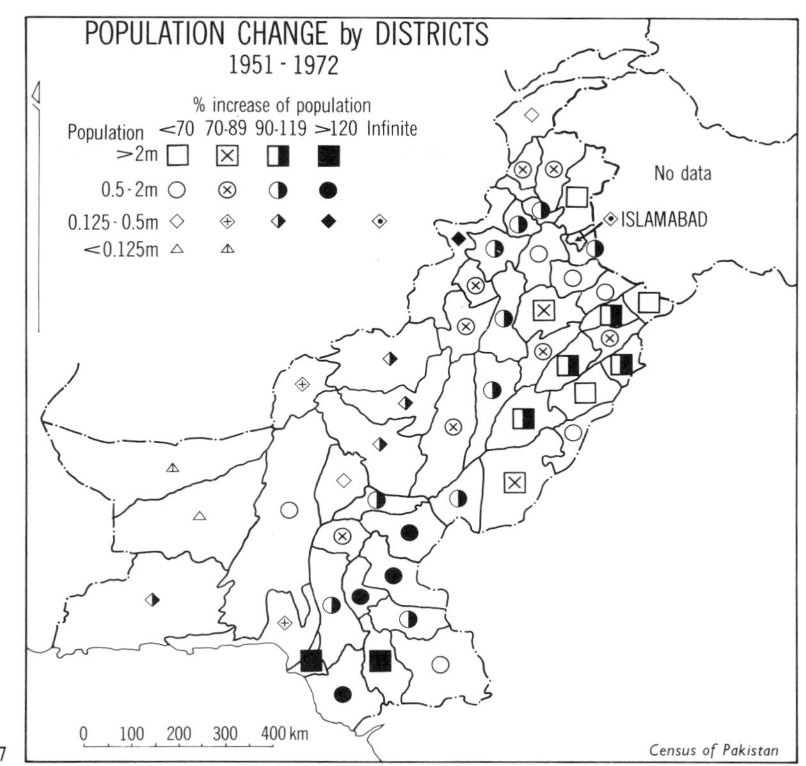

FIG. 12.7

Bahawalnagar suffer perhaps from proximity to the Indian frontier, though this has not thwarted Lahore overmuch. Sialkot shares with another group of slow developers, all of them rainfed districts, Gujrat, Jhelum, Attock and Hazara, the misfortune is not having mainly canal-based agriculture. In the age of the Green Revolution this is a drawback. Tharparkar is another laggard, and has always suffered from its position on the desert fringe and from salinity also. Thatta, which used similarly to compare unfavourably with its neighbours Hyderabad and Karachi, in this instance shows very high growth due to the spread effects of Karachi. In Baluchistan limited resources hamper growth, though the districts east of Quetta do better than average probably on account of mining development (though census error may not be absent).

URBANIZATION AND URBAN GROWTH

Urbanization is the process by which the proportion of the total population living in towns increases, and the level of urbanization is the point reached in this process. Ultimately, as is very nearly the case in Hong Kong or Singapore, the finite limit is 100 per cent urban. Generally speaking the developed countries of the world have a high level of urbanization, for instance Australia 85.6 per cent, England and Wales 78 per cent, USA 73.5 per cent. The more agriculturally-based and underdeveloped a country, the more likely it is to have a low level of urbanization, and to some extent improvement in that level is a measure of development. This is true only, however, if the cause of the shift from rural to urban living by such a proportion of the population as will bring about a change in the level, is a *real* change in the economic structure producing work in the towns to which unemployed rural labour can go. Unfortunately it is only too true of many countries of the Third World, among them Pakistan, that an appreciable part of the rural-urban migration stream merely changes the locale of its unemployment or underemployment.

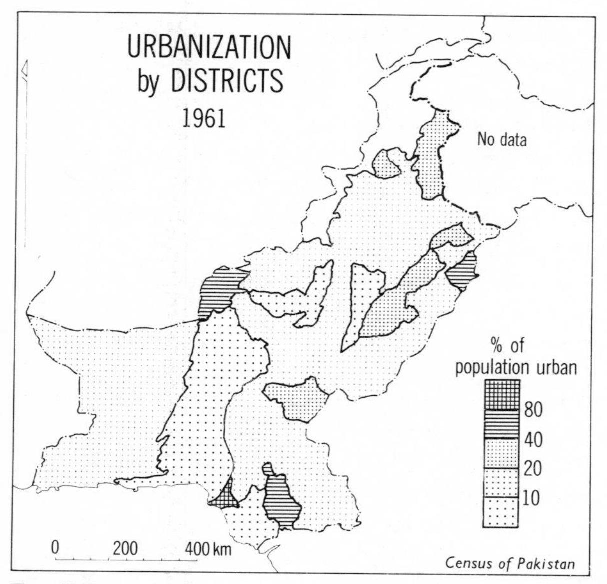

FIG. 12.8

In 1972, 26.3 per cent of the population was recorded as urban. Fig. 12.8 shows the wide disparity between districts in the level of urbanization. Only nine districts out of 45 exceeded the average, indicating that a few large cities account for the national level being as high as it is. The median value was 15 per cent.

Urban growth is simply the increase in population of the cities treated individually or by classes of one kind to another. The term carries no imputation as to how the growth took place, whether mainly by natural increase or by migration. The percentage of the total population living in Pakistan's urban areas at each census since 1901 is given in Table 12.1.

The static situation in the first three census periods reflects the stagnating economy of colonial time when urban and industrial development were not encouraged. After the world economic depression of the 1930s there were signs of modernization even in British India and the level of urbanization reached 14.2 per cent in 1941, though this may be a somewhat inflated figure as urban dwellers were the more politically aware of the possibilities of enhancing their community's chances at a possible partition if they maximized their ennumeration in the census. After independence urbanization has increased apace. The intercensal increase in urban population is not, of course, all due to urbanization. The natural increase of the urban dwellers has differed little from that of the rural population.

There is always a problem to decide what is and is not 'urban'. The meaning of the term as understood by the Census of India which was responsible for the first five decennial censuses listed, allowed small places to be counted as urban provided they had urban functions, a somewhat difficult concept to apply in practice. Since independence, India found that (in 1961) an excessive enthusiasm to see places as being urban led to some distortion of reality, and a more careful assessment was made in 1971 based mainly on size class. This is also the practice of United Nations demographers, and is used here in conjunction with the findings of the Pakistan census. The UN defines urban as an agglomerated settlement of more than 20,000 people. Probably few such agglomerations lacking urban characteristics (not necessarily *modern* urban characteristics) exist now in Pakistan,

TABLE 12.1
Urban population, 1901–1972

Census year	*Total population (millions)*	*Urban population (millions)*	*Percentage urban*	*Intercensal increase in urban population (%)*
1901	16.6	1.6	9.8	–
1911	19.4	1.7	8.7	4.3
1921	21.1	2.1	9.8	21.8
1931	23.5	2.8	11.8	34.5
1941	28.3	4.0	14.2	45.0
1951	33.8	6.0	17.8	49.9
1961	42.9	9.7	22.5	60.4
1972	64.9	17.0	26.3	76.6[a]

Source: Burki, S. J., 'Rapid Population Growth and Urbanization', *Pakistan Economic & Social Review*, VII, 1973, pp. 239–276.
[a] 59.8 at 1971.

though if they do they are likely to be in the NWFP hills where settlements are tightly nucleated and often large. The Census of Pakistan takes a broader view of what is urban, and is seems that in 1951 52 per cent of the population then designated as urban was in 'towns' of less than 20,000. Many of these ultimately made the grade, since by 1972 only 18 per cent were in such small towns, and the number involved had actually fallen by 60 per cent. Table 12.2 and Fig. 12.11 set out some salient features of urban growth since independence, i.e. between 1951 and 1972. For ease of comparison with other countries the size ranges used by United Nations are used, and the proportion of urban population in each range is related to the total in towns of over 20,000. The proportions related to the total urban population as defined by the census are also shown.

The number of towns overall increased almost three times from 43 to 107 and remarkably consistently within size classes (lines 1 and 8 in the table). The number in the over 500,000 class increased 300 per cent from 2 to 6 while in the other three classes the increase was 257,257 and 256 per cent. Lines 2 and 3 compare the population of the towns in the size classes of 1951 with the population those same towns had achieved in 1972, and the percentage change is given in line 4. Growth has been stronger in the towns starting with over 100,000 than in the two lower classes. The two cities over 500,000 in 1951, Karachi and Lahore, moved into the multimillion class, Karachi from 1.138 to 3.607 million (261 per cent growth), Lahore from 0.849 to 2.165 million (155 per cent). The range of growth (line 5) is considerable. Of the seven starting at over 100,000 in 1951, only Sialkot (22 per cent) and Peshawar (77 per cent) failed to double in size; the strongest growth was Faisalabad's (359 per cent), refugees playing an important part. The other cities in this group of 1951 are Rawalpindi (159 per cent growth), Gujranwala (198 per cent), Multan (185 per cent) and Hyderabad (160 per cent).

TABLE 12.2
Urban growth, 1951–1972

Size class	*Over 500,000*	*100,000 to 499,999*	*50,000 to 99,999*	*20,000 to 49,999*	*Less than 20,000*	*Total*
1. No. of towns 1951[a]	2	7	7	27	–	43
2. Population 1951 (millions)	1.987	1.289	0.482	0.903	– (4.993)	4.661[a] (9.654)
3. Population of *these* towns in 1972 (millions)	5.772	3.439	0.931	1.925	–	12.067
4. Percentage growth 1951–1972[a]	290	267	193	213	–	259
5. Range of growth values for individual towns within group (per cent)	155–261	22–359	24–157	29–1050	–	–
6. Median growth value (per cent)	208 (average)	160	85	171	–	–
7. Population of *all* towns 1972 (per cent)	8.379	2.253	1.274	2.136	– (2.998)	14.042[a] (17.040)
8. No. of towns 1972	6	14	18	69	–	107
9. Percentage of total urban population 1951 (towns > 20,000 only)	43	28	10	19		100
10. ditto 1972	60	16	9	15		100
11. Percentage of total urban population (census definition) 1951	21	13	5	9	52	100
12. ditto 1972	49	13	7	13	18	100

Sources: Census of Pakistan. Burki, S. J., *op. cit.*
[a]Exclusive of towns less than 20,000.

The relative importance of the different size classes in the total urban picture is given (for the UN definition of urban, i.e. over 20,000 population) in lines 9 and 10. The top class has grown at the expense of the lower classes all of which lost ground relatively, particularly the class over 100,000 (mapped in Fig. 12.9).

The effect of relating the population of each class to the larger urban total as given by the census is seen in lines 11 and 12. All classes gain or remain with the same percentage (but of a larger total–17.040 million in 1972 compared with 9.654 in 1951) except the lowest class of all, the less than 20,000 group. This indicates that it was from this lowest group of 1951 that the recruits were taken to advance into the higher classes, without however being replaced from aspiring villages. This suggests that the Green Revolution brought such small towns into prominence as rural service centres supplying the essential inputs. It did not require great numbers of rural dwellers to move into the nearby market town to raise its population to urban status. Of 55 places that moved across the 20,000 boundary between 1951 and 1972, a fifth increased their population by 320 per cent or more, two-fifths by 180 per cent or more. In Fig. 12.11 discussed below almost half of them appear in the highest category of growth rate. Only one-fifth of them failed to double their population. Fig. 12.10 compare the rank-size distribution graphs of urban centres of over 20,000 population for 1951, 1961 and 1972. Although Pakistan started independence with two cities of metropolitan status and not dissimilar in size – Karachi just over 1 million, Lahore 0.849 million – because of Karachi's unique combination as the country's only port, its relative security from the only enemy, India, and its direct links by air with the world at large, it was chosen as capital. Lahore had more grace and tradition but for a capital it was too exposed to any threat from India. This fact has inhibited its growth and development, thus encouraging industrial investment in relatively nearby regional cities such as Faisalabad and Gujranwala. These factors bid fair to place Karachi incontestibly at the head of a rank-size hierarchy

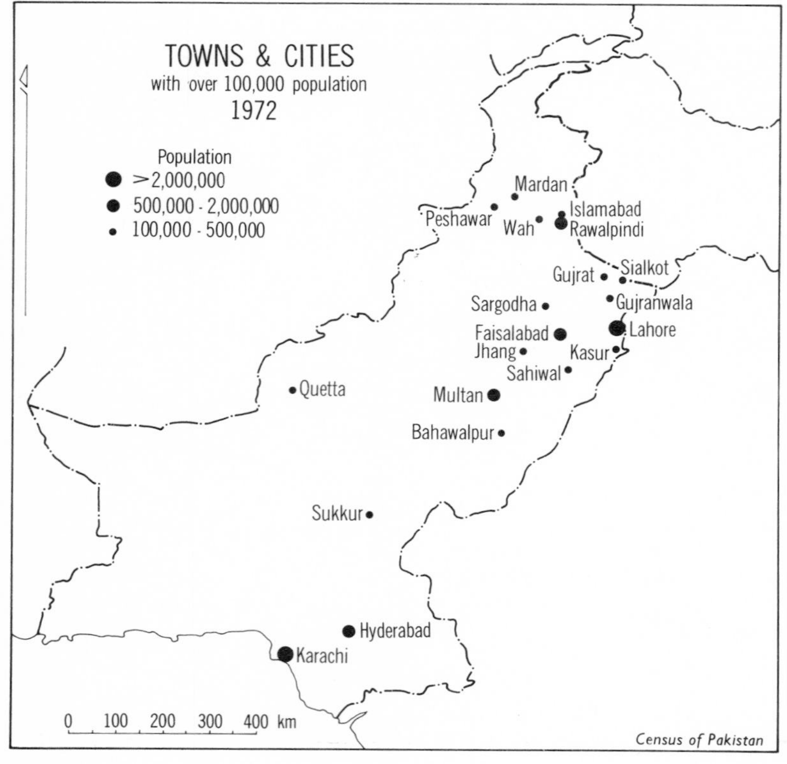

FIG. 12.9

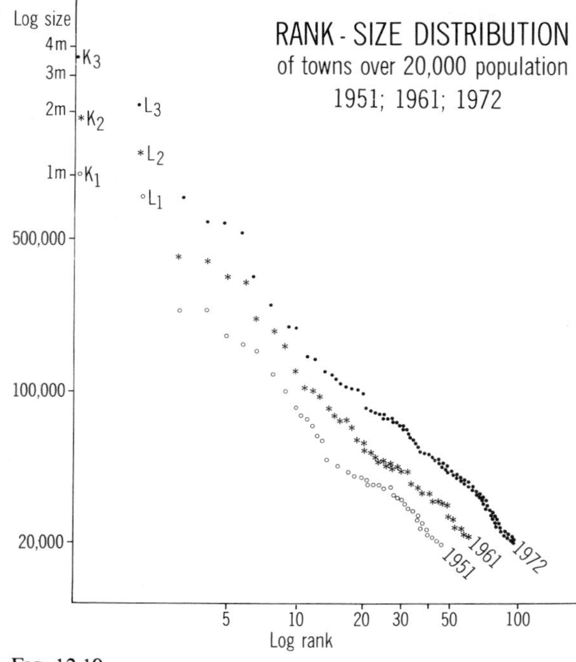

FIG. 12.10

and its faster growth rate would guarantee it doubling Lahore's population in the near future.

The map of urban growth, 1951 to 1972 (Fig. 12.11) is simultaneously a distribution map of towns by size classes in 1972. Considering first the distribution pattern, it is possible to see an urban corridor from Peshawar through Rawalpindi to Lahore – the Grand Trunk Road – and then southwest along the main railway line through Okara and Sahiwal to Multan and Bahawalpur from where it sweeps down the Indus valley past Sukkur and Hyderabad to Karachi. The towns threaded like beads on the string of the main railway line include five of the six cities in the 'over half million' class, and eight out of 14 towns in the 'over 100,000' class. Three others in this latter group are on short off-set branches from the main line: Mardan, Sialkot and Kasur. Of the four major towns and cities remaining, three form a group in central Punjab, Faisalabad, Sargodha and Jhang and the fourth is Quetta, strategic outpost above the Bolan Pass to Afghanistan and Iran. Many of the towns in the two smaller size classes form lesser beads to the

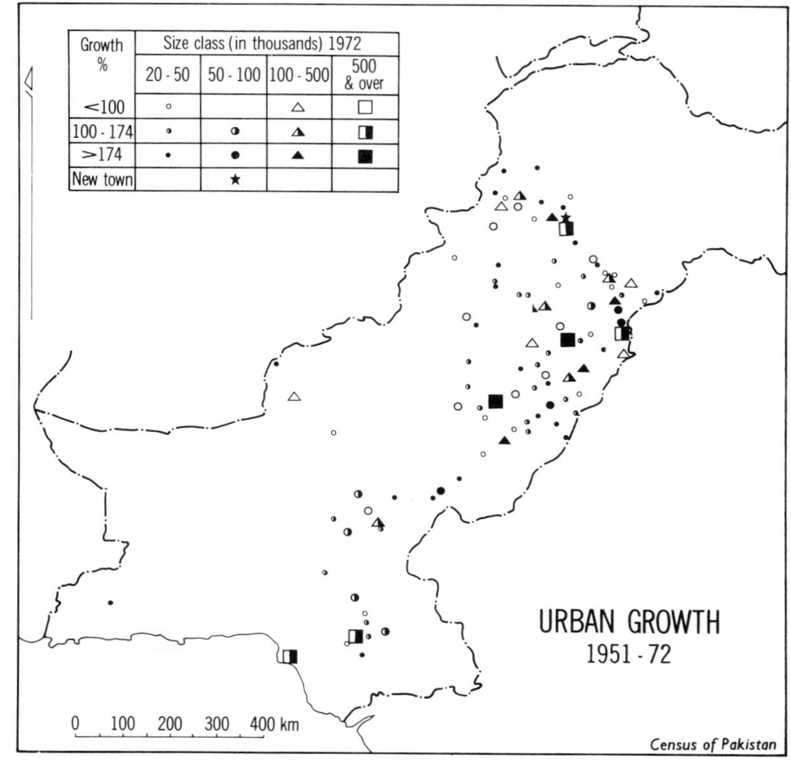

FIG. 12.11

string along the main corridor, but there are some notable exceptions. In southeast Punjab two secondary strings of smaller towns follow major canal systems on either side of the Sutlej. Another two mark out the alignment of the canals that have colonised the Thal region since independence from Daudkhel and Mianwali to Khushab along the foot of the Salt Range scarp, and a more extended string southwards from Mianwali along the Indus left bank through Bhakkar, Leiah and Kot Addu to Muzaffagarh and Multan. Across the Indus, Kohat, Bannu, Dera Ismail Khan and Dera Ghazi Khan cannot yet be seen as part of a north-south corridor, but rather are the railheads and bridgeheads of past and present strategic control of the long frontier beyond the Sulaiman Range. In Sind, the right-bank canal systems from the Gudu and Sukkur barrages indirectly support a cluster of towns (Shikarpur, Larkana, Jacobabad and Warah) while bifurcations in the Rohri and Eastern Nara canal systems on the left bank and the canals from the Ghulum Mohammed barage at Kotri, account for the urban centres in a semi-circle east of Hyderabad from Nawabshah through Sanghar, Mirpur Khas to Tando Allahyar and Tando Mohammed Khan. To the far west, Turbat in Makran (with 680 per cent growth to 27,642) points to strategic rather than economic interest in this remote corner of Baluchistan. Similarly frontier control and of course some trading functions, account for Chaman, north of Quetta.

Growth tends to follow the same patterns sketched for urban distribution, though a regional rather than linear approach provides a better basis for explanation. Apart from the Rawalpindi–Wah–Islamabad complex of the new capital and Gujrat, the barani lands in general, including the submontane piedmont of the Punjab plains from Sialkot to Jhelum, the Peshawar basin, Kohat and Bannu, show relatively low urban growth rates. Similarly the trans-Indus towns above the 50,000 mark have tended to lag.

The Punjab irrigation towns show a stronger growth than those of Sind. Agriculture has been more buoyant and the Green Revolution more successfully achieved here with consequential stimulus for agro-towns, agricultural engineering and, at Faisalabad, a fully fledged Agricultural University. In Sind (apart from Karachi) growth has been comparatively modest and of course based on smaller and fewer centres. The intermediate region of southernmost Punjab (Rahimyar Khan District) and northern Sind has a cluster of vigorously growing towns stimulated by canal development and natural gas.

COMMUNICATIONS

As an economy expands and with it the towns that are the key points linking the agricultural sector to urban markets and processing plants, and to the centres of secondary and tertiary industrial development, so also increases the need for interconnection between centres to facilitate the flow of commodities, goods, people and ideas.

For linking the widely spread population of Pakistan still for the most part living at a pre-aviation level, the railway holds pride of place for inter-provincial travel, though within the provinces road transport of both passengers and goods is now dominant.

Railways

The railway network is shown in Fig. 12.12. Routes total 8809 km, mostly in broad gauge. From Karachi to Bahawalpur (with the exception of the Kotri-Hyderabad bridge section over the Indus, which is presently under conversion from single track) the line is double track, and there are small sections of double track near Lahore, Rawalpindi and Quetta. Elsewhere the single line has constantly to be upgraded to increase its carrying capacity to the maximum. Diesel locomotives are favoured, but the change from coal-burning steam engines is not complete.

Within the present network, little altered since independence, the dual purposes of British Indian railway construction can be seen. An overall strategic design links Karachi to Lahore, the headquarters of the former North Western Railway. Lahore was tied by several lines to Delhi and Agra (and so to Bombay) and through Delhi or Saharanpur to Calcutta. From Lahore northwestwards the mainline paralleled the Grand Trunk Road to Rawalpindi and Peshawar. A strategic rather than an economically justifiable line tied Rawalpindi to Multan along the left bank of the Indus. Within the Indus plains of the Punjab and Sind the network

70 West of the Indus there are a few stretches of narrow gauge railway like this one between Bannu, Lakki and Kalabagh. Coal-burning steam engines persist, though on the main lines, they have been replaced by diesel units.

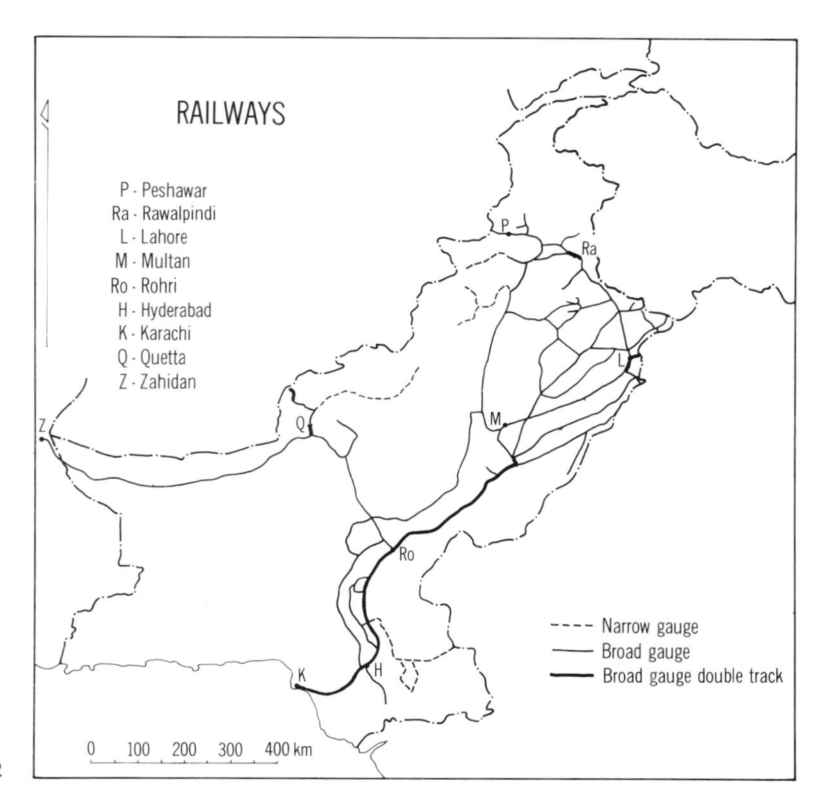

FIG. 12.12

developed for economic reasons. West of the Indus, a number of spur lines, some of them narrow gauge, penetrated the mountains to carry military forces and supplies to the frontier regions. Thus the Khyber, Kohat and Bannu in the north and Fort Sandeman in the Zhob District (linked to Quetta) were railheads of strategic importance only. The long and magnificently engineered line from Sibi through Quetta to the Afghan frontier at Chaman and to Zahidan just inside the Iranian border far to the west could not be accounted an economic line, though Quetta District has benefited greatly from being able to send fresh temperate fruit and vegetables expeditiously to Karachi and Multan.

In 1976–77, 141.8 million passengers were moved an average of 91.6 km (12,992 passenger kilometres in all) and 10.9 million tonnes travelled an average of 718 km per tonne (7804.6 million tonne/km).

Road Transport

Since independence the use of motor vehicles has greatly increased. Roads of a high standard (sealed) totalling 8130 km in 1947–48 now total 27,152 km; those of low quality shingle went from 14,108 km to 22,774 km in the same period. Another 56,300 km of earth roads link many villages to the main routes. There has been more than an equivalent expansion in vehicle numbers to judge from the traffic one now encounters to the main highways, let alone in the congested cities. In the five years 1960 to 1975 powered road vehicles increased by more than four times, from 1.1 million to 5 million. To this must be added untold thousands of rubber-tyred bullock, buffalo, camel and donkey carts, horsedrawn rehras and passenger tongas, and pack animals under cumbersome loads that solemnly plod their way along the same roads where trucks and buses hurtle past with horns blaring in full cry. Federal and provincial governments strive to keep abreast of the traffic boom. The road network is shown in Fig. 12.13. Of recent interest are roads connecting Pakistan to its neighbours. The existing links to Iran and Afghanistan carry a heavier load of international traffic every year. A new road, the Karakoram Highway, is under construction by the Chinese and Pakistan military engineers to link Swat and Hazara to China through the Hunza valley. The terrain is extremely difficult for road building as land slides and screes present ever-moving surfaces. Also in the northern mountains, an all-weather road is being built to Chitral, formerly cut off by snow for a period each winter. A tunnel under the Lowari Pass will solve the problem.

Little economic justification can be made for the Karakoram Highway. Its strategic significance to China probably justifies the Chinese expenditure on the project. Some Pakistani planners have been arguing for the construction of an Indus Super Highway of 1187 km to link Karachi and Peshawar along the west bank of the Indus at present followed by a road of variable quality. This road again has more strategic than economic justification, and hence has not found the lavish support desired of foreign donors who no doubt see other more pres-

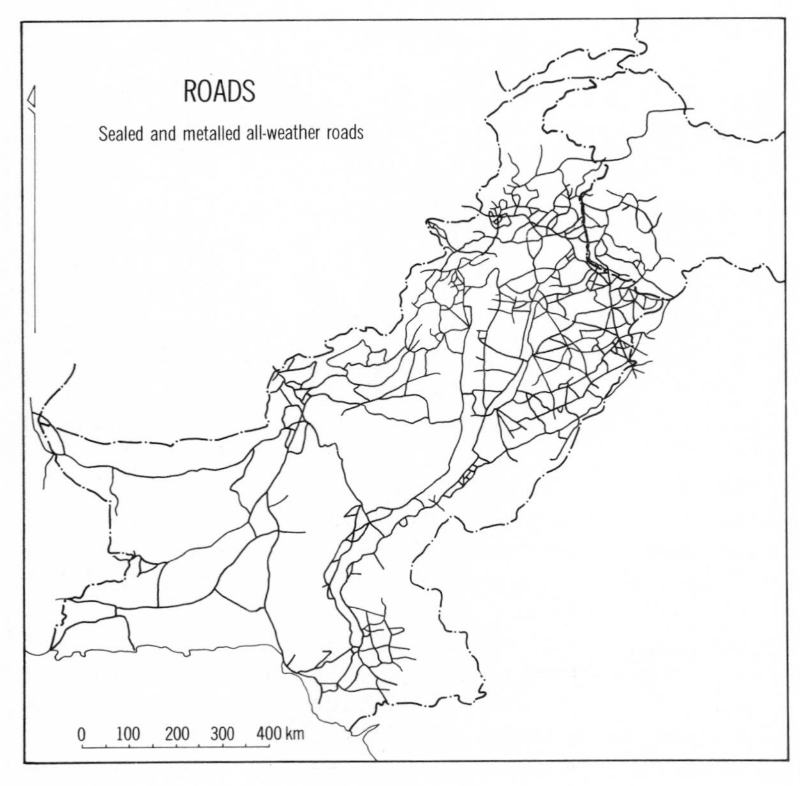

Fig. 12.13

sing development opportunities for their limited funds. Apart from the section of the road up to Larkana and Shikarpur in Sind the Indus Highway would traverse relatively undeveloped country lacking in the kind of resources that only await a road for them to be exploited. However, the Highway has become something of a matter of prestige and may well divert funds that could usefully be spent urgently to duplicate the Grand Trunk Road and other overcrowded routes in the more closely settled districts of Punjab.

Communication linkages designed for specialised use are the electricity grid and the pipelines for natural gas and for oil which have been mentioned in Chapter 11.

Air Transport

With its great linear extent separating Karachi from Lahore, and Rawalpindi-Islamabad and Peshawar, and the national capital from the provincial capitals, air services perform an essential task in carrying administrators and business men. Increasingly too, the affluent Pakistani worker who goes overseas from his frontier village runs the gamut of transport media from horse rehra and bus to take to the air in Fokker Friendship, Boeing 727 and finally Jumbo Jet.

Fig. 12.14 shows the passenger carrying capacity of the internal network, clearly designed to meet the inter-city traffic between the widely separated major cities. A more modest range of services help fill in the intervening spaces and to penetrate remote areas like Gilgit, Chitral and Baluchistan.

International services operate through Karachi for the most part; traffic westward far exceeding that to the east. For a decade or so after independence, Karachi was one of the nodal points in the airlines system linking Europe and America with South, Southeast and East Asia, and with Australia-New Zealand. As aircraft increased their range, and sought more rapid turn-round to maximise the use of capital invested, Karachi began progressively to be over-flown by services going direct from a Middle Eastern or even European airport to Bangkok for example. Lahore carries some international traffic and Rawalpindi even less.

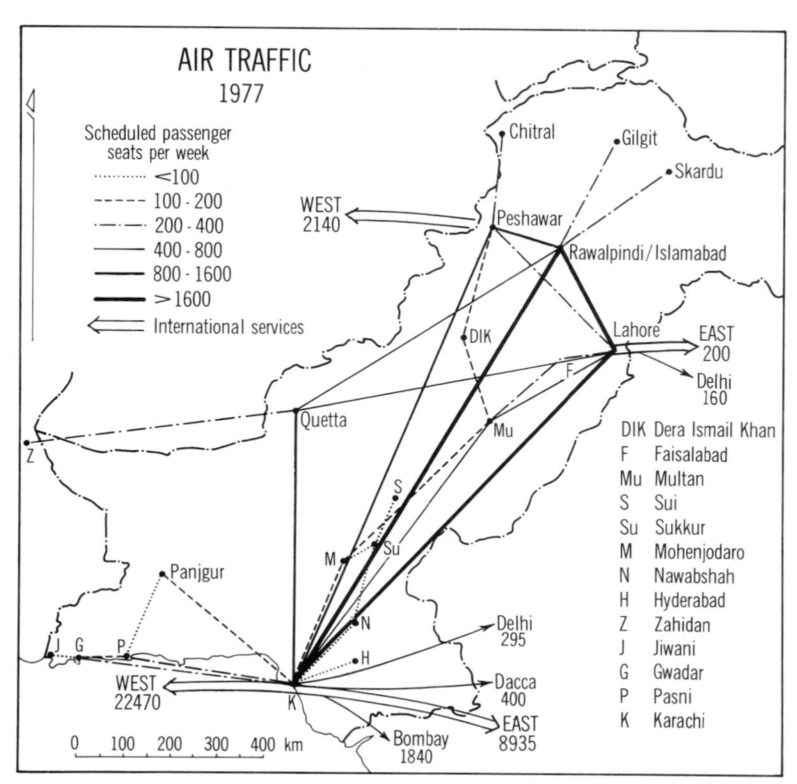

Fig. 12.14

It may be mentioned in passing that the national air line, Pakistan International Airlines Corporation, has built up an enviable record as an international carrier. P.I.A. runs at a profit and expands its turnover of passengers annually. Passenger/kilometres have increased from 1592 million in 1971–72 to 2896 million in 1975–76. Domestic services are usually fully booked, and the passenger load factor stands at 56.6 per cent of capacity.

Sea Transport

Although country craft have not entirely disappeared from the Indus, the increasing number of barrages and the uncertainty of flow deep enough for boats in the dry season make it unlikely that inland water transport will develop.

Maritime shipping was brought under national control in 1974, involving 53 ships with a deadweight tonnage of 620, 643. Port facilities at Karachi are overcrowded, and the completion of Port Qasim in the near future will ease the pressure as well as taking the new traffic to be generated by the Karachi Steel Mills at Pipri. Port Qasim will have an oil terminal, and bulk terminals for phosphate rock, fertilizers, sulphur, and wheat imports in addition to the special provision for coal and iron ore vessels.

Along the Makran coast, Ormara, Pasni, Gwadar and Jiwani are small fishing ports whose markets lie in Karachi.

CHAPTER THIRTEEN

VILLAGE, TOWN AND CITY

SUMMARY

In the introductory section some of the fundamental traits characterising settlements in Pakistan are discussed: the influence of Islam on traditional patterns of behaviour; the rudimentarily simple needs of the self-sufficient village; the elaboration of those needs as the agricultural bases of the economy becomes commercialized and technologically more sophisticated; the imprint of the dual economy, and, particularly clear in urban settlements, of the still living vestiges of the administrative structure of colonial rule.

The urban form of several of the larger cities is reviewed to show some consistent functional patterns differing only in the detail imposed by site and by historical heritage. Closer study is made of Lahore, Karachi and of the national capital 'in transition' Islamabad and its foster parent city, Rawalpindi.

INTRODUCTION

The dualism exemplified in the economy of Pakistan – the co-existence of traditional agriculture and craft industry with roots in a formerly subsistence way of life alongside their modernized and commercialized counterparts — is carried into the forms and structures of settlements. These reflect also the social fabric of a dualistic society and, in a somewhat different dimension, the persistence of patterns and structures deriving from the political administration by alien imperial rulers.

The preponderance of Islamic cultural mores throughout the country guarantees a certain basic similarity in a domestic architecture designed to ensure privacy to the family, particularly for its womenfolk. Extremes of summer heat and general dryness permit the simple flat-roofed house, which has the added advantage of an open air 'bedroom'

71 This village in the NWFP south of Peshawar is characteristic. Privacy and the seclusion of the womenfolk are assured by the surrounding wall which also encloses the livestock shed and fodder stacks though some of the latter are on the rooftops. The taller structures are towers in which to sleep in the hot weather and from which to watch out for feuding enemies (at least in the past).

72 An Afridi village in the Khyber Pass, NWFP. The high towers have a defensive purpose enabling a watch to be kept against enemies pursuing blood feuds which used to continue from generation to generation. The sparse vegetation of the Khyber Hills can be seen behind the village, and a rail of the Khyber line in the foreground.

on the roof in summer. Economics and the resources found in the local environment dictate the building materials. Only in the mountains are stone and sometimes timber freely available for walls. In the plains a few sturdy beams of timber 30 cm in square cross section have to be purchased for the roof, but the rest of the house can be constructed by the owner and his family from mud bricks, reeds and straw, the finish of walls and floors being rendered in mud plaster by the women. With increasing affluence, 'pucca' burnt bricks are replacing 'kachha' sun-dried mud bricks, one great advantage being their ability to withstand the occasional cloudburst or flood that can dissolve the kachha dwelling back into mud. Reed screens, matting and thatch are variable adornments depending on the local vegetation. Even if the main living room is 'pucca', often the kitchen, cowbyre and outhouses are of 'kachha' brick and cob (mud with kankar or straw binding moulded in place), and the stores for grain are usually of smoothed mud plaster. The whole family compound is usually surrounded by a mud wall within which they and their livestock, farm equipment and grain store are accommodated.

Flying over the villages one can readily see how man's basic needs are met. The surrounding fields provide food, the mud-walled compounds give shelter and privacy, and between fields and village a zone of lower ground, often foul with village drainage and always flooded in the rains, marks the 'borrow pit' from which fresh mud has been dug to repair and renew the houses for generations. A few compounds have their own well, but many housewives must draw water from a communal well at the centre of the settlement, though the traditional procedure of drawing up the water in a pot on a rope is fast giving way to collecting it from the gushing outlet of a tube well pumped by electricity. Standing out against the monotonous khaki of the mud houses an occasional splash of whitewash indicates the village mosque, or the ostentatious two storeyed house and grainstore of a wealthy landlord. In Sind the houses are often topped with wind collecting funnels, while in the Frontier hills, the profile is varied by the watch towers rising three storeys high above the general level of compounds, here more highly walled than in the longer more peaceful plains. Within the village narrow winding lanes barely passable by bullock carts connect houses to the surrounding fields. When roads eventually reach the village they stop at the outskirts, developing around themselves a new alignment of structures, but leaving undisturbed the original compact village of pedestrian movement, foot and hoof. At a grander scale this same distinctive differentiation between ancient and modern is repeated in the older cities.

Such is the pattern of the traditional village and there are many thousands of them throughout the country. In the canal colonies established by British fiat particularly in the Punjab plains, and by

73 On the margins of the Indus in Sind the abundance of trees and reeds allows villagers to construct houses like this one.

the continuing tradition of the irrigation departments since independence, regularity of form continues into the village the rectilinear geometry of canal distributaries and the cadastral survey pattern of land holdings. At the crossing of main streets the village well is sited, and along the roads each compound has its fixed allotted space. In the Thal development areas, a 'pucca' brick house formed the nucleus of the family compound, around which mud sheds and walls could be constructed by the occupier to suit individual tastes and needs. The sense of a collection of people having been imposed upon the landscape rather than of a community having evolved within and from it is heightened by the primary designation of the village being a reference number not a name. Chak 80–5–R six kilometres from the town Sahiwal is village no. 80 on Right Bank Distributary No. 5 of the Lower Bari Doab canal (Fig. 13.4). Within the same radius of Sahiwal, eighteen such Chaks have no other name on the map, though about half that number have achieved some personal identity in the names they carry as sub-titles to their official designation. Thus Delhiwala may denote the antecedents of an early landlord gifted the land by the British, Gurmukh Singhwala (no doubt a name since superseded) points to a former Sikh settlement, Murshidabad, Islamabad and Shah Madar are quite clearly Muslim. The canal system, the survey and the administration of the planning of the Canal Colonies were projections of the imperial industrial scientific and political might of an alien power, the village in most respects remained 'pre-industrial' and culturally indigenous.

74 The atta (flour) market of a small Punjabi town. Note that the shoppers are men. Atta is used to make the staple food, chapattis, a form of unleavened bread.

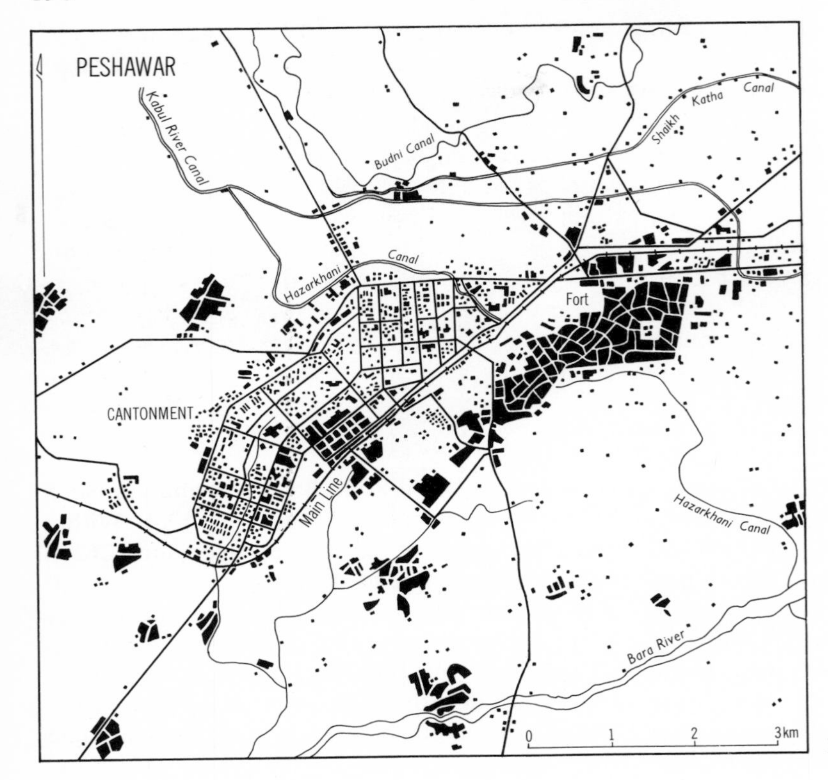

FIG. 13.1

The degree of rural village self-sufficiency is rapidly dwindling, and the urban functions essential to an exchange economy are percolating downwards through the settlement hierarchy. The larger settlements that used to be distinguished as market centres for the surplus grain, fruit and livestock of basically subsistence villages, provided in return necessities like salt and near-luxuries such as kerosene, metal pots and mill cloth. Today a more sophisticated commercial economy obtains and the more affluent among the farmers have many needs that have to be supplied from outside: spare parts for tractors and tube wells, finer fabrics in gaudier designs for wives, electric light bulbs and maybe a cooking ring, if power has reached the village.

Superimposed on this pattern of economic functions of the urbanised village are the administrative functions of an external government – external in the sense that decisions come from outside through the bureaucratic machines of departments responsible for agriculture, irrigation, revenue, police, education, and so on. These functions bestow status on the settlement, as a tehsil headquarter town, or maybe as a sub-divisional or district headquarter town in the administrative hierarchy. Outsiders to the district recruited generally from the urbanised elites of the cities who are perhaps a generation or more remote from village life, people the bureaucracy at its executive levels at any rate with officers aspiring to progress upwards through the system with a metropolitan posting as its crowning reward.

In that the administrative structure is a little changed inheritance from British rule, it tends to be peopled by educated social elites of semi-independent means derived from property ownership, who inhabit simultaneously two different worlds: that of the (at best) semi-sophisticated populace their task is to administer, and that of a westernized urban society, seeking for their children entry to professions that will rarely take them near a village. At the small tehsil town level, such administrators are among the very few residents who have absorbed much of western culture. Some among the business men and professionals like lawyers and doctors, more numerous in district headquarter towns, share similar backgrounds and outlook. In business, many are the agents in the field of commercial networks organised from the metropolitan cities and even internationally. The dealings at the small town level of all these people with the local population, its landlords-cum-grain merchants and the like, represent the interface between the two parts of the dual economy and society.

In the bigger towns and cities the dual economy persists, but both parts operate at a more sophisticated level and with a greater measure of interpenetration. The marketing of petroleum products provides a simple illustration. Petrol stations little different from those of developed countries service the motorised customers, while trotting through the narrow alleys of the pre-industrial city, donkey-drawn 'tanker wagons' deliver kerosene to the urban housewife. The traditional economy penetrates the homes of the westernized elite directly through the hawkers of cottage-made mats and brushes, and more subtly through the linkage between the craftsmen in backstreet workshops who adapt their skills to new designs and materials to produce the stylish shoes and ready-made garments displayed in the shops.

The urban fabric changes slowly, however, and in towns and cities of every size an increasingly uncomfortable co-existence between traditional and modern ways persists, most evident in transportation. As the study of Lahore below shows, the confrontation of old and new produces problems insuperable within the present forms and structure of the city. Trucks and buses cannot physically penetrate the Old City, and the question of how to modernize without destroying its character is very real and urgent.

URBAN STRUCTURE

Six 'thumbnail' sketch maps of Peshawar, Sargodha, Faisalabad, Sahiwal, Hyderabad and Multan (Figs. 13.1, 13.2, 13.3, 13.4, 13.5, 13.6) and several illustrating aspects of Lahore (Figs. 13.7, 13.8, 13.9 and 13.10) and Karachi (Figs. 13.11, 13.12, 13.13) serve to demonstrate the broader features of Pakistani cities, features they share with many

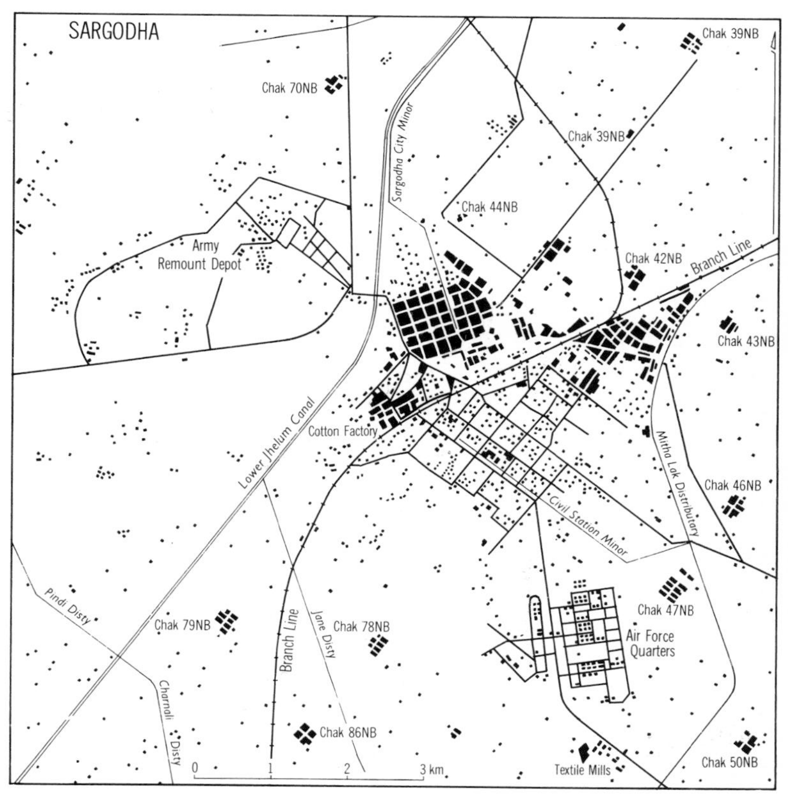

FIG. 13.2

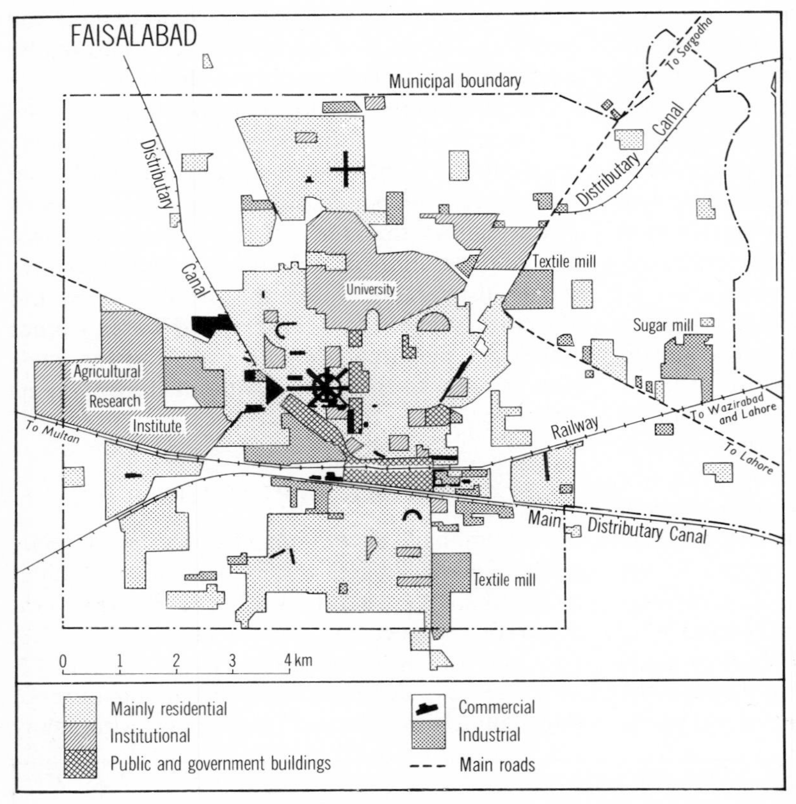

FIG. 13.3

others in former British India. Four of the seven cities Peshawar, Lahore, Multan and Hyderabad can claim ancient lineage, and their early nuclei are clearly discernable in the present pattern of lanes and alleyways, which serve the warren-like congested housing of their 'old cities'. In each an old fort dominates one corner of the formerly walled city. The railway passes close outside the city walls and widened 'circular' roads trace their circumference. In one sector of the external city, the regular rectilinear street pattern marks out the cantonment, where British administrators and their army depots and hospitals were located, and where now their Pakistani successors continue the tradition. Between these two elements are found the monuments of civil administration: law courts and municipal offices and regional branches of higher administrative bodies like the irrigation department. A separate railway station serves the cantonment. Again beyond the old city walls in a different sector or along the major roads approaching the city are found the larger units of modern industry. Mainly post-independence housing suburbs spread in other sectors. Commercial detail is indecipherable at the scale of most of these sketch maps but it may be noted that the 'old city' and cantonment had their own retail shopping centres, and sometimes their own vegetable markets. In the new suburbs a planned shopping centre is provided. Some of these details are pursued below in the case of Lahore.

Of the other four cities, three are the product of canal colonization in the Punjab: Sargodha on the Lower Jhelum canal, Faisalabad, formerly Lyallpur on a branch of the Lower Chenab canal, and Sahiwal, formerly Montgomery, on the Lower Bari Doab canal. The fourth, Karachi, while it had some earlier settlement in the eighteenth century as a small fortified port serving a hinterland in Sind, Punjab, Afghanistan and Baluchistan by camel train, achieved real importance as a city under the British. Thus in all four cases the pre-British element can scarcely be discerned except in the somewhat irregular street pattern of Old Karachi close to the present docks which presumably were the commercial nucleus of the pre-British settlement.

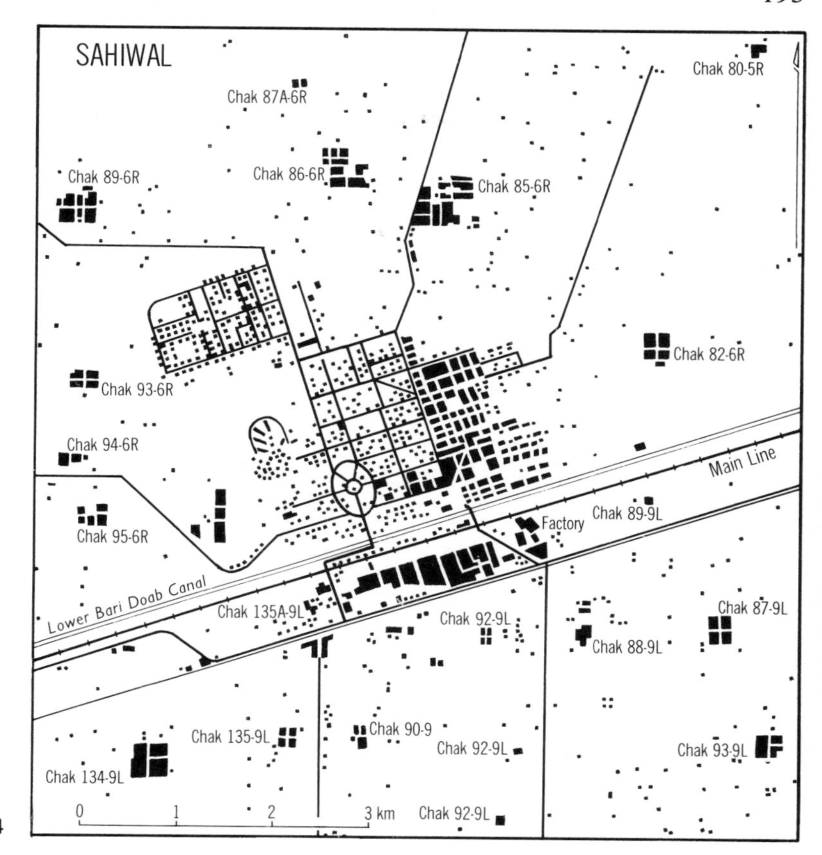

Fig. 13.4

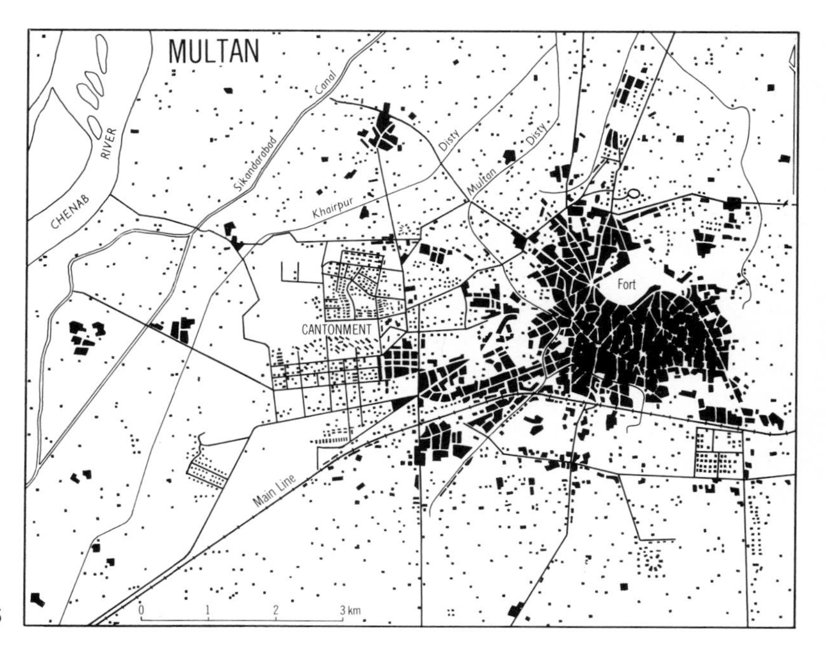

Fig. 13.5

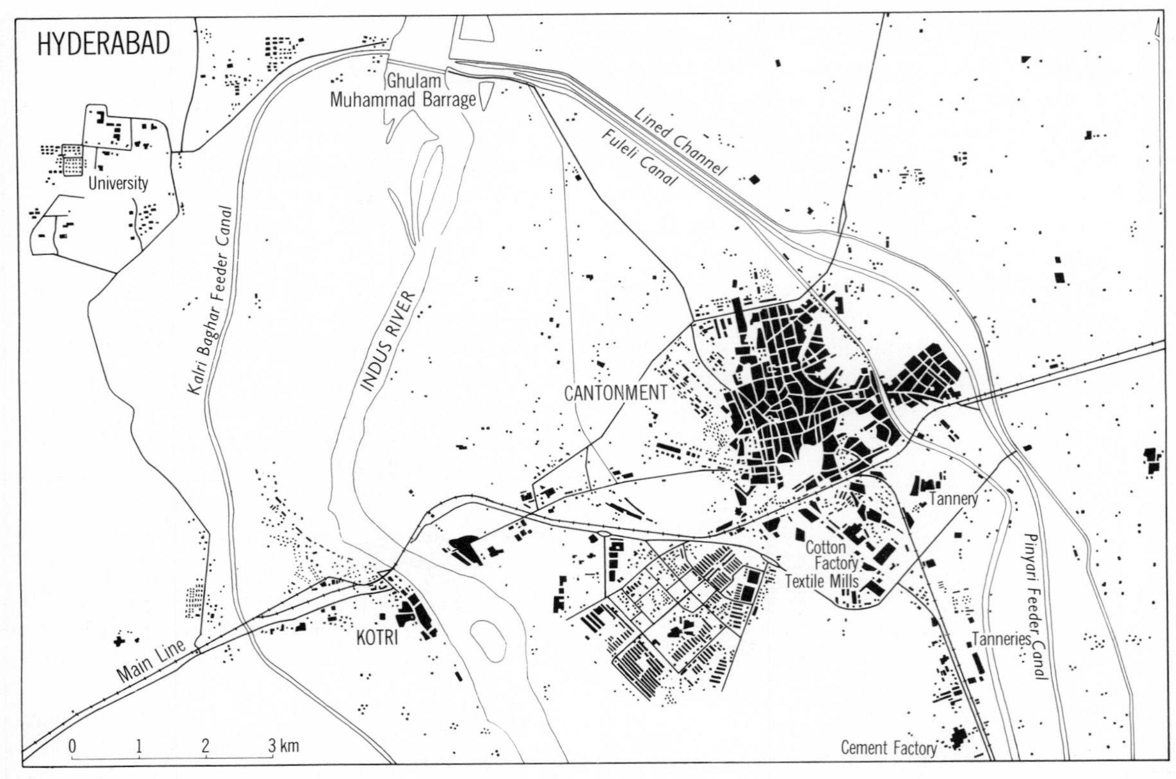

FIG. 13.6

For the rest, the imprint of colonial rule, civil, military and (in Karachi) commercial is seen in the regularity of the street plan and the elements of the cantonment and civil administration present outside the old cities in the preceding four examples. A contrast in residential densities between the 'garden city' environment of the cantonments and the closer but still regulated pattern of the one time 'native quarter' of the indigenous population is quite apparent. In Sahiwal and Faisalabad civil administration dominates the scene, and the military presence, if any, was and remains subdued. Sargodha on the other hand prides itself on having an important Air Force base as well as an Army Remount Depot with extensive areas for breeding healthy horses. In Karachi the former British military presence in Cantonment and 'Soldier Bazar' and the rifle ranges have in part at least been overrun by modern urban growth. The pattern of the Civil Lines where administrators lived between the Courts and the Sadr Bazar (the Cantonment's shopping centre) is still to be seen. An outlying air-force cantonment at Drigh Road has evolved to serve a much larger airport. Karachi as it was on the eve of Pakistan's independence is shown in Figure 13.12. Sahiwal and Faisalabad were planned cities in every respect and industry was located in particular blocks close to the railway up to independence. Subsequent expansion and motor transport has led to industry spreading further afield.

LAHORE*

In all its diverse ingredients Lahore provides a classic example of the South Asian city. (See Fig. 13.7)†.

*This section is heavily indebted to Muhammed Mushtaq Chaudhury', *'Lahore: a Geographical Study'*, University of London Ph.D. thesis, 1965. Chapters have appeared in *Pakistan Geographical Review*, vols. 22, 23, 27 in 1967, 1968 and 1972. Recent data is courtesy of Lahore Development Authority. Fig. 13.9 is based on an original courtesy Prof. Samuel V. Noe.

†The present author in *India; Resources and Development*, Heinemann Educational Books, London 1978 elaborated a 'model' of the Indian city which could equally apply to Lahore.

Its character and importance cannot be appreciated without some knowledge of its past. The Old City of Lahore stands on a terrace bluff of the River Ravi, a part of the left-bank cover flood plain, with the fickle Ravi washing and cutting into its northwestern corner as it flows on its braided meandering course towards the southwest. If present-day planners bemoan the threat the Ravi presents and have had to protect Lahore by massive bunds along its northern and western flanks, the site of the Fort at the extreme northwest corner of the city suggests that the defensive possibilities of the site facing northwest were well-appreciated in the past. It was from the northwest that invaders have repeatedly entered the sub-continent, travelling down the Grand Trunk Road from the Khyber Pass and Peshawar to Lahore, and then heading for Delhi and the Ganges plain.

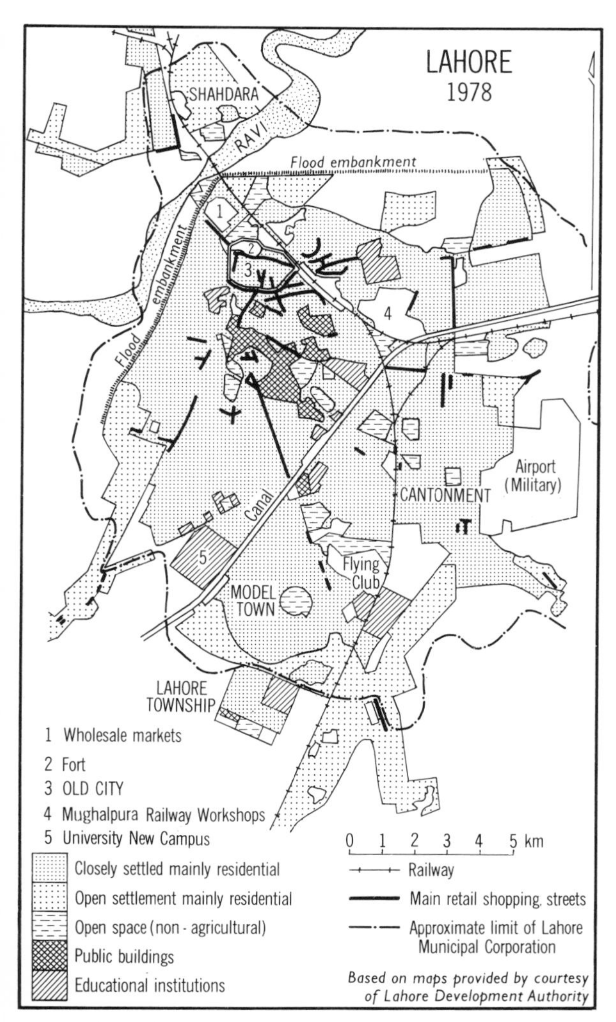

FIG. 13.7 Lahore: land uses.

In the twelfth century Lahore became the capital of the Ghazni Empire that stretched westwards beyond the Indus to its mother country in eastern Afghanistan. As this empire expanded eastwards, the capital of its ruling sultans was moved to Delhi, leaving Lahore a regional capital. The Moghuls were next to invade, and Babar laid the city waste before adopting it in 1526 as the Moghul capital, a position it shared from time to time with Delhi, Agra (and Fatepur Sikri, briefly). Under Babar, Akbar, Jehangir and Shahjahan, Lahore lived out its golden age, acquiring much of its magnificent architecture. Akbar built the Fort in 1566, to which Jehangir and Shahjahan added embellishments in the early half of the seventeenth century. Under Shahjahan, Jehangir's Tomb and the Shalimar Gardens were built, but the crowning glory of the Moghuls came towards the last quarter of the century with the Badshahi Masjid or Mosque built by Aurangzeb.

In the declining years of the Moghul Empire, Lahore was once again exposed to invasion from the northwest by the Afghans seeking to struggle with the Maharattas for control of Delhi. From 1767 to 1840 when they in turn were dislodged by the British, the Sikhs ruled in Lahore, then but a desolate shadow of its former glory. Under the British, Lahore regained something of its earlier importance. Once more it outshone Delhi as the British chose Calcutta as their Indian capital; Lahore became the cultural leader of the rich northwestern provinces until displaced again by Delhi becoming the Imperial capital in 1912. Lahore became the fulcrum of British political and economic strategy in the Indus basin. With the arrival of the railway from Calcutta, Bombay and Delhi via Amritsar in 1862 and from Karachi in 1865, Lahore's strategic position as the rear base for the Northwest Frontier region was assured. Although rivalled and overtaken by Delhi in imperial rank, Lahore remained capital of the Punjab, the most prosperous and best-run province of British India. Partition in 1947 temporarily shook the Punjab, leaving Lahore too dangerously exposed close to the Indian border to become capital of Pakistan, (and too thoroughly Punjabi to find favour as the capital of a new nation in which Punjabis, Sindhis, Pathans and Baluchis had to be united, not to mention the Bengalis in distant East Pakistan). For a while, between 1956 and 1970, West Pakistan was administered as 'one unit' with Lahore as its capital.

The national capital was in Karachi, moving nearer to Lahore (and the Punjabi sphere of influence) when it was transferred to Rawalpindi in 1959 with its ultimate location planned for nearby Islamabad. Today Lahore continues as provincial capital of the Punjab, but still maintains its status as the intellectual and cultural centre of Pakistan.

The Lahore Development Authority estimates the present population at 2.63 million (1977). The course of population growth since the census is set out in Table 13.1

TABLE 13.1
Lahore: population, 1881–1977

Year	*Population (millions)*	*Average annual growth (%)*	
1868	0.125		
1881	0.149	1.9	
1891	0.177	1.2	
1901	0.203	1.5	
1911	0.229	1.3	
1921	0.296	2.3	
1931	0.449	5.3	
1941	0.672	5.6	
1951	0.849	2.7	
1961	1.296	5.3	
1972	2.422	8.0	6.4[b]
1977[a]	2.630	1.7	

[a]Estimate.
[b]In view of the dubious status of the 1972 census total, it is appropriate to take the 16 year period from 1961 to the 1977 estimate as the basis for estimating the average annual increase.

Overmuch reliance cannot be placed on the figures for 1931 and 1941 for political reasons already mentioned in Chapter 2. 1931 was a year of under-enumeration, 1941 of over-enumeration. 1951 – the first census after independence – recorded that 45.5 per cent of Lahore's population were refugees who numbered 386,000. It is probable that they replaced at least as many Sikhs and Hindus who fled to India. Between 1951 and 1961 about half the increase was due to migration (mainly from within Pakistan) and the proportion of the increase accountable to migration has if anything increased between 1961 and 1972.

Lahore suffers many of the constraints characteristic of a city standing on a promontory projecting into the ocean. The physical threat of the Ravi prevents any urban expansion northwards into the wedge of left-bank terrace left to Pakistan. Eastwards the Indian border-post is only 25 km from the city centre so the hinterland in that direction is practically nil. Growth has perforce to go southwards parallel to the Ravi and the frontier, but at a strategic distance from both.

THE URBAN REGIONS OF LAHORE

Fig. 13.8 shows in simplified fashion the major regions discernable in the morphology of Lahore today. Six different basic morphological types can be identified:

1. the historic core: the Old City;
2. an intermediate zone south and southeast of the core;
3. the Cantonment in the southeastern sector;
4. the suburban zone surrounding the core and intermediate zone;
5. the arterial industrial ribbons;
6. the 'urban' wedges and fringe.

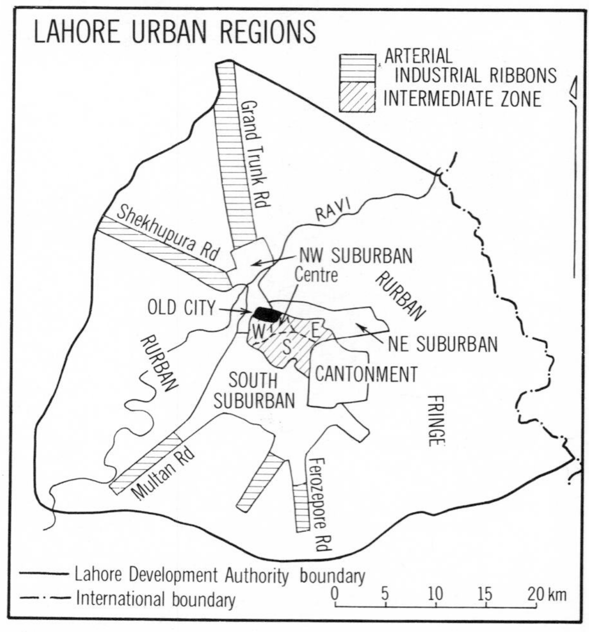

FIG. 13.8

1. The Old City

This is the most distinctive area of Lahore (Fig. 13.9). Formerly a walled city dominated by the fort at its northwestern corner, overlooking the Ravi (now an abandoned channel at this point), the line of the walls is now marked by gardens and broad streets. Covering about 2.6 km² the Old City contains some third of a million inhabitants living obviously at extremely high density. One quarter, Waksowali, is estimated to have 194,224 per km², that is 786 people per acre or 503,040 per mile². To achieve such density the minimum space is given for movement among the tiny plots which often carry four-storey buildings, occupied at ground level by a shop or workshop, with family living quarters above, topped by a flat roof for summer sleeping and a latrine. Some efforts have been made to open up sections of the Old City but it retains its essential character as a pedestrian enclave, only its wider 'streets' being accessible to slow moving bullock or horse drawn carts tongas and auto-rickshaws. Away from such lanes, only head porterage penetrates the alleys. Mushtaq classifies the streets as (a) 'bazars' where some motor traffic can penetrate from one of the thirteen city gates, lined with shops and 5–7 m wide at most; (b) 'streets' are about 2 m wide, sinuous and dark; with (c) 'lanes' 1–1.3 m wide leading off them to house plots. Open drains run down the side of the streets which themselves quickly become fouled with rubbish despite regular sweeping. Some of the lanes as well as the bazar streets, are occupied by specialized retail quarters, each trade clustering together. Fig. 13.9 shows the street pattern of part of the Old City. The Old City confronts the town planner schooled in western philosophies of urban life and speedy and efficient transport flow with a living fragment of the classic pre-industrial city. Until he attunes himself to the traditions of the place and realizes that all is not squalor with-

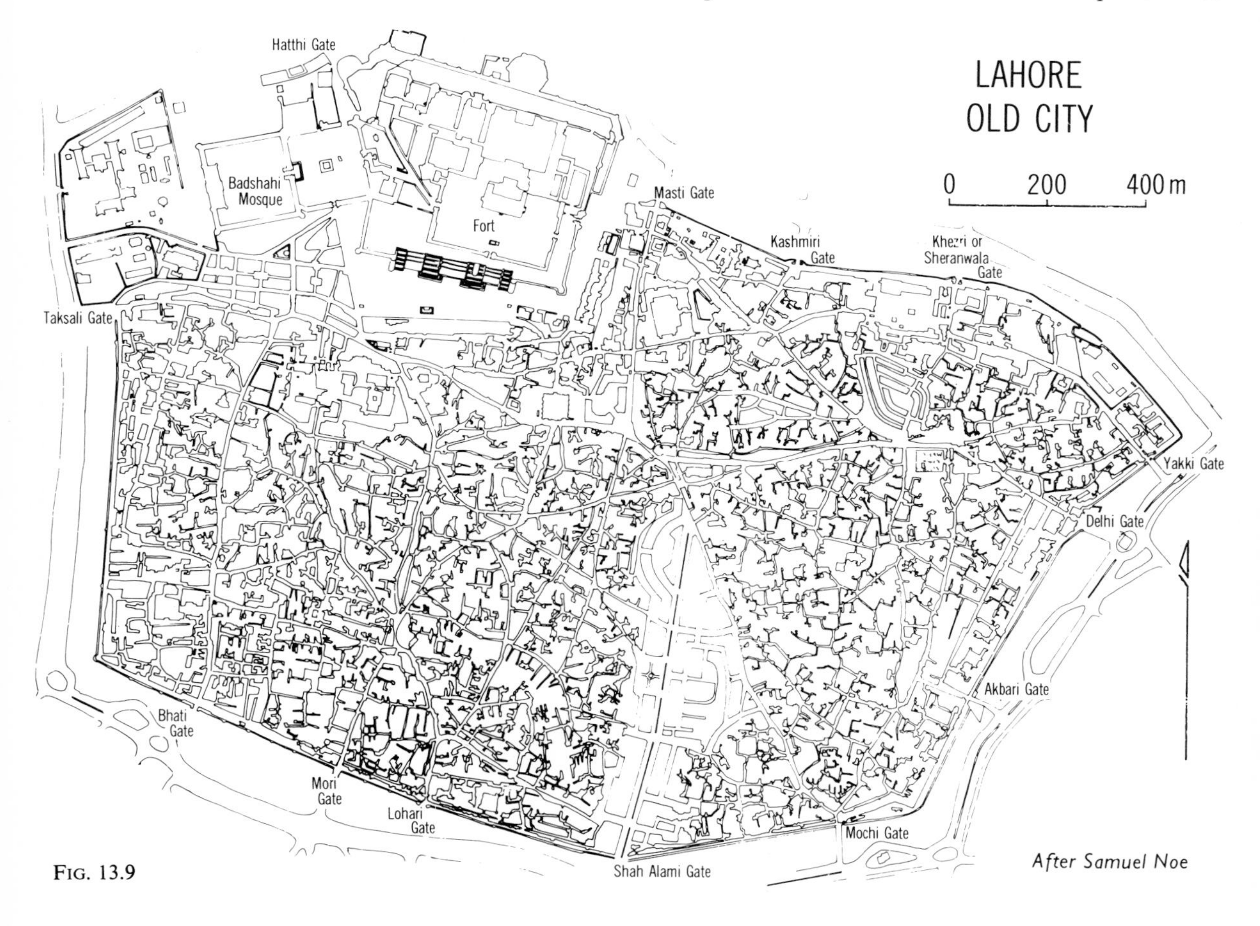

FIG. 13.9

in, and finds the corners of great wealth and industry among its alleys, he may well think that the only solution is to raze it to the ground and build anew a modern C.B.D. in glass steel and concrete parking lots and wide streets agog with perilous traffic.

That its drains, its light, its water supply – all its 'amenities' – can be much improved there is no doubt, but it is most improbable that the living organs that constitute the Old City would stand surgery and transplantation. To this observer the answer seems to be to leap the many evolutionary stages that have shaped the western city and to come in one gentle shuffle, as it were, to the city of pedestrian precincts so applauded by the avant-garde in planning.

2. The Intermediate Zone

This zone lies south and southeast of the Core. The British administration in the nineteenth century established its offices in the western part of this zone, into which some of the commercial and retail functions for the expanding Lahore spilled out from the Old City into Anarkali. Banks, Law Courts, the G.P.O., the Punjab University (Old Campus), and the Museum are concentrated in this part. The centre of the Intermediate Zone is more closely settled with a residential retailing and small industrial workplaces. Foundries, hardware shops, markets for wool, metals and fruit make this a very busy quarter in which the roads are fortunately reasonably wide. To the east of it is the distinctive area of the railway stations, passenger and goods, the railway workshops, and the residential quarters of its staff. The southern part of the zone contains the Civil Lines, the well laid out residential quarter for the British administrators, with cheek-by-jowl, sharing the security and prestige of the imperial rulers, the huge residences of the Indian princes – Bahawalpur House, Kashmir House, Patiala House, and Chamba House. Government offices now occupy them. The Upper Mall provided high class shopping facilities and a link to the administrative centre in the west of the zone: exclusive colleges catered for the elites and botanical gardens for their recreation. Altogether density was low and remains so, though infilling and intensification of land use is now visible.

3. The Cantonment

The Cantonment in the southeastern quadrant of Lahore is the military reflection of the administrative Civil Lines and a functional centre. From the 1850s it was developed to house the army, its troops in barracks with parade grounds and rifle ranges, hospitals and bazars, the officers and their families in widely spaced bungalows in gardens. Independence has changed the occupants but neither the settlement pattern nor the way of life.

4. The Suburban Zone

This zone contains a range of residential areas which vary in density of settlement inversely with socio-economic status. Cramped for space closer to the centre, industry has found a foothold here and there, but has been overwhelmed by residential sprawl before much in the way of an industrial region could develop.

75 Anarkali Bazar in Lahore leads to the crowded Old City. Note the concentration of cloth shops side by side.

In the northwest an area of very mixed and generally poor development occupies the badly drained ground of the former bed of the Ravi and its sand bars. It extends beyond the Ravi in Shahdara. Flood risks make it a 'less desirable' residential area on both sides of the river and is thus relegated to poor low class dwellings. The strategic economic position of the left-bank portion close to the Old City and the bustle of the western parts of the Intermediate Zone make it important as a wholesale market area for fodder vegetables and livestock.

The mirror image of this area to the north and east of the Old City and Intermediate Zone is similarly lowlying and only protected from flooding by the bund. Parts of the Ravi's cutoff billabongs still hold stagnating water, and occasionally (as in 1955) excessively high floods may outflank the defences. The whole area has something of a 'beyond the tracks' character about it. Although the Engineering University has been established here development has been a mixture of industry and housing clusters for the working class population, separated from Lahore by the 6m embankment of the railway. In the east of this sector, between the Grank Trunk Road leading to Shalimar Garden and the Lahore Branch of the Upper Bari Doab canal is more mixed development characterized by crowded huttments. Proximity to the Indian border has no doubt inhibited more profitable use of this area which lies beyond the Lahore Municipal Corporation's boundary.

To the south of the Intermediate Zone, flanked by the Ravi Bund to the west and Cantonment to the east, suburban growth has had unfettered opportunity to sprawl. To western eyes, backed by a consciousness of high land values, such sprawl seems at first sight an excessively lavish use of resources. Furthermore it tends to put people at a considerable distance from the job opportunities of the inner city areas and the industrial ribbons to be discussed below. This south suburban sector contains several planned better class suburbs, some like Model Town dating from before independence and using as do indeed the later Gulberg suburbs, plans reminiscent of British garden cities. At Kot Lakhpar, beyond the Municipal boundary, Lahore Township is another example of a planned integrated residential suburb. Travelling through these suburbs it is often hard to believe one is in Pakistan, and what one is seeing is the homes of the urbanized and westernized elite infinitely small in relation to the population as a whole, but wielding great political and economic power, a fact that helps explain the glaring contrasts in housing standards within the society.

Not all the south suburban zone is high class residential. Within Gulberg, space at an appropriately higher density is provided for the menials who serve the luxury residences. Some Government departments, like Posts and Telegraphs, have built staff colonies in the southwest, the University has its New Campus (designed by Doxiades) in the south and there are several colonies for lower paid government servants and others. One cannot, however escape the sense of occupational 'caste' in the residential patten.

5. The Arterial Industrial Ribbons (Fig. 13.10)

The justification for strong planning laws and their firm application is too often seen in the disasters that result in their absence. Despite the abundant examples from western experience of laissez-faire development in the inter-war period, industrial growth around Lahore has been allowed to take place along several of the major highways that bring traffic to the city, in particular from the north along the Grand Trunk Road from Gujranwala, and from the west along the Sheikhupura road. For many kilometres right up to the limits of the Lahore Development Authority's jurisdiction and beyond, these roads are lined intermittently with industrial sites to which as a rule, the labour force has to be transported by 'company bus' from the city 20 km and more distant. There is a tendency for similar ribbon development along the Multan and Ferozepore Roads, southwest and southwards from the city. The opportunity seems to have passed to lay out a consolidated industrial sector nearer to the homes of potential workers.

6. The 'Urban' Wedges and Fringes

Finally, between these ribbons and the other roads that radiate from the city, beaded with clusters of settlements and various kinds of service and commercial activity, remain wedges of mainly agricultural land dotted with villages in form typical of any Punjabi rural settlement, but in function poised

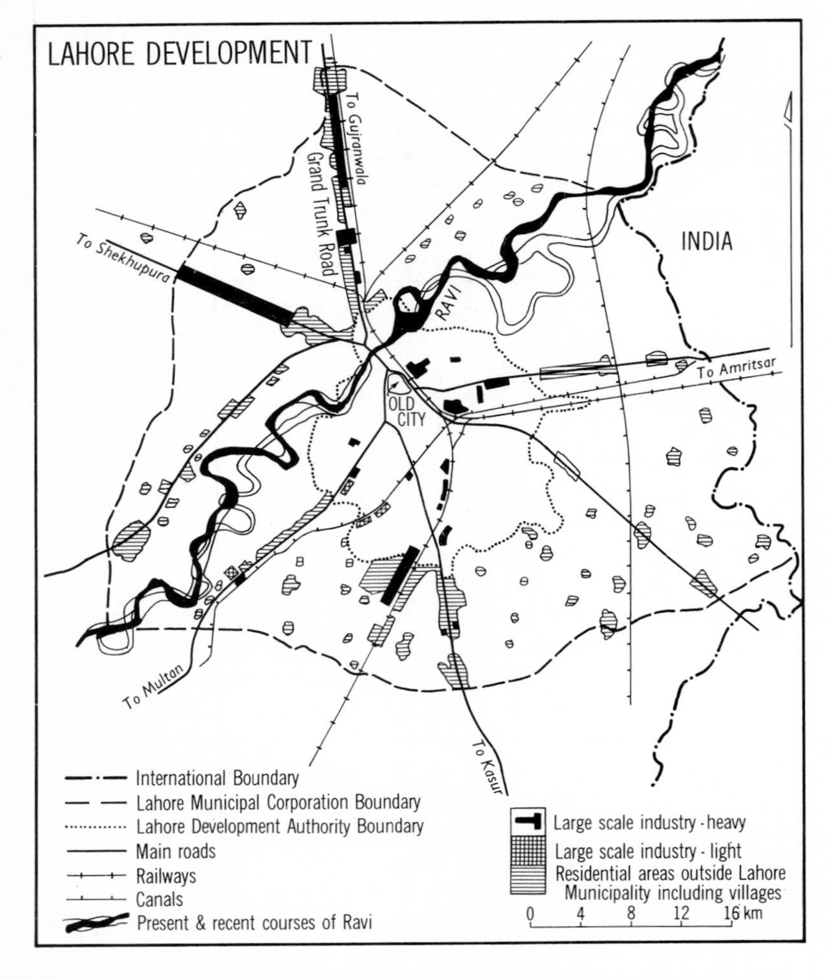

FIG. 13.10

76 On the banks of an abandoned branch of the Ravi on the outskirts of the Old City, Lahore, squatters pitch their tents and dairy herds browse under the trees.

between town and country. The farmers grow some vegetables for the Lahore market, but increasingly urban sprawl is devouring the traditional market gardening areas, or they are giving way to fodder crops, bulkier and less able than vegetables to stand the costs of carriage from further away. Day and night bullock and buffalo carts and 'rehras' ply the radial roads to bring cut fodder into the city to feed its milk herds and its draught animals. The urban villages have become semi-urban in their employment base, the workers either cycling to work or riding in the numerous small buses that serve the outer fringe of the city's labour catchment. Some villages have already lost all their lands to become 'urban' villages, mud houses, farm yards and all, overwhelmed in the tide of urban expansion, but adapting to the situation by capitalizing on their ability to supply fresh milk to the citizens, albeit becoming themselves dependent on distant supplies of fodder.

Housing in Lahore

The rapid growth of population in Lahore creates a constant demand for housing. The influx of the rural poor, the dispossessed and the redundant from the march of modernization, has led to much squatter settlement, often in neglected corners, or on unviable strips of land along drains and railways. L.D.A. policy is now to accept their presence as inevitable and to help the dwellers in these 'jhuggis' or 'kachha abaddis' to make the best of their environment by providing water and drainage. They have even been granted rights of occupance where they are on government land. Many are located where building projects first attracted them as labourers, and there they have remained. There are some 120 jhuggi settlements housing about 80,000 people in Lahore.

More formal housing is undertaken by the Lahore Development Authority and by private enterprise. By 1976–77 the L.D.A. had built 6500 residential flats and 3500 'quarters' (single-storey small home units); 1000 flats were built in the public sector in that year alone, and 3000 more in the private sector. The L.D.A. acquires land, provides it with urban services and either builds homes for sale or sells land for private development. Its function is to act as a catalyst in the housing business rather than to become a super-landlord itself.

77 Utensils in copper, aluminium and tin in a street of the Old City, Lahore. The large pots are used for milk, and the trays for serving food.

Retail Trade

The bazars of the Old City have already been mentioned. Fig. 13.7 shows the main shopping areas. Apart from the concentrations in Anarkali and the adjacent Upper Mall and Gwalmandi areas, the pattern of retailing tends to be linear along Multan and Ferozepore Roads within the south suburban sector, and in small shopping and market centres in the several Gulberg suburbs, Model Town, etc. In every shopping area, informal commerce in the shape of itinerant peddlars and pavement hawkers seems to occupy as much of the roadway as it dares. The open stalls of the Old City and the markets gradually give way to the glass fronted shops of the suburban areas, more familiar to the western visitor, but not designed for the casual haggling commerce of the bazars.

Transport

A major problem facing the planner of Lahore in the last quarter of the twentieth century is the persistence therein of traditional transport modes competing for road space with modern motorized vehicles. In 1976 the vehicles registered in Lahore were as in Table 13.2 with the 1964 data for comparison.

Some major city streets have been banned to animal drawn traffic, but away from these confusion breeds confusion as tongas, rickshaws, buses and buffaloes try to sort themselves out at cross roads. In addition to vehicles there are the animals themselves: donkeys with saddle sacks of bricks and sand proceeding on their lawful occasions, the occasional camel train loaded with firewood or fodder and the stately half-tonne buffalo cows going down from their dairy plot to the canal to drink, or herds of prime goats on the way to market and ultimately the butcher. Amongst it all, unlike in the developed western city, people in their thousands walk and jostle under head bundles and baskets, and cyclists wend their way among the crowds.

And the more the planners strive to open up the roads, the more they come, as the enormous increase in vehicles between 1964 and 1976 indicates, and the more they will come if living standards rise to bring a motor cycle into every home where a push bike now carries the breadwinner to work.

TABLE 13.2
Lahore: registered vehicles, 1964–76

	1964	*1976*
Motorized		
Motor cycles/scooters	11,548	52,283
Motor cars	9,274	22,832
Jeeps	1,201	1,109
Buses	2,298	4,331
Taxis and auto-rickshaws	2,120	9,251
Trucks and delivery vans	3,509	5,578
Tractors, etc.	859	4,720
Animal drawn		
Tongas	3,974	4,041
Rehras	3,728	5,400
Bullock carts		1,200
Hackney carriages (human-hand carts)		9,383
Others	?	5,000

Note: The categories of non-mechanized vehicles are not strictly comparable between the years. Bicycles are thought to number one for every other household – say 250,000.

Urban Farms

One last feature that distinguishes Lahore in its pre-industrial phase from the industrial city of the developed world is the continued presence within its limits of great numbers of dairy cattle, particularly buffaloes. While they ensure a fresh (if not an undiluted or an unpolluted) supply of milk within the city, they put a great load on the transport system to maintain them in fodder. The herds of ruminating buffaloes may be seen along the canal drains easements, on almost every vacant lot, even in the heart of the Old City. Their dung, a precious by-product, is plastered in neat round cakes to dry on the walls and to be used or sold as fuel for the kitchens of the poor.

KARACHI*

Situated on a coast of low platforms, mangrove swamps, sandspits and beaches, and backed by semi-desert etched with the courses of intermittent torrents, Karachi presents a very different picture from that of Lahore (Fig. 13.11). Its history prior to British rule was undistinguished. The British capturing the town in 1839 found a small fortified settlement in the bend of the River Lyari about 2 km northwest of the present railway station which had been protected as a trading post by the Khans of Kalat in the eighteenth century and from 1797 by the Talpurs who built a fort on the Manora headland to give protection against sea raiders. The port traded westwards to the Persian Gulf and south to Bombay with which it had strong links. With their conquest of Sind in 1843 the British made Karachi its capital, then a town of about 14,000 people. Until connected by railway to Kotri in 1861 and via Multan to Lahore and Delhi in 1878 Karachi served a limited hinterland by river. The Indus Flotilla Steamship Co. towed flats as far as Multan (whence a railway later went to Lahore in 1865) saving British troops the long march from Calcutta in the pre-railway days. Land transport by camel train penetrated the remote interior of Baluchistan and Afghanistan.

It was the opening of the Suez Canal in 1869 that gave the west coast ports of British India – Bombay and Karachi – their major boost. Karachi had an economic advantage over Calcutta and Bombay for the hinterland westwards from Ambala (in East Punjab). With canal colonization of the Punjab, its trade expanded greatly and processing industries to prepare crushed bones, tanned hides, pressed cotton bales, washed wool and flour were established. In the twentieth century the three major fillips to growth were the opening up of Upper Sind to more intensive cultivation with the construction

*The major sources for this section are Mohd. Zafar Ahmad Khan, '*Karachi – a Pre-industrial City in Transition*', Univ. of London Ph.D. thesis, 1968, chapters of which appear in *Pakistan Geographical Review*, vols. 23, 24, 26, 1968, 1969 and 1970, and maps and material supplied by Karachi Development Authority.

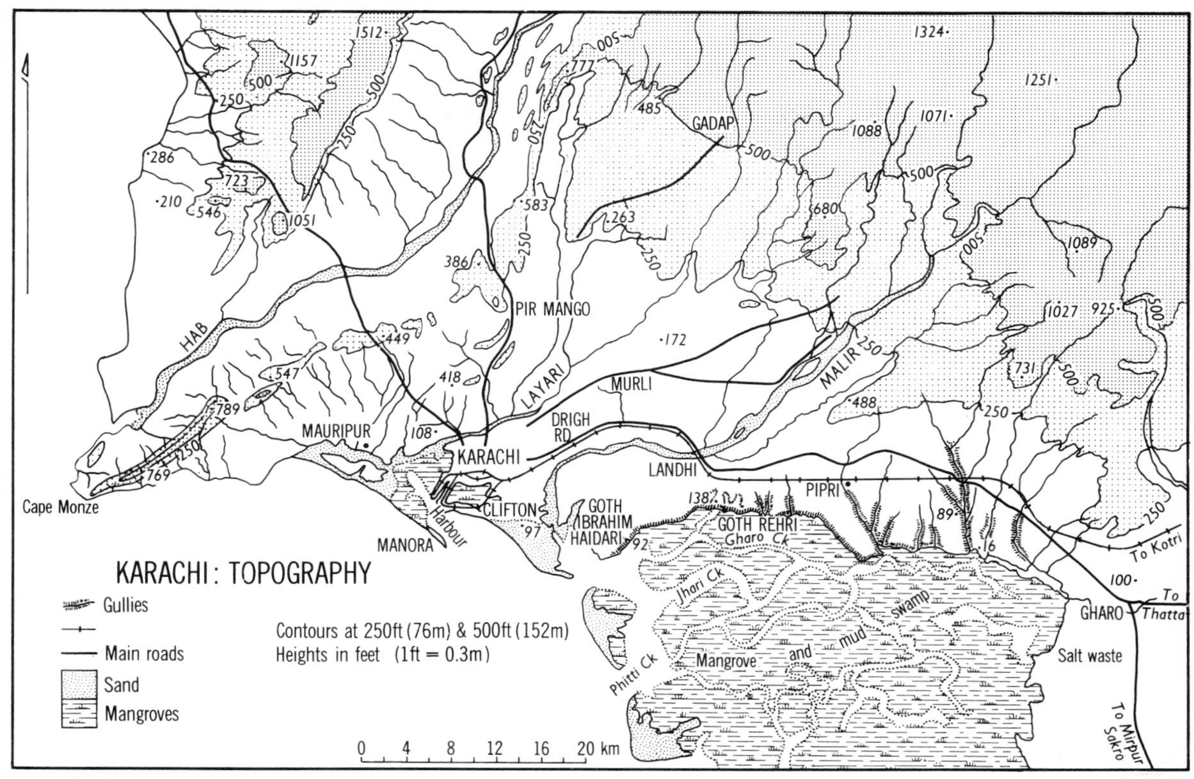

FIG. 13.11

of the Sukkur barrage in 1932; World War Two which brought new strategic significance to a western outlet for supplying British armies in North Africa and Mesopotamia, and to the major enlarged airforce base; and lastly partition, as a consequence of which Karachi not only became the capital of Pakistan, and its only port, but also had to shoulder a major part of the burden of refugees from India. Since then in-migration from inland Pakistan has brought an unabated increase in population over and above the natural increment. Table 13.3 shows the growth of population from 1843 to 1972.

At the last census before partition Karachi had a majority of Hindus – 51 per cent of its population compared with 42 per cent Muslim. By 1951 the Muslims had reached 96 per cent. Some 120,000 Hindus left the city to be replaced by 150,000 refugee Muslims in the space of two years. Almost the only resource available in Karachi to meet the problems brought by such an influx of refugees, plus the immigration of the Pakistani born, has been land. Water and food have to be brought in from a distance: water from the Indus, food from upcountry Sind. Jobs had to be created, shelter provided, not to mention hospitals, education and general welfare services. Although Karachi lost its national capital status in 1959 it became Sind's capital again.

TABLE 13.3
Karachi: population 1843–1972

Year	*Population (million)*	*Average annual increase (%)*
1843	0.014	–
1872	0.057	–
1881	0.074	2.7
1891	0.105	4.2
1901	0.117	1.1
1911	0.152	3.0
1921	0.217	4.3
1931	0.264	2.2
1941	0.387	4.7
1951	1.138	19.4
1961	1.913	6.8
1972	3.607	8.0

Fig. 13.11 shows the physiography of the site of Karachi and its immediate hinterland, the mouth of the River Lyari where the trading port grew up and the low mangrove coast to the east where modern port development is taking place. Fig. 13.12 shows the city on the eve of independence, and Fig. 13.13 the major features of its subsequent growth. Reference has already been made to the lay out of the city in British times. With the arrival of thousands of refugees housing was the first task.

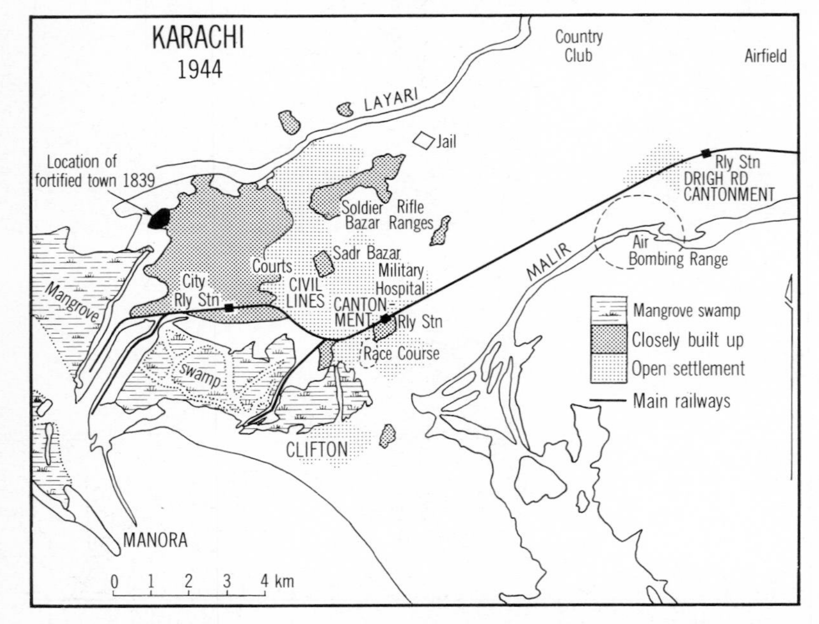

Fig. 13.12

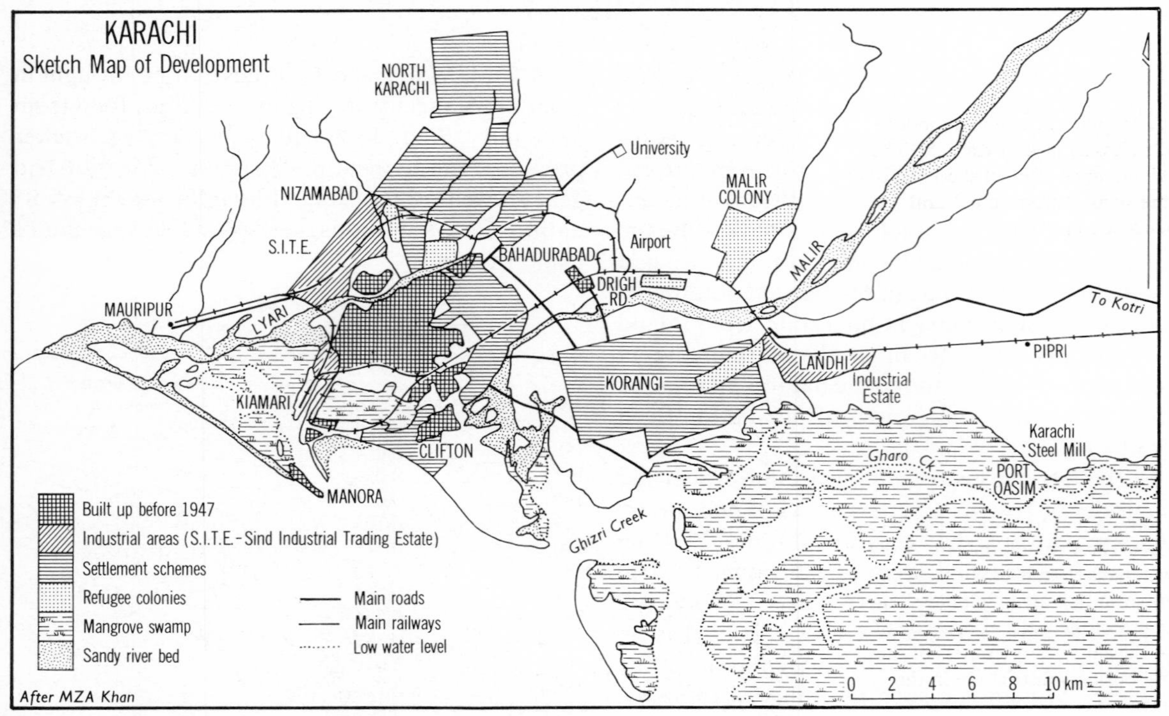

Fig. 13.13

78 Refugee housing in a Karachi colony. These houses are of good quality compared with many that had to be occupied soon after partition. Note the variety of materials and styles. Many of the roofs are weighed down with rocks to prevent the corrugated iron sheeting being blown away in a storm.

79 A shopping centre in a modern middle class suburb of Karachi. The traditional open-fronted shop persists but the signs of western cultural influence are apparent.

80 Karachi from the east. The Thole Produce Yards are on the left. A diesel drawn goods train is moving out from the main station area. Squatters' huts make use of the walls as one side of a lean-to dwelling. Skyscrapers of the central business district are in the distance.

Many families lived for a decade in kachha jhuggis of indescribable squalor. (Visiting one such settlement in 1957 when it was being evacuated to a pucca colony, the writer found the crowding of huts and people was such as to make photography of a single home almost impossible.) In 1959 it was estimated that 750,000 people were still living or rather clinging on to life in such jhuggis. The major unplanned refugee clusters have now mostly given place to resettlement areas spread over many square kilometres and provided with the minimum of amenities. Often all a family came to was a pucca brick one-roomed box, to which they perforce added kachha rooms and lean-to shelters. Many jhuggis settlements sprang up in the sandy beds of the nalas which seldom flow but when they do can create great havoc as in 1977 (see Chapter 6 above). For the middle and upper income groups cooperative housing schemes have provided spaciously laid out suburbs with modern shopping centres and schools. The better suburbs are reasonably close to the fringe of the pre-independence city: Clifton, Sind Muslim Housing Society, Defence Officers' Housing Society, etc. The problems of transport to work for the residents of the resettlement areas 12 km from industrial employment can readily be appreciated when the Chief Planner tells of buses carrying up to 142 passengers in the rush hour.

Industry, originally concentrated near the docks and scattered through the pre-independence city, has been provided with two major industrial estates: S.I.T.E. (Sind Industrial Trading Estate) lies west of the River Lyari served by the circular railway; the Landhi Industrial Estate lies 25 km east of the city centre finding its labour in the huge Korangi housing area. Further east again is the new developing industrial complex around Pipri and Port Qasim (see Fig. 11.5) where the Karachi Steel Mills are under construction.

The industries in the unplanned areas of the old town are a hotch potch of hovels, huts and workshops with noxious tanneries, wool washeries and smoking foundries jostling with housing between Lawrence Road and the Lyari. South of this small industry tends to cluster big trades, motor repairers, jewellery, handloom and dyeing, for example. S.I.T.E. by contrast is spaciously planned for almost half the industrial workers in Karachi. Textiles, metal working and chemical industries are important, but there is great variety: tyres, matches, bone crushing, pharmaceuticals, etc. Near the docks, at West Wharf at their northern side and Kiamari to the south, the industries are linked to the port: motor vehicle assembly plants using imported components, ship building and repair, chemicals. Landhi deals rather with resources from the hinterland, cotton textiles, food products as well as metal trades, but nearby the oil refineries have found adequate space and terminal facilities for tankers (but with some hazard from the flooding River Malir). Karachi with its singular advantages as a port and its unrestricted availability of level land is likely to dominate the industrial scene in Pakistan for the foreseeable future. Water for its factories and its population will be a recurrent problem, as also adequate finance properly to house its burgeoning population.

ISLAMABAD

Largely on account of its importance as Pakistan's military headquarters, with the appropriate ordinance factories, training establishments and regimental depots, Rawalpindi after independence was the fourth largest city in the country with 237,000 inhabitants (including 95,000 – 40 per cent – refugees). It had become the regional centre for not only its own district and the others in the Potwar plateau, but also for the valleys leading north into the hills past Taxila to Abbotabad and Manshera, for the oil fields of the plateau, and for Azad (Free) Kashmir. Peshawar was now too far from the strategically important areas to compete, and furthermore it was not in the Punjab. In 1959 under the military dictatorship of General (President) Ayub Khan, the capital was transferred from Karachi, exposed too much to exotic cultures and too removed from what was felt to be the heart of Pakistan national identity – Punjab, the Army and the North. Rawalpindi became the foster parent for a new capital to be built at Islamabad to the plans of the Greek designer Doxiades. By the 1972 census Islamabad had reached a population of 77,318 and Rawalpindi 615,374. Fig. 13.14 shows Rawalpindi and Islamabad as they are at present, the one an old district town onto which the British grafted large military installations which came completely to dominate its character and

81 Rawalpindi's main street carries a variety of transport: bullock cart, horse-drawn tonga, motorcycles, bicycles and cars. A mixture of scripts is evident in the street signs, as are Pakistan's increasing links with the Middle East – Saudia Airways, Cafe Iran.

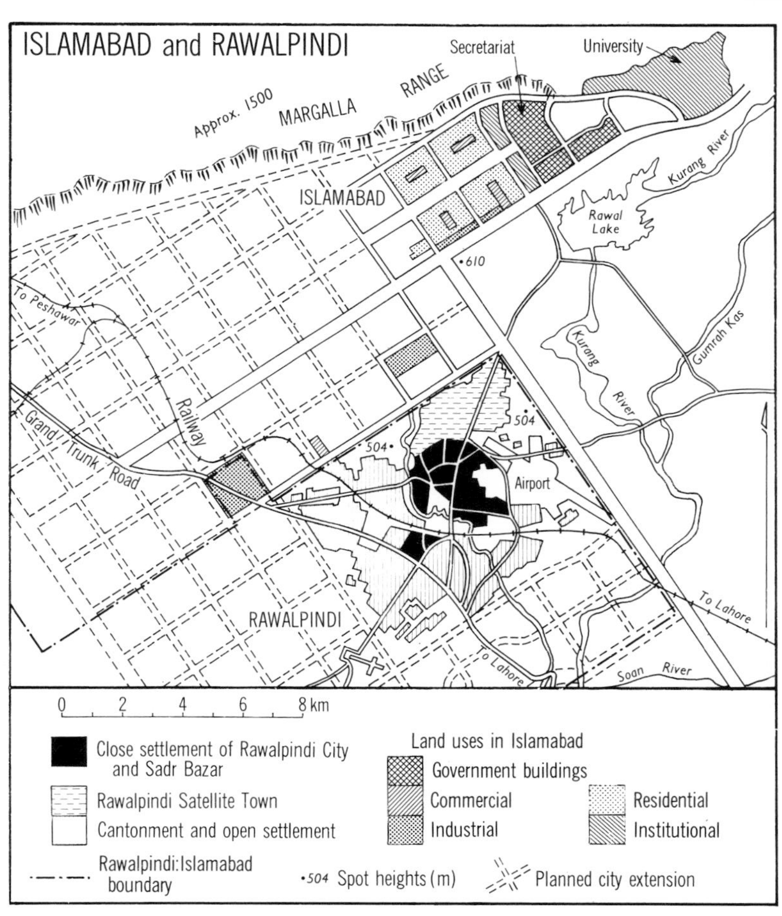

FIG. 13.14

functions, the other brash and new and barely yet an independent city. The twice daily chariot race between the mini-buses that link the two cities, carrying commuting clerks from the relative cheapness of 'Pindi to work in the huge Secretariat buildings, are clear evidence of their interdependence. The plan for Islamabad envisages, somewhat cavalierly, the ultimate incorporation of Rawalpindi almost as an ancient relic into the spreading capital. The Punjab Provincial Government, still master in its own home, has an alternative plan for 'Pindi, independent of that of the Federal Capital planners!

Rawalpindi evidences in its form the several characteristics of the dualism seen in other cities that gained favour under British rule (Fig. 13.14).

82 Islamabad, the national capital, is dominated by the white blocks of the Secretariat which house government offices. The Murree Hills rise in the background.

These have been discussed above: the traditional city and exotic cantonment, unplanned bazars in the Old City, tidy rectilinear regimentation for the civil lines and the military. Islamabad is exotic in a quite different sense. (One may comment somewhat ruefully in parentheses, that at least the British lived in the cantonments they designed: it is sometimes hard to imagine for whom the linear form of Islamabad was conceived as being the home.) In its grand design Islamabad will ultimately stretch almost 20 km along the foot of the splendid Margala Hills with the Government Secretariat and the University at the extreme northeastern corner. Foreign embassies have begun to cluster near the Secretariat and some five blocks of residential quarters, neatly graded according to the hierarchical status of their occupants in the civil service, represent the first completed in 38 planned for the areas north of the Murree Highway, and 52 more for the areas south, excluding Rawalpindi and the airport left like an island, three parts surrounded. On the fourth side, in the southeast third of the master plan, a national park 15 km square is envisaged, in which Lake Rawal has already been completed to supply water to the city. In the linear scheme, commercial and industrial zones, and a continuous belt of open space parallel the residential blocks. The latter, each 0.6 km square, are already two deep, and will become four deep when the capital reaches its full extent to the southwest, and consequently the green belts will be enjoyed more by the passing commuter than by the residents three, four or six kilometres distant. Islamabad is the linear city, conceived originally to cope with urban expansion in a highly mobile society, translated prematurely into a community who can barely afford the paisa for a bus ride and very few of whom have motorized transport of any kind. For the elites in the upper echelons of the Secretariat and for the foreign embassy staffs, Islamabad may be a tolerably comfortable and increasingly green and beautiful city in which to live but it is a far greater conceptual distance from the peasant ploughing his fields behind a pair of bullocks than Canberra, Brasilia, New Delhi or other planned national capitals are from their respective citizens. In a way it epitomises a recurrent and significant problem for Pakistan: for whom is the country to be built?

INDEX

Entries thus: 22 refer to text; *22* refer to figures or tables; **22** refer to photographs; n indicates a footnote. All are page numbers.